Theoretical Ecology

PRINCIPLES AND APPLICATIONS

Theoretical Ecology

PRINCIPLES AND APPLICATIONS

EDITED BY

ROBERT M. MAY

Biology Department
Princeton University

companion volume
= Behavioral Ecology
Krebs + Davies eds. 1978
Oxford —

Second Edition

SINAUER ASSOCIATES, INC. ● PUBLISHERS

SUNDERLAND MASSACHUSETTS

First published 1976
Second Edition 1981

Distributed in the U.S.A. by
Sinauer Associates, Inc., Publishers
Sunderland, MA 01375

Library of Congress Cataloging in
Publication Data

Main entry under title:
Theoretical ecology
    Bibliography: p.
    Includes index.
    1. Ecology.   2. Ecology—mathematical models.
I. May, Robert M.
QH541.T49   1981   575.5   80-19344

ISBN 0-87893-514-2
ISBN 0-87893-515-0 (pbk.)

Printed in Great Britain

Contents

List of Authors

ROY M. ANDERSON   Department of Zoology, Imperial College of Science and Technology, London, SW7 2AZ, England

GRAEME CAUGHLEY   Department of Biology, University of Sydney, Sydney, N.S.W., 2006, Australia

COLIN W. CLARK   University of British Columbia, Vancouver, British Columbia, V6T 1Y4, Canada

GORDON CONWAY   Centre for Environmental Technology, Imperial College of Science and Technology, London, SW7 2AZ, England

JARED M. DIAMOND   Physiology Department, U.C.L.A. Medical Center, Los Angeles, California 90024, U.S.A.

STEPHEN JAY GOULD   Department of Paleontology, Harvard University, Cambridge, Massachusetts 02138, U.S.A.

JOHN L. HARPER   School of Plant Biology, University College of North Wales, Bangor, Wales

MICHAEL P. HASSELL   Imperial College Field Station, Silwood Park, Ascot, Berks SL5 7PY, England

HENRY S. HORN   Biology Department, Princeton University, Princeton, New Jersey 08544, U.S.A.

JOHN H. LAWTON   University of York, Heslington, York, YO1 5DD, England

ROBERT M. MAY   Biology Department, Princeton University,
Princeton, New Jersey 08544, U.S.A.

ERIC R. PIANKA   Department of Zoology, University of Texas,
Austin, Texas 78712, U.S.A.

T. R. E. SOUTHWOOD   Department of Zoology, University
of Oxford, South Parks Road, Oxford OX1 3PS, England

Acknowledgements

In a multi-authored volume such as this, the number of people to whom the authors are obliged for stimulating conversation and helpful criticism is unmanagably large. A short list includes Robert Campbell (who suggested both the original version of the book and its substantial revision and alteration in this second edition), J.R. Beddington, J.T. Bonner, W.H. Drury, T. Fenchel, L.R. Lawlor, B.R. Levin, S.A. Levin, the late R.H. MacArthur, J. Maynard Smith, G.F. Oster, Ruth Patrick, J. Roughgarden, T.W. Schoener, J.M.A. Swan, J. Terborgh, M. Williamson and all the graduate students in the population biology group at Princeton University. B. DeLanoy and C.L. Hewitt typed most of the manuscripts, and helped produce order out of chaos.

In choosing the topics to be included in this book, I have exhibited a certain amount of bias and caprice; the guilt is entirely mine.

Princeton University   R.M.M.
August 1980

Introduction

An increasing amount of study is being devoted to mathematical models which seek to capture some of the essential dynamical features of plant and animal populations. Some of these models describe specific systems in a very detailed way, and others deal with general questions in a relatively abstract fashion; all share the common purpose of helping to construct a broad theoretical framework within which to assemble an otherwise indigestible mass of field and laboratory observations.

The present book aims to review and to draw together some of these theoretical insights, to show how they can shed light on empirical observations, and to examine some of the practical implications. In so doing, the book seeks to occupy a useful niche intermediate between the compendious and general text (of which there are an increasing number of excellent examples) and the often highly technical journal and monograph literature on theoretical ecology (which many people will find impenetrable). The book is directed to an audience of upper level undergraduates, graduate students, or general readers with an educated interest in the discipline of ecology.

Attention is focussed on the biological assumptions which underlie the various models, and on the way the consequent mathematical behaviour of the models explains aspects of the dynamics of populations or of entire communities. That is, the approach is descriptive, with emphasis on the biological inputs in constructing the models, and on the emergent biological understanding. The intervening mathematical details are, by and large, glossed over; this is a book for people who did not get beyond a freshman course or A levels in calculus. Those readers who dislike ex cathedra pronouncements, or who wish to savour the detailed mathematical development, will find signposts to guide them to the more technical literature. Other people may be content to follow the advice given by St Thomas Aquinas (concerning technical details of proofs of the existence of God): 'Truths which can be proved can also be

known by faith. The proofs are difficult, and can only be understood by
the learned; but faith is necessary also to the young, and to those who,
from practical preoccupations, have not the leisure to learn philosophy.
For them, revelation suffices' (from Russell, 1946, p. 46).

As will be seen from the chapter headings, the first two-thirds of the
book (chapters 2 to 11) deals with plant and animal ecology as such:
theoretical and empirical aspects of the dynamics of single populations, of pairs of interacting populations, and of whole communities of
different species. The last third of the book (chapters 12 to 16) is devoted
to various subjects which are having fruitful reciprocal interaction with
theoretical ecology. Although primarily aimed at review and synthesis,
the book does contain a significant amount of new material. The second
edition differs from the first in that some chapters have been extensively
revised to bring them up to date (this revision has, for example, almost
doubled the length of chapter 9), other chapters (7, 12 and 14) have been
completely rewritten by new authors, and two new chapters (4 and 16)
have been added.

Chapter 2 outlines the dynamical behaviour of single species in which
population change takes place either in discrete steps, or continuously,
subject to density dependent regulatory mechanisms with time-delays;
the environment may be constant, or it may vary in time. The consequent population behaviour may be a stable equilibrium point (with
disturbances damped in a monotonic or an oscillatory manner), or
stable cycles, or even apparently random fluctuations, depending on the
relations among the various natural time scales in the system. In chapter
2, these ideas are applied narrowly to the dynamical trajectories of
particular laboratory and field populations, and in chapter 3 they are
applied much more widely to discuss the way an organism's bionomic
strategy (size, longevity, fecundity, range and migration habit) is
fashioned by its general environment. Broadly speaking, plants tend to
differ from animals in having modular patterns of development and
growth; arguably the leaf, rather than the tree, is the relevant unit of
population. Chapter 4 explores these ideas, aiming to stimulate further
work on the subject; the chapter deliberately raises more questions than
it answers.

Chapter 5 gives a brief survey of the basic dynamical features of two
species interacting as prey–predator, as competitors, or as mutualists.
This serves as a background to the next three chapters, which focus
upon two special cases of the general prey–predator relationship, and
upon competition. In chapter 6, mathematical models are combined

with field and laboratory studies to elucidate the components of predation in arthropod systems. Chapter 7 does a similar thing for plant–herbivore systems. Competition is addressed in chapter 8, which shows how theory and empirical observation can illuminate such questions as the meaning of the ecological niche or the limits to similarity among coexisting competitors.

Quantitative understanding of the populations' dynamics, of the sort which enlivens some of the earlier chapters, is rarely feasible for complex communities of interacting species. Here the search is more for broad patterns of community organization. Chapter 9 discusses some of these patterns: energy flow (where much progress has accrued from the work of the International Biological Program); relative abundances of the different species in the community; convergence in the structure of geographically distinct communities in similar environments; and the relation between the complexity of an ecosystem (measured by the number of species or the richness of the web of interactions among them) and its ability to withstand natural or man-made disturbances. Some of these notions can be developed further and applied to the biogeography of islands, and thence to the design and management of floral and faunal conservation areas; this is done in chapter 10. Chapter 11 presents a somewhat revisionist account of succession, arguing from theory and observation that ecosystems require occasional (neither too frequent, nor too infrequent) disturbance back to the zeroth successional state, if they are to maintain their potential diversity.

The final five chapters deal with areas on the edge of mainstream ecology. Chapter 12 looks at some ways in which theoretical ecology has advanced our understanding of the structure and organization of animal societies ('sociobiology'). The way recent ecological advances may shed light on such long debated riddles as the great waves of extinction which mark the end of many of the conventional geological epochs is the subject of chapter 13, devoted to palaeobiology. Chapter 14 pursues the population biology of infectious diseases, discussing how parasites (broadly defined to include viruses, bacteria, protozoans, helminths and arthropods) are transmitted and maintained, and how they can in some circumstances regulate their host populations. These ideas are applied to hookworm, malaria and other infections in human populations, to viral and bacterial infections in populations of laboratory mice, and to population cycles of forest insects. Chapter 15 draws together ecology and economics in a discussion of optimal strategies for pest control. This theme is developed further in chapter 16, which shows how resource economics, when applied to biological resources, can give insights into what has actually happened in the management of forests and of fish and whale populations. It seems useful to have the mathematical underpinning set out more fully in chapters 14 and 16 than elsewhere in the book; the essential biological assumptions and messages are, however, spelled out in plain language, and the equations can be skipped in reading these two chapters.

I hope that the selection of topics has provided a representative range of examples where theoretical models have been successfully intermeshed with real world observations. More than this, I hope that the collection may convey a sense of excitement, and may, to some small degree, serve to indicate unanswered questions and future directions for research.

2
Models for Single Populations

ROBERT M. MAY

2.1 Introduction

One broad aim in constructing mathematical models for populations of plants and animals is to understand the way different kinds of biological and physical interactions affect the dynamics of the various species. In this enterprise, we are relatively uninterested in the algebraic details of any one particular formula, but are instead interested in questions of the form: which factors determine the numerical magnitude of the population; which parameters determine the time scale on which it will respond to natural or man-made disturbances; will the system track environmental variations, or will it average over them? Accordingly, attention is directed to the biological significance of the various quantities in the equations, rather than to the mathematical details; to do otherwise is to risk losing sight of the real wood in contemplation of the mathematical trees.

In the use of mathematical models to grasp at general principles, it is helpful to begin with models for a single species. Models of this kind seek to elucidate the behaviour of a single population, $N(t)$, as a function of time, t.

Many isolated populations, carefully maintained in a controlled environment, may realistically be modelled by such a single equation.

On the other hand, there are few, if any, truly single species situations in the natural world. Populations will tend to interact with their food supply (below them on the trophic ladder), with their competitors for these resources (on the same trophic level), and with their predators (above them on the ladder). In addition, populations will be influenced by various factors in their physical environment. Even so, it is often useful to regard these biological and physical interactions as passive parameters in an equation for the single population, summarizing them as some overall 'intrinsic growth rate', 'carrying capacity', or the like.

Section 2.2 discusses models where generations completely overlap and population growth is a continuous process (first order differential equations), and section 2.3 treats models where generations are non-overlapping and growth is a discrete process (first order difference equations). Some of the emergent insights are applied to field and laboratory data in section 2.4, and extended to encompass time-varying environments in section 2.5. Section 2.6 briefly discusses the more complicated case of many distinct but overlapping age classes. Section 2.7 concludes the chapter by reechoing the major themes.

2.2   Continuous growth (differential equations)

In situations where there is complete overlap between generations (as in human populations), the population changes in a continuous manner. Study of the dynamics of such systems thus involves differential equations, which relate the rate at which the population is changing, dN/dt, to the population value at any time, $N(t)$.

2.2.1   *Density independent growth*

The simplest such model has a constant per capita growth rate, r, which is independent of the population density:

$$dN/dt = rN. \tag{2.1}$$

This has the familiar solution

$$N(t) = N(0) \exp{(rt)}. \tag{2.2}$$

There is unbounded exponential growth if $r > 0$, and exponential decrease if $r < 0$. In either event, the characteristic time scale for the 'compound interest' growth process is of the order of $1/r$.

2.2.2   *Density dependent growth*

Such unbounded growth is not to be found in nature. A simple model which captures the essential features of a finite environment is the logistic equation:

$$dN/dt = rN(1 - N/K). \tag{2.3}$$

Here the effective per capita growth rate has the density dependent form $r(1-N/K)$: this is positive if $N < K$, negative if $N > K$, and thus leads to a globally stable equilibrium population value at $N^* = K$. K may be thought of as the carrying capacity of the environment, as determined by food, space, predators, or other things; r is the 'intrinsic' growth rate, free from environmental constraints.

In any such dynamical system, it is useful to christen a 'characteristic return time', T_R, which gives an order-of-magnitude estimate of the time the population takes to return to equilibrium, following a disturbance (for a more formal discussion, see May *et al.*, 1974, and Beddington *et al.*, 1976a). In eq. (2.3), this characteristic time scale remains $T_R = 1/r$. To elaborate this point, we rewrite eq. (2.3) in dimensionless form by introducing the rescaled variables $N' = N/K$ and $t' = rt = t/T_R$. This gives the parameter-free equation

$$dN'/dt' = N'(1-N').\qquad(2.3a)$$

Such rescaling arguments are of general usefulness in disentangling those factors which influence the *magnitude* of equilibrium populations from those factors which bear upon the *stability* of the equilibrium. In this particular example, it is clear that the magnitude of the equilibrium population depends only on K, whereas the dynamics—the response to disturbance—depends only on r. This fact underlies the metaphor of r and K selection, developed in the next chapter.

It must be emphasized that the specific form of eq. (2.3) is not to be taken seriously. Rather it is representative of a wide class of population equations with regulatory mechanisms which biologists call density dependent, and mathematicians call nonlinear. A plethora of other such models, taken from the ecological literature, is catalogued in May (1975a, pp. 80–81). All share with eq. (2.3) the essential property of a stable equilibrium point, $N^* = K$, with any disturbance tending to fade away monotonically. One way of justifying eq. (2.3) is to regard it as the first term in the Taylor series expansion of these more general density dependent models.

2.2.3 *Time-delayed regulation*

In eq. (2.3), the density dependent regulatory mechanism, as represented by the factor $(1 - N/K)$, operates instantaneously. In most real life situations, these regulatory effects are likely to operate with some built-in time lag, whose characteristic magnitude may be denoted by T. Such time lags may, for example, derive from vegetation recovery times or other environmental effects, or from the time of approximately one generation which elapses before the depression in birth rates at high densities shows up as a decrease in the adult population. A very rough way of incorporating such time delays is to rewrite eq. (2.3) as

$$dN/dt = rN[\text{I} - N(t - T)/K]. \tag{2.4}$$

This delay-differential equation was first introduced into ecology by Hutchinson (1948) and Wangersky and Cunningham (1957),and by now it enjoys an extensive literature (for a brief guided tour, see, e.g., May, 1975a, pp. 95–98). One way of deriving it is as a crude approximation to a fully age-structured description of a single population, in which case T is the generation time. As discussed below, the detailed form of the equation is not to be taken too literally; amongst other things, in more realistic treatments the regulatory term is likely to depend not on the population at a time exactly T earlier, but rather on some smooth average over past populations (for a more mathematical discussion, see May, 1973a). Nonetheless, elucidation of the dynamic properties of eq. (2.4)—many of which are representative of wider classes of models—has helped to advance our understanding in recent years.

The qualitative nature of the solutions of eq. (2.4) follow from precepts familiar to engineers. If the time delay in the feedback mechanism (namely, T) is long compared to the natural response time of the system (namely, T_R or I/r), there will be a tendency to overshoot and to overcompensate. For modest values of the time delay this overcompensation produces an oscillatory, rather than a monotonic, return to the equilibrium point at $N^* = K$. As the time delay becomes longer (as T/T_R or rT exceeds some number of order unity), there is a so-called Hopf bifurcation, and the stable point gives way to stable limit cycles. These stable cycles are an explicitly nonlinear phenomenon, in which the population density, $N(t)$, oscillates up and down in a cycle whose amplitude and period is determined uniquely by the parameters in the equation. Just as in the case of a stable equilibrium point, if the system is perturbed it will tend to return to this stable cyclic trajectory. Such stable limit cycle solutions are a pervasive feature of nonlinear systems, for which conventional mathematics courses (with their focus on linear systems) give little intuitive appreciation.

Specifically, eq. (2.4) has a monotonically damped stable point if $\text{o} < rT < e^{-1}$, and an oscillatorilly damped stable point if $e^{-1} < rt < \frac{1}{2}\pi$. For $rT > \frac{1}{2}\pi$, the population exhibits stable limit cycles, the period and amplitude of which are indicated in Table 2.1. These numerical details (e^{-1} and $\frac{1}{2}\pi$) are peculiar to eq. (2.4), but the character of the solution, with a stable equilibrium point giving way to stable cycles once T/T_R exceeds some number of order unity, is generic to a much wider class of models with time-delayed regulatory mechanisms.

Table 2.I. Properties of limit cycle solutions of eq. (2.4).

rT	$N(\text{max})/N(\text{min})$	Cycle period$/T$
1·57, or less	1·00	—
1·6	2·56	4·03
1·7	5·76	4·09
1·8	11·6	4·18
1·9	22·2	4·29
2·0	42·3	4·40
2·1	84·1	4·54
2·2	178	4·71
2·3	408	4·90
2·4	1,040	5·11
2·5	2,930	5·36

In particular, it is worth noting that once stable limit cycles arise in equations of the general form of eq. (2.4), their period is roughly equal to $4T$. A qualitative explanation of this fact is as follows: In the first phase of the cycle, the population continues to grow $(dN/dt > 0)$ until the earlier population value in the time-delayed regulatory factor attains the potential equilibrium value $(N(t - T) = K)$; at this point, population growth ceases $(dN/dt = 0)$, and the population begins an accelerating decline from its peak value. Thus the first phase, where the population grows from around K to the cycle maximum, takes a time T. Similar arguments applied to the subsequent phases of the cycle suggest an overall period of roughly $4T$. The exact results in Table 2.1 show that this rough rule remains true, even as the amplitude of the cycle (population maximum/population minimum) increases over several orders of magnitude.

More realistic equations for describing the dynamics of species in which population growth is, to a good approximation, a continuous process have the form:

$$dN/dt = -\mu N + R[N(t - T)]. \tag{2.5}$$

Here losses due to deaths (at a constant per capita mortality rate μ) depend simply on $N(t)$, but recruitment into the adult population [described by some nonlinear function of the population size, $R(N)$] at time t depends on the population, $N(t-T)$, a time T earlier; T is the typical time taken to attain maturity. The equations used, for example, by the International Whaling Commission, IWC, in setting quotas for whaling are essentially of the form of eq. (2.5), with $R(N)$ having the explicit form

$$R(N) = N[p + q\{\mathrm{I} - (N/K)^z\}]. \tag{2.6}$$

Here p is the pregnancy rate (discounted by the probability to survive the T years to adulthood) in the pristine equilibrium state at $N = K$, $p+q$ is the 'intrinsic' pregnancy rate at low stock densities, and z is a phenomenological parameter describing the steepness of the density dependence as stock are depressed below K. For whales, the time taken to attain maturity varies from $T \simeq 6$ years for the relatively small minke whales in recent years, to $T \simeq 25$ years for male sperm whales. Other differential-delay and integro-differential equations of a form broadly similar to eq. (2.5) with eq. (2.6) have been discussed in physiological contexts by Glass and Mackay (1979) and MacDonald (1978), and in human demography by Frauenthal and Swick (1980), Swick (1980) and others.

   Equation (2.4) turns out now to be a somewhat unfortunate basis for studying the effects of time delays, because the more general eq. (2.5) can, in fact, manifest a wider and more astonishing range of dynamic behaviour than can eq. (2.4); eq. (2.4) happens to be a rather special case. More generally, eq. (2.5) exhibits a stable point, which as time delays lengthen (increasing T) and density dependent effects become severe (increasing z) gives way first to a simple limit cycle, and then to a cascade of further cycles (each with period roughly twice its predecessor), and eventually to apparently random or 'chaotic' fluctuations, with no discernible periodic structure (even though the underlying dynamic equations are simple and rigidly deterministic). Finally, for very severe nonlinearity ($z \gg \mathrm{I}$), the system collapses back to a single simple cycle. This bewildering array of behaviour is not yet fully understood, but in broad outline it closely parallels the phenomena recently elucidated for difference equations, which will be discussed in some detail in the next section. For recent reviews, see May (1980a,b).

   Incidentally, although I gave the IWC equations as examples, this was a bit of a tease; the values of the biological parameters p, q and z used in the whaling models put all the species well within the domain of stable point behaviour (May, 1980a).

In short, equations such as (2.4) and (2.5) constitute minimally realistic models for a single population, in which the density dependent regulatory effects (derived from maturation times, or food supply limitations, or crowding, or whatever) operate with a time delay. The consequent population dynamics can be monotonic damping to an equilibrium point, or damped oscillations, or sustained patterns of stable cycles, or apparently chaotic fluctuations, depending essentially on the ratio between T and T_R. A variety of population data can be surveyed in this light, and this is done in section 2.4 and in chapter 3.

2.3 Discrete growth (difference equations)

At the opposite extreme from section 2.2, many populations are effectively made up of a single generation, with no overlap between successive generations, so that population growth occurs in discrete steps. Examples are provided by many temperate zone arthropod species, with one short-lived adult generation each year. Periodical cicadas, with adults emerging once every 7 or 13 or 17 years, are an extreme example.

In these circumstances, the appropriate models are difference equations relating the population in generation $t+1$, N_{t+1}, to that in generation t, N_t. In contrast to section 2.2, time is now a discrete variable.

2.3.1 *Density independent growth*

The difference equation analogue of eq. (2.1) is the simple linear equation

$$N_{t+1} = \lambda N_t. \tag{2.7}$$

Here λ (conventionally misnamed the 'finite rate of increase') is the multiplicative growth factor per generation; the 'compound interest' growth rate is* $r = \ln \lambda$. Equation (2.7) describes unbounded exponential growth for $\lambda > 1$ $(r > 0)$, exponential decline to extinction if $\lambda < 1$ $(r < 0)$.

---

* Throught this volume, we follow the conventional practice of using ln to denote natural logarithms (to the base e), and log to denote logarithms to the base 10.

2.3.2 *Density dependent growth*

More generally, and more realistically, we will have a density dependent relation of the form

$$N_{t+1} = F(N_t), \qquad (2.8)$$

where $F(N)$ is some nonlinear function of N. A fairly complete catalogue of the many forms which have been proposed as discrete analogues of the logistic eq. (2.3), along with their biological provenances, has been given by May and Oster (1976). These forms $F(N)$ all share the essential features of a propensity to population growth at low densities, to population decrease at high densities, and a parameter (or parameters) which measures the severity of this nonlinear response.

Table 2.2 gives a short list of four such expressions for $F(N)$. Each case is complete with the value of the possible equilibrium point, N^*,

$$N^* = F(N^*), \qquad (2.9)$$

and of the characteristic return time, T_R, which describes how quickly the system tends to return to equilibrium following a perturbation.

The first of these expressions, form A, is that preferred by mathematicians, because it is *the* simplest nonlinear difference equation. Indeed it is startling that so simple an equation possesses the bizarre range of dynamical behaviour discussed below. However it has the biologically ugly feature that if the population ever exceeds $K(1+r)/r$, it becomes negative (i.e., extinct) in the next generation: the other three forms all have the more attractive property that population fluctuations are bounded above and below, and that the stability properties are global. Form B has an extensive pedigree in the biological

Table 2.2. Specific formulae for the function $F(N)$ in eq. (2.7).

Label	Form* for $F(N)$	Equilibrium point, N^*, from eq. (2.8)	Characteristic return time, T_R
A	$N[1 + r(1-N/K)]$	K	$1/r$
B	$N \exp [r(1-N/K)]$	K	$1/r$
C	$\lambda N(1 + aN)^{-\beta}$	$(\lambda^{1/\beta} - 1)/a$	$[\beta(1 - \lambda^{-1/\beta})]^{-1}$
D	λN^{1-b}; for $N > \varepsilon$	$\lambda^{1/b}$	$1/b$
	λN; for $N < \varepsilon$		

* For a catalogue of the original sources for these various forms, see May and Oster, 1976.

literature, and corresponds to modifying the simple eq. (2.7) with a mortality factor, $\exp(-aN)$, which becomes exponentially more severe for large N; this is a plausible model for populations which at high density are regulated by epidemics. Form C is included because it is the basis of a study of field and laboratory data, discussed in section 2.4.3 (see Fig. 2.6). Form D is the model which corresponds to the empirical method of analysing data for density dependence by plotting $\log(N_t/N_{t+1})$ against $\log(N_t)$: the slope of the consequent regression line is b.

Unlike eq. (2.3), the forms A–D and their friends and relatives do not contain any explicit time lags. However, there is a time delay implicit in the structure of eq. (2.7), in the one generation time step between the expression of the density dependent regulatory effects in generation t, and their manifestation in the census data in generation $t+1$. As above, we then expect exponential damping, or oscillatory damping, or sustained oscillations, or chaos, depending on the ratio between the time delay (which is now 'T' $=1$) and the natural response time (T_R, as catalogued in Table 2.2).

This analogy between delay-differential equations and ordinary difference equations is developed in more detail by May $et\ al.$ (1974). It sheds light on the behaviour of the general eq. (2.8) and the specific examples in Table 2.2, namely a monotonically damped stable point for $T_R > 1$ ($1 > r > 0$ in forms A and B), an oscillatorily damped stable point for $1 > T_R > 0.5$ ($2 > r > 1$ in forms A and B), and sustained but bounded oscillations for $T_R < 0.5$ ($r > 2$ in A and B), but it gives no hint of the bewildering richness of the spectrum of dynamical behaviour which has recently been uncovered in this oscillatory regime ($T_R < 0.5$, $r > 2$).

2.3.3 Stable points, stable cycles, chaos

A full mathematical and biological account of this spectrum of behaviour for nonlinear difference equations has been given by May and Oster (1976, see also May, 1974a, 1975b, 1976c), with emphasis on the generic character of the phenomenon. What follows is a brief summary, emphasizing the implications for population biology. To be definite, we will refer to the forms A and B of Table 2.2 (see also Fig. 2.6), with numerical illustrations drawn from form B.

So long as the nonlinearities are not too severe, the time delay built into the structure of the difference equation tends to be short compared to the natural response time of the system, and there is simply a stable equilibrium point at N^* [determined by eq. (2.8)].

What happens when this equilibrium point ($N^* = K$ in forms A and B) becomes unstable, as it does for $r > 2$? As this point becomes unstable, it bifurcates to produce two new and locally stable fixed points of period 2, between which the population oscillates stably in a 2-point cycle. With increasing r, these two points in turn become unstable, and bifurcate to give four locally stable fixed points of period 4. In this way there arises, by successive bifurcations, an infinite hierarchy of stable cycles of period 2^n—a cascade of period doublings— as illustrated in Fig. 2.1.

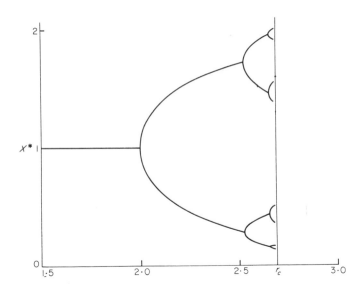

Fig. 2.1. The hierarchy of stable fixed points (X^*) of periods 1, 2, 4, 8, ..., 2^n (corresponding to stable cycles of periods 2^n), which are produced from eq. (2.8) with the form B of Table 2.2 as the parameter r increases. Each pair of points arises by bifurcation as a previous point becomes unstable. The sequence of stable cycles of period 2^n is bounded by the parameter value r_c; beyond this lies the chaotic region.

This sequence of stable cycles with period 2^n converges on a limiting parameter value, r_C say, beyond which the system enters a regime which has aptly been termed 'chaotic' (Li and Yorke, 1975). For any parameter value in this domain there are an infinite number of different periodic orbits, as well as an uncountable number of initial points for which the population trajectory, although bounded, does not settle into any cycle. In particular, Li and Yorke have proved the surprising theorem that, for *any* first order difference equation, once there exists a 3-point cycle (which happens for $r > \sqrt{8} = 2.828$ in form A, and for $r > 3.102$ in form B) then there necessarily also exist cycles of every integer period, along with an uncountable number of asymptotically aperiodic trajectories. There will, however, usually be one cycle that attracts almost all initial points (Guckenheimer *et al.*, 1976). But these cycles give way to each other kaleidoscopically with very tiny changes in the parameter values, so that the overwhelming impression is of irregular, chaotic dynamics, for all the world like the sample function of some random process.

In other words, the underlying mathematical structure of the 'chaotic' regime is exquisitely intricate, and rich in fascinating details (to the point where it has constituted a fashionable growth industry in pure mathematics in the past few years; for reviews, see May, 1976c; Guckenheimer, 1979). But the impression given by numerical simulations or casual observation of the dynamics is that a stable point gives way first to regular cycles (of periods 2, 4, 8, etc.), and then to irregular fluctuations (where initially the population alternates raggedly between 'high' and 'low' values, and eventually—for extreme linearity—shows effectively random fluctuations).

Figure 2.2 aims to illustrate this range of dynamical behaviour for various values of r in the form B of Table 2.2. Table 2.3 summarizes the various regimes for both the forms A and B, and Fig. 2.6 illustrates these regimes, as functions of the two pertinent parameters λ and β, for the form C. The form D is interesting in that (by virtue of the non-analyticity of this function) there is an abrupt transition from a stable point if $b < 2$ directly into chaos for $b > 2$, with no intermediate regime of stable cycles.

Once in the chaotic regime, the population dynamics of these deterministic models is best described in probabilistic terms. This has been done for the form B of Table 2.2, where it is seen that for largish r the trajectories are almost periodic, with an approximate period $[\exp(r-1)]/r$, as illustrated by the top trajectory in Fig. 2.2 (May, 1975b). The task of finding general methods of probabilistic description of the population dynamics in the chaotic regime has received some attention, but much remains to be done before this is in a form where it is a useful tool in the analysis of population data.

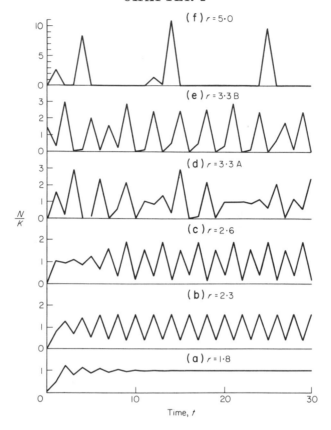

Fig. 2.2. The spectrum of dynamical behaviour of the population density, N_t/K, as a function of time, t, as described by the difference equation (2.8) with the form B of Table 2.2, for various values of r. Specifically: (**a**) $r = 1\cdot8$, stable equilibrium point; (**b**) $r = 2\cdot3$, stable 2-point cycle; (**c**) $r = 2\cdot6$, stable 4-point cycle; (d to f) in the chaotic regime, where the detailed character of the solution depends on the initial population value, with (**d**) $r = 3\cdot3$ ($N_0/K = 0\cdot075$), (**e**) $r = 3\cdot3$ ($N_0/K = 1\cdot5$), (**f**) $r = 5\cdot0$ ($N_0/K = 0\cdot02$).

Quite apart from their intrinsic mathematical interest, the above results raise very awkward biological questions. They show that simple and fully deterministic models, in which all the biological parameters are exactly known, can nonetheless (if the nonlinearities are sufficiently severe) lead to population dynamics which are in effect indistinguishable from the sample function of a random process. Apparently chaotic population fluctuations need not necessarily be due to random environmental fluctuations, or sampling errors, but may reflect the workings of some deterministic, but strongly density dependent, population model. This question is pursued further in section 2.4.3.

As a postscript, I cannot resist remarking that production of such apparently random dynamics by simple deterministic models is of interest to mathematicians (as a relatively tractable example of a multiple bifurcation process) and to physicists (where it provides a metaphor for the phenomenon of turbulence). Indeed, Feigenbaum has recently elaborated some aspects of the analysis outlined above, and applied his insights to make some predictions about the period-doubling phenomenon and the spectrum of fluctuations observed at the onset of turbulence in fluids (Feigenbaum, 1978, 1979; May and Oster, 1980). Feigenbaum's predictions seem to be confirmed by recent experiments (Libchaper and Maurer, 1980). To see mathematical ecology informing theoretical physics is a pleasing inversion of the usual order of things.

Table 2.3. Dynamics of a population described by the difference eq. (2.8) with the forms A and B of Table 2.2.

Dynamical behaviour	Value of the growth rate, r	
	Form A	Form B
stable equilibrium point	$2 \cdot 000 > r > 0$	$2 \cdot 000 > r > 0$
stable cycles of period 2^n		
2-point cycle	$2 \cdot 449 > r > 2 \cdot 000$	$2 \cdot 526 > r > 2 \cdot 000$
4-point cycle	$2 \cdot 544 > r > 2 \cdot 449$	$2 \cdot 656 > r > 2 \cdot 526$
8-point cycle	$2 \cdot 564 > r > 2 \cdot 544$	$2 \cdot 685 > r > 2 \cdot 656$
16, 32, 64, etc.	$2 \cdot 570 > r > 2 \cdot 564$	$2 \cdot 692 > r > 2 \cdot 685$
chaotic behaviour		
[Cycles of arbitrary period, or aperiodic behaviour, depending on initial condition]	$r > 2 \cdot 570$	$r > 2 \cdot 692$

In all, the message emerging from this section reinforces that of section 2.2. Depending on the ratio between the characteristic return time, T_R, and the time delay in the regulatory mechanism (here 'T' = 1), single populations exhibit either a stable equilibrium point (damped monotonically or oscillatorily) or patterns of self-sustained fluctuations (which can be periodic or chaotic).

2.4   Applications to field and laboratory data

2.4.1   *Nicholson's blowflies*

The first half of this century saw several classic experiments, in which laboratory populations in a constant and limited environment exhibited growth in accord with the logistic eq. (2.3). These experiments, on the protozoan *Paramecium*, on yeast cells, on bacteria such as *E. coli*, on *Drosophila*, and on various flour beetles, are reviewed, for example, by Krebs (1972, pp. 190–200).

Other similar single species experiments have shown sustained patterns of large-amplitude oscillations. The most notable of these is Nicholson's study, illustrated in Fig. 2.3, of the sheep-blowfly *Lucilia cuprina*.

An accurate model for this population would need to include both the details of the reproductive physiology of blowflies, and the fact that there are several distinct but overlapping age-classes. However, the essential dynamics of the system may be captured by the naive eq. (2.4), which includes the three major features: an intrinsic rate of increase, r; a resource limitation, K, set by the amount of the constant supply of ground liver; and a time delay, T, in the action of this limitation, roughly equal to the time for a larva to mature into an adult. After the x and y axes in Fig. 2.3 are scaled, the model gives a 1-parameter fit to the data. This single parameter, namely rT, is chosen to

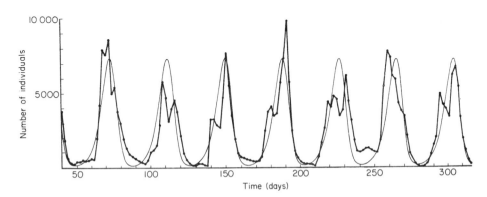

Fig. 2.3. From the one-parameter family of limit cycles generated by the time delayed logistic eq. (2·4), we display that which best fits the oscillations in Nicholson's blowfly populations (namely $rT = 2\cdot 1$: see Table 2.1). After May (1975a).

fit the observed ratio between minimum and maximum populations (see Table 2.1): the result is the theoretical curve illustrated in Fig. 2.3. From this particular choice of rT, one can infer an egg-to-adult development time of 9 days, in fair agreement with the actual figure of 11–14 days.

Although this crude and phenomenological fit to Nicholson's data has been superseded by more detailed and age-structured models (Oster and Ipaktchi, 1978; Gurney *et al*, 1980), it is encouraging to see how the simple ideas developed in sections 2.2 and 2.3 can give a good caricature of the dynamics.

2.4.2   *Responses to temperature change*

Other sets of field and laboratory data show populations with monotonic damping at one temperature, and oscillations (either damped or persistent) at another temperature. These include, *inter alia*, McNeill's field studies of the grassland bug *Leptoterna dolobrata* (see May *et al.*, 1974); Pratt's (1943) laboratory observations on population growth in the water flea *Daphnia magna*, which appears to show a stable point at 18°C and sustained oscillations at 25°C; and the laboratory studies by Beddington (1974; also Beddington and May, 1975) on the Collembola *Folsomia candida*.

Such qualitative changes in the population dynamics may plausibly be associated with changes in the ratio between T and T_R (i.e., with changes in rT) produced by the interplay between external temperature and metabolic rates.

One particularly elegant illustration is provided by Fujii's (1968) laboratory data on three different strains of the stored-product beetle *Callosobruchus*. As one goes from the population illustrated in Fig. 2.4(a) to that in Fig. 2.4(c), the intrinsic growth rate r increases and T_R decreases. Consequently the ratio T/T_R (or rT) increases, and the dynamics goes from the monotonically damped stable point of Fig. 2.4(a), to the damped oscillations of Fig. 2.4(b), to the stable cycle shown in Fig. 2.4(c).

Another nice example is given by Halbach's (1979) study of the population dynamics of rotifers; here we have a system where population growth is effectively continuous, so that differential-delay equations are the best description. As illustrated in Fig. 2.5, the dynamical behaviour appears to go from damped oscillations to sustained cycles of increasingly severe amplitude as the temperate increases. Another

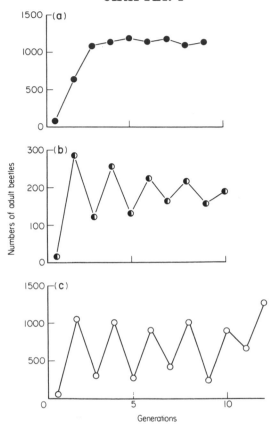

Fig. 2.4. Population changes in laboratory cultures of three different strains of stored-product beetles, displayed as number of adult beetles in successive generations: (**a**) *C. chinensis* (after Fujii, 1968); (**b**) *C. maculatus* (after Utida, 1967); (**c**) *C. maculatus* (after Fujii, 1968).

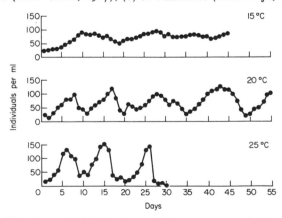

Fig. 2.5. The dynamical behaviour of laboratory populations of rotifers are shown as functions of time, for three different temperatures (from Halbach, 1979).

series of studies of a similar kind, with similar results, is by Goulden
and Hornig (1980).

2.4.3   *Insect population parameters*

From the available literature, Hassell *et al.* (1976a) have culled data on
28 populations of seasonally breeding insects in which generations do
not overlap. In each case, the data have been fitted to the difference eq.
(2.8) with the particular form C of Table 2.2 for $F(N)$, and estimates
made of the parameters λ, β and a. (The intrinsic growth factor, λ, is
first estimated independently, and then β and a are found by fitting the
census data: the above reference should be consulted for details.)

Figure 2.6 shows the theoretical domains of stability behaviour for
this difference equation: depending on the values of λ and β, there is a

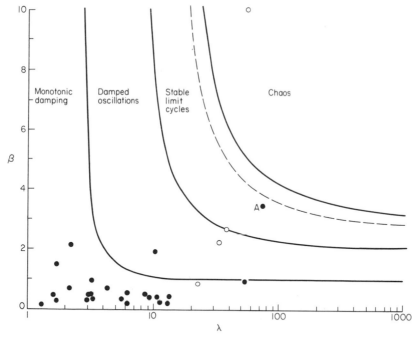

Fig. 2.6. Dynamical behaviour of eq. (2.8) with the form C of Table 2.2.
The solid curves separate the regions of monotonic and oscillatory damping to a stable point, stable limit cycles, and chaos; the broken line indicates where 2-point cycles give way to higher order cycles. The solid
circles come from the analyses of life table data on field populations, and
the open circles from laboratory populations. After Hassell *et al.*
(1976a), where details are given.

stable point, or stable cycles in which the population alternates up and down, or chaos. The points show the parameter values for the actual populations, with solid circles denoting field populations and open circles laboratory populations.

Note that there is a tendency for laboratory populations to exhibit cyclic or chaotic behaviour, whereas natural populations tend to have a stable equilibrium point. The laboratory populations are maintained in a homogeneous environment, and are free from predators and many other natural mortality factors, which may well make for exaggeratedly nonlinear behaviour, and give a misleading impression as to their population dynamics in the outside world.

In short, Fig. 2.6 can be interpreted as indicating a tendency for natural populations to exhibit stable equilibrium point behaviour. It is perhaps suggestive that the most oscillatory natural population (labelled A in Fig. 2.6) is the Colorado potato beetle, *Leptinotarsa*, whose contemporary role in agroecosystems lacks an evolutionary pedigree.

These sweeping generalizations should, however, be approached with extreme caution. Quite apart from the biases in data selection, and inadequacies in data reduction, which may be inherent in Fig. 2.6, it must be kept in mind that there are *no* truly single population situations in the real world. To subsume the population's biological interactions with the world around it in passive parameters such as λ and β may do violence to the multi-species reality. Add to this the fact that it requires less severely nonlinear behaviour to take multi-species models into the chaotic regime (May and Oster, 1976; Guckenheimer *et al.*, 1976), and Fig. 2.6 falls far short of a proof that natural populations have stable point behaviour.

2.4.4 'Wildlife's 4-year and 10-year cycles'

The natural world exhibits many instances of 3-to-4 year population cycles, particularly among small mammals in boreal regions. These have attracted much attention, from the time of Elton (1942) onward (see, e.g., Chitty, 1950; Pitelka, 1967; Dajoz, 1974).

Without attempting to explain the biological mechanisms responsible for density dependent regulation in such populations, it is plausible to assume that in strongly seasonal environments such mechanisms operate with built-in time lags of slightly less than, or of the order of, one year. For beasts with a relatively large intrinsic capacity for population growth, r (so that rT significantly exceeds unity), the upshot will tend to be stable limit cycles. As observed in section 2.2.3, regardless of the amplitude, these cycles will have periods of the order of $4T$, which is to say approximately 4 years.

Here is a detail-independent explanation of 'wildlife's 4-year cycle'.

Figure 2.7 shows Shelford's (1943) data for a population of the almost legendary lemming, along with a theoretical curve obtained from eq. (2.4) with $rT = 2.4$ (chosen from Table 2.1 to fit Shelford's numerical data on the amplitude of the cycle) and $T = 9$ months (very roughly the time from the end of one summer to the beginning of the next). Equation (2.4) is, of course, a very crude model. In particular, it necessarily gives uniformly spaced cycles with a period intermediate between 3 and 4 years, whereas the seasonal northerly environment compels the real cycles to be lock-stepped to an integral number of years, which is sometimes 3, sometimes 4 years.

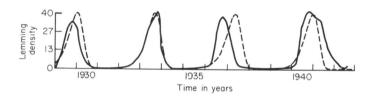

Fig. 2.7. Shelford's (1943) data on the lemming population in the Churchill area in Canada (expressed as numbers of individuals per hectare), compared with a naive theoretical curve (dashed line) obtained from the simple time delayed logistic eq. (2.4); the time delay T is taken to be a little under one year ($T = 0.72$ yr.). For further details, see the text.

This explanation is independent of the biological mechanism(s) producing the time delay. Needless to say, elucidation of these biological details remains a fascinating and relevant task.

Even more familiar to those conversant with introductory ecology texts, are the 10–11 year cycles shown by many mammals in the boreal regions of North America. The classic cases, which show pronounced and regular oscillations extending back nearly 200 years, are the snowshoe hare and the lynx, but similar cycles are exhibited by other fur bearers (muskrat, fisher, marten and others). In deliberate evocation of the title of Elton's (1942) book, Keith (1963) has surveyed much of this data under the title *Wildlife's Ten-Year Cycle*. Most of these population cycles are inferred from the recorded cycles in pelts traded by fur trappers, and there is some dispute about the extent to which the eighteenth and nineteenth century cycles in pelt counts reflect population density, as opposed to the foraging practices of the native hunters. An excellent recent discussion of the matter is by Winterhalder (1980), who concludes the population cycle in the snowshoe hare is real (though those inferred for some other animals may indeed be complicated by the foraging strategy of the Cree-Ojibwa hunters).

In this context, it is interesting that Pease *et al.* (1979) and Bryant (1980) have shown that many of the deciduous plants browsed by snowshoe hare in Alaska manifest defensive responses to heavy browsing, by producing resins or other compounds that are repellent or toxic to hares, and that the time lag in mounting this response is around 2 to 3 years. As emphasized by Bryant, such a time lag would, for the general reasons outlined above, tend to produce population cycles with a period of around 10 years. Species predatory upon the hare would then track this cycle (and may tend to amplify it).

2.5   Time-varying environments

In all the models discussed so far, the biological and environmental parameters (such as r and K) were taken to be constants. We now examine some of the effects of letting r and K vary in time, in either periodic or random fashion.

There is no difficulty in solving the logistic differential equation (2.3) for arbitrarily time-dependent $r(t)$ and $K(t)$: see, e.g., Poluektov (1974), Kiester and Barakat (1974), May (1976a). In the particular case when r is constant, the population $N(t)$ tends, after a sufficiently long time, to the asymptotic solution

$$N(t) = \left\{ r \int_0^t [1/K(t')] \exp [r(t - t')] \, dt' \right\}^{-1}. \tag{2.10}$$

The population's characteristic response time is again $T_R = 1/r$. Equation (2.10) says that $N(t)$ is the inverse of some weighted harmonic average over past values of the carrying capacity, with this average reaching a typical distance T_R into past time.

Consider the situation when the carrying capacity $K(t)$ varies periodically:

$$K(t) = K_0 + K_1 \cos (2\pi t/\tau). \qquad (2.11)$$

Here we require $K_0 > K_1$ if $K(t)$ is not to go nonsensically negative during its cycle. The behaviour of the population, $N(t)$, will now depend on whether its characteristic response time ($T_R = 1/r$) is long or short compared to the period (τ) of the environmental oscillations.

In the limit when T_R greatly exceeds τ ($r\tau \ll 1$), it may be shown that

$$\cdot \quad N(t) = \sqrt{K_0^2 - K_1^2} \, [1 + o(r\tau)]. \qquad (2.12)$$

The expression in square brackets specifies that the correction terms are of relative order $r\tau$. In this limit the population averages out the environmental variations: note, however, that this average population value is not simply the average value of $K(t)$, namely K_0, but is less than it. This approximately constant population trajectory in the limit $T_R \gg \tau$ is illustrated in Fig. 2.8(a).

Conversely for T_R short compared with τ ($r\tau \gg 1$), we have approximately

$$N(t) = K_0 + K_1 \cos (2\pi t/\tau) \, [1 + o(1/r\tau)]. \qquad (2.13)$$

The population now tends to 'track' the environmental variations, as illustrated by Fig. 2.8(b).

Similar considerations apply to situations where $K(t)$ has a more general time dependence, with stochastic components (Roughgarden, 1975a). Reechoing the themes which pervade this chapter, we find that populations with relatively large r (short response time T_R) are condemned to track environmental fluctuations, whereas those with relatively small r may average over essentially all fluctuations. More specifically, populations will tend to average over high frequency (short time scale) components of the environmental noise spectrum, and to track low frequency (long time scale) components; the transition zone occurs for frequency components $\omega \sim r$ (time scales of the order of T_R).

An investigation of the many other aspects of population behaviour in a stochastic environment is largely beyond the scope of this book (for a good review, see Turelli, 1977). Even though the population may average over environmental variations, to maintain an approximately steady value, results such as eq. (2.12) or Fig. 2.8(a) make it plain that this steady value will depend on the relative magnitude of the variations (i.e., on the ratio of K_1 to K_0), and that larger environmental fluctuations will make for lower average population values. Studies of the relation between the magnitude of environmental fluctuations and the

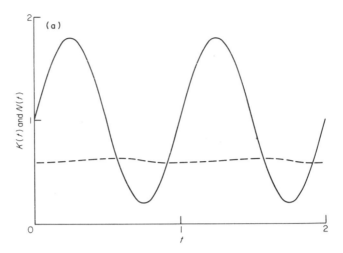

Fig. 2.8a. Illustrating the way a population [$N(t)$, dashed line] varies if it obeys a logistic equation with a time dependent carrying capacity [$K(t)$, solid line]. This figure is for the case when the population's natural response time ($1/r$) is long compared to the periodicity in the environment (τ), i.e., for relatively small r. Specifically, the figure is for eq. (2.3) with eq. (2.11), and $r = 0.2$, $\tau = 1$, $K_0 = 1$, $K_1 = 0.8$; i.e., $r\tau = 0.2$.

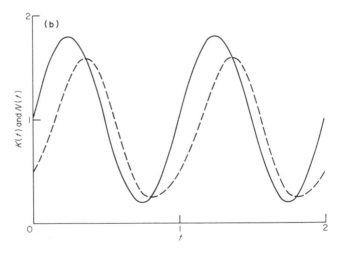

Fig. 2.8b. As for Fig. 2.8a, but in the opposite limit when the natural response time is short (relatively large r) so that the population tracks environmental variations. The details are as for Fig. 2.8a, except that $r = 10$, whence $r\tau = 10$.

probability of the population becoming extinct were initiated by Lewontin and Cohen (1969) and Levins (1969); recent developments are reviewed, e.g., by May (1975a, ch. 5 and pp. 229–231). As a one sentence summary, it may be said that temporal variations in the environment are a destabilizing influence.

Many of these considerations are relevant to the harvesting of natural populations. Earlier discussions of harvesting tended to focus on static estimates of the maximum yield that could be taken in a sustainable way, that is, on estimating the maximum sustainable yield, MSY (see ch. 16). More recently, however, several people have stressed that the ability of a population to recover from random or periodic environmental disturbances will be affected by harvesting (Gulland, 1977; Doubleday, 1976; Sissenwine, 1977; Beddington and May, 1977). In the simplest case, the characteristic return time T_R lengthens as populations are depressed below their pristine levels, and consequently stocks tend to take longer to recover from perturbations about the new, lower equilibrium levels created by 'sustainable' harvesting. It has been suggested that such effects are the cause of observed tendencies for yields of fish and whales to show greater and greater variability as harvesting efforts become more intense (May et al., 1978). The discussion has become more complicated as it has become more realistic (Horwood et al., 1979; Beddington, 1978; Beddington and Grenfell, 1979), and there are systematic dynamic differences between most fish ('r-selected species') and most marine mammals ('K-selected species'): for recent reviews, with emphasis on implications for the management of real fisheries and whaling, see Shepherd and Horwood (1979), May et al. (1978) and May (1980a).

2.6 Populations with age structure

The models considered in sections 2.2 and 2.3 pertain to the two opposite circumstances where generations overlap in a continuous manner (leading to first order differential equations) or do not overlap at all (leading to first order difference equations). But many populations have the intermediate structure of several distinct but overlapping age classes. Such is the case, for example, for most commercial fish populations, and most mammals in temperate zones. The consequent description in terms of m interlinked equations for the m age classes has many of the complicating features of multi-species situations, particularly once density dependent effects are included [i.e., the multiple age class analogues of eqs. (2.3), (2.4) and (2.8)]. The hope is that basic elements of such systems are laid bare by considering the opposite extremes of complete and of no overlap, the broad dynamical similarities between which have been the main theme of this chapter.

One result for density independent multiple age class models should, however, be made explicit. This result concerns the relation between the population's actuarial 'life tables' and its intrinsic rate of increase, r. For a single age class model [i.e., eq. (2.7)] this relation is immediate: $r = \ln \lambda$, where λ, the 'finite rate of increase', is the nett number of surviving offspring produced each year, per capita. More generally, in the density independent multiple age class case [the multiple age class analogue of eq. (2.1) or (2.7)], we need to know the effective value of r in terms of the fecundity (m_x) and survival probability (l_x) of the various age classes (x).

The exact form of this relation is well known (see, e.g., Krebs, 1972):

$$\sum_x e^{-rx} l_x m_x = 1. \tag{2.14}$$

Although this implicit equation for r may be readily solved on a computer, its biological meaning is not at all transparent, and it is ill suited to the sort of general discussion undertaken in chapter 3. An excellent approximation, the biological significance of which is much clearer, is

$$r \simeq (\ln R_0)/T_c. \tag{2.15}$$

Here R_0 is the expected number of offspring (or females) produced by the average individual (or female) over its lifetime,

$$R_0 = \sum_x l_x m_x, \tag{2.16}$$

and T_c is an average generation time, defined in straightforward statistical fashion as

$$T_c = \sum_x x \, l_x \, m_x / \sum_x l_x \, m_x. \tag{2.17}$$

The error introduced by using eq. (2.15) to approximate the exact solution of eq. (2.14) is of relative magnitude $\frac{1}{2}(\ln R_0) (\sigma^2/T_c^2)$, where σ^2 is the variance in the generation time. The approximation is good either if $R_0 \simeq 1$ (as in human populations), or if the coefficient of variation of the generation time is not too large; a more detailed justification for the use of eq. (2.15) is given by May (1976b).

2.7 Summary

Regardless of whether population growth is a continuous or a discrete process, simple density dependent models show that single populations may exhibit either a stable equilibrium point (damped in a monotonic or an oscillatory manner), or sustained fluctuations (as stable cycles or as chaos), depending on whether the natural periodicities or time-delays in the regulatory mechanisms are short or long compared with the characteristic response time of the system.

Since a population's characteristic response time is often of the order of $1/r$, these considerations introduce elements of self-consistency into the evolution of population parameters. In an unpredictable environment, there will be an advantage in having a largish r, to recover from bad times and to exploit the good times. But large r (short T_R) condemns the population to track environmental fluctuations, and makes for the sort of overcompensation which is inimical to population regulation, thus exacerbating the perceived unpredictability of the environment. Conversely, relatively small r implies a long response time, with the advantage that the population may maintain steady values and may average over environmental variations, but with the disadvantage of slow recovery from traumatic disturbances. Chapter 3 expands and applies these ideas, which are further developed elsewhere by Horn (1978).

This chapter has neglected the effects of spatial heterogeneity, which often has an important stabilizing effect. Such effects are discussed in chapters 3, 6, 10 and 11.

3

Bionomic Strategies and Population Parameters

T. R. E. SOUTHWOOD

In the preceding chapter it has been shown that many of the essential features of the growth of populations of single species in limited environments may be described by simple models. When population growth is a continuous process, an approximate model is

$$dN/dt = rN[\mathrm{I} - N(t - T)/K].$$

(3.1)

In the discrete case one has the general eq. (2.6), with, for example, the form B of table 2.2,

$$N_{t+1} = N_t \exp [r(\mathrm{I} - N_t/K)],$$

(3.2)

or the form D,

$$N_{t+1} = (\lambda N_t^{-b})N_t.$$

(3.3)

The infinite variety of nature may therefore be capable of being caricatured by the few parameters contained in these expressions. Each organism will have a bionomic strategy (i.e., size, longevity, fecundity, range and migration habit) that is summarized by the parameters of these models; this strategy will evolve to maximize the fitness of the organism in its environment. Hence the organism's habitat may be viewed as a templet against which evolutionary forces fashion its bionomic or ecological strategy (Southwood, 1977a). This chapter establishes the general form of the relationship between habitat type and bionomic features, as expressed by the impact of habitat on the parameters in the above expressions.

3.1   The parameters

Three parameters may be distinguished:

(1) N^*, the population at equilibrium. Normally in a single species situation this will be $N^* = K$, the carrying capacity.

(2) Time delays, T (the time delay in the response due to some

environmental lag) and τ (the generation time). These time delays have equivalent effects (May *et al.*, 1974), but only τ is directly responsive to evolutionary pressures on the species itself.

(3) The finite rate of increase of the population, $\lambda = \exp r$, itself the difference between gains and losses. This determines the system's return time or natural response time, T_R, which is the time it takes to return to equilibrium, following a disturbance (May *et al.*, 1974). It is a property of the system: for eqs. (3.1) and (3.2), $T_R = 1/r$, and for eq. (3.3), $T_R = 1/b$ (see Table 2.2).

3.2   Habitats

Habitats may be classified according to a number of characteristics:

(1) *Duration stability*, the length of time the particular habitat type remains in a particular geographical location. Thus, for example, a large tree in a climax forest may last for hundreds of years, a herb in successional vegetation for only a handful of seasons, a dung pat for a few weeks. Clearly the significance of this duration stability depends on the relationship between the organism's generation time (τ) and the length of time the habitat remains favourable (H).

(2) *Temporal variability*, the extent to which the carrying capacity (K) of a habitat varies during the time that site is tenable by the organism (i.e., temporal heterogeneity). Variations in K may be predictable, like the spring flush of plant growth in northern deciduous forests, or *un*predictable, like desert rainstorms and are due to changes in the physical environment. When such unfavourable periods become lengthened the habitat can be regarded as adverse (Whittaker, 1975a). Diapause and other mechanisms involving resistance to water loss or temperature extremes are characteristic bionomic strategies for species subject to adversity selection [or 'stress-selection' (Grime, 1977)].

(3) *Spatial heterogeneity*, continuity versus patchiness. In a tropical rainforest any particular tree species will have a very patchy, scattered distribution; but in northern coniferous forests the same one or two species may continuously cover thousands of square miles.

For any particular animal, the habitat may be defined as that area accessible to the trivial movements of the food-harvesting stages. The range of the animal's movements will therefore determine the scale of the habitat; for the *Drosophila* larva the ripe fruit of a tropical forest tree is a temporary habitat and if it survives to adulthood it will migrate to another site, but for the orang-utan the whole forest is a stable and permanent habitat. Great longevity will tend to reduce the significance of temporal variability and increase the degree of predictability of a given location.

An initial discussion of habitat characteristics may be simplified by confining attention to duration stability; that is, by limiting attention to the value of the ratio τ/H for the habitat.

In those species where τ/H approaches unity, one generation cannot affect the resources of the next; there will be no evolutionary penalty for overshooting the carrying capacity of the habitat. These species are then exploiters, opportunists and (using the terminology of MacArthur and Wilson, 1967) may be referred to as r-strategists.

Conversely, for those animals that occupy long-lived habitats where the carrying capacity (K) is fairly constant, significant overshooting will lower K, and will adversely affect subsequent generations. Many other species will have colonized such stable habitats, and hence interspecific competition in all its forms, including predation, is likely to be intense. Such species are referred to as K-strategists. They are selected for harvesting food efficiently in a crowded environment (MacArthur and Wilson, 1967); their greater competitive ability is associated with reduced rates of population growth (Luckinbill, 1979).

3.3 Size and its implications

A number of extremely significant bionomic characters are closely linked with the size of an organism. In responding to the evolutionary pressures arising from its habitat, a modification of the organism's size will inevitably move it one way or another on the r–K continuum of strategies.

A key parameter is clearly r, the per capita rate of increase. As discussed in chapter 2, r is effectively dependent on the nett reproduction rate, R_o, and the generation time, T_c:

$$r \simeq (\ln R_o)/T_c. \tag{3.4}$$

There is a strong positive correlation between size and generation time in organisms ranging from bacteria to whales and redwoods (Bonner, 1965; see Fig. 3.1). This relationship is probably due to longevity being inversely proportional to total metabolic activity per unit of body weight. At the cellular level the rate of cell growth and the length of the cell cycle seems to be positively correlated with the DNA content of the nucleus (its 'C-value') (Cavalier-Smith, 1978).

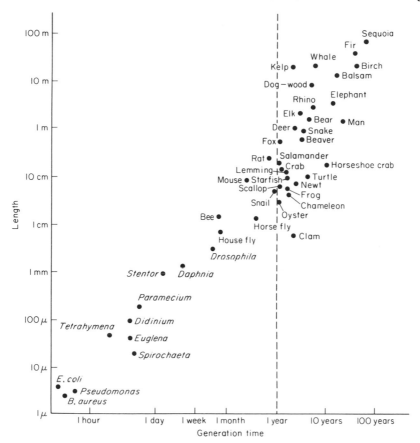

Fig. 3.1. The relationship of length and generation time, on a log-log scale, for a wide variety of organisms (from Bonner, 1965). From John Tyler Bonner, *Size and Cycle: An Essay on the Structure of Biology.* (© 1965 Princeton University Press). Fig. 1 p. 17. Reprinted by permission of Princeton University Press.

A consideration of eq. (3.4) immediately shows that r tends to be more sensitive to changes in the generation time than to changes in R_o. Halving the generation time will double r; but doubling R_o (e.g., by doubling fecundity) will only increase r by the difference of the values of the natural logarithms (i.e., R_o from 5 to 10 would lead to changes in r from 1·6 to 2·3). It is therefore not surprising that the relationship of r to T_c for a wide range of organisms falls into a narrow straight band, with slope −1, when plotted on a log-log scale (Heron, 1972; see Fig. 3.2). This train of consequences of size change may be expressed as

$$r \propto \frac{1}{T_c} \propto \frac{1}{\text{size}} \propto \text{metabolic rate per unit weight.}$$

Indeed Fenchel (1974) has shown that the relationship of r, and of metabolic rate/unit weight, to organism size are strikingly parallel (Fig. 3.3), both falling to three groups; unicellar, poikilotherms and homeotherms. Each major evolutionary step slightly increases both the metabolic rate and r for a given size.

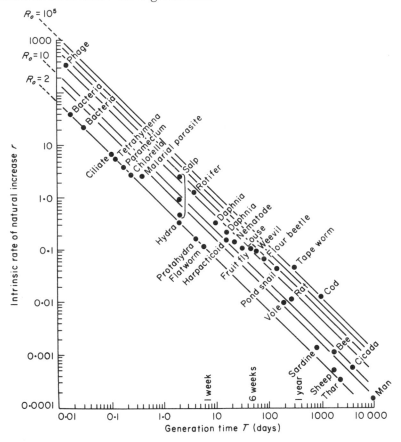

Fig. 3.2. The relationship of the intrinsic rate of natural increase and generation time, with diagonal lines representing values of R_o from 2 to 10^5, for a variety of organisms (from Heron, 1972).

Size change has other implications for the ecology of the species. Larger species are at an advantage in interspecific competition (lions may drive hyaenas from their kill), and in defence from predators (May, 1978a). The allometric growth of offensive or defensive appendages will enhance this ability and, combined with longevity, allows the possibility of a high level of parental care and protection.

The size of an animal will influence the scale of its habitat (May, 1978). To return to the example of two species living on tropical fruits, the size of the orang-utan means that the range of its trivial movements encompasses many types of tree, so that the habitat is predictable and stable. The *Drosophila* larva, on the other hand, is limited to a few centimetres around the oviposition site. The adult stage of *Drosophila*

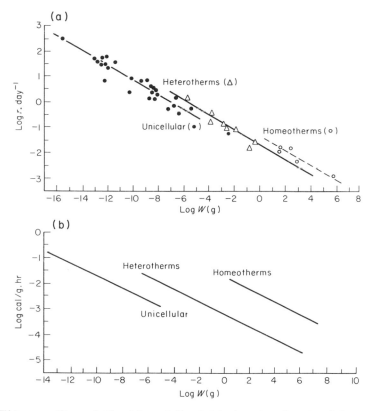

Fig. 3.3. The relationships of the intrinsic rate of natural increase (**a**) and metabolic rate (**b**) to weight for various animals (from Fenchel, 1974).

migrates to new food sources, and indeed many small insects can travel great distances on air currents (Johnson, 1969). But small wingless animals are perforce restricted in the scale of the migrations they can undertake on land. The regular long-distance migrations of the wildebeest on the savanna cannot be emulated as an ecological strategy by a small rodent.

3.4   Fecundity versus longevity

The energy available to an organism capable of reproduction may be directed towards the survival and growth of that organism, or towards the production of offspring, or towards some partitioning between the two. As with size, any change in the allocation of energy will influence the species' bionomic strategy, and hence the population parameters.

The problem can be illustrated by reference to a hypothetical and simplified organism, the parthenogenic 'block-fish' (Fig. 3.4). Each summer the productivity of each block fish is two blocks. These may be added to itself or used in reproduction. Any fish that does not add to itself dies. Every winter each 'block-fish' gains one block; breeding does not occur at this season. As shown in Fig. 3.4, there are two mutations. Mutation 1 puts all its summer productivity into reproduction, and at the end of three years is represented by eight 1-block-fish. Mutation 2 only puts half its summer productivity into reproduction, and at the end of three seasons is also represented by eight individuals; but these show an age and size range from a 7-block-fish to four 1-block-fishes.

From this information we cannot tell which mutation will be successful. Let us consider various possible conditions imposed by the environment.

(1) Carrying capacity limited to a total of 8 blocks. Mutation 1 would remain as shown, but in mutation 2 only the large 7-block-fish and its small 1-block daughter would survive. The population of mutation 2 is thus more vulnerable to extinction, whilst the population of mutation 1 could easily 'bounce back'; even the loss of half its individuals would not affect its size next year.

(2) Half the juveniles (1-block-fish) are killed by predators each year. If this happens, mutation 1 remains as one 1-block-fish, whereas mutation 2 has a population of three (7-, 3- and 1 block).

Various other conditions can easily be envisaged and their outcome determined. For example: (a) block-fish over a certain size are able to exploit additional resources (K measured in standing crop of blocks is increased); (b) larger block-fish become less efficient, and so their productivity falls below three blocks per year; (c) the productivity is a proportion of size; (d) mortality is higher on larger block-fish; (e) the block-fish has a strict size limit (which is realistic for the majority of organisms other than fish and plants). In this last case (e), surviving individuals simply become older; often as in most higher vertebrates, this is accompanied by increased competitive advantage (for example in holding territory).

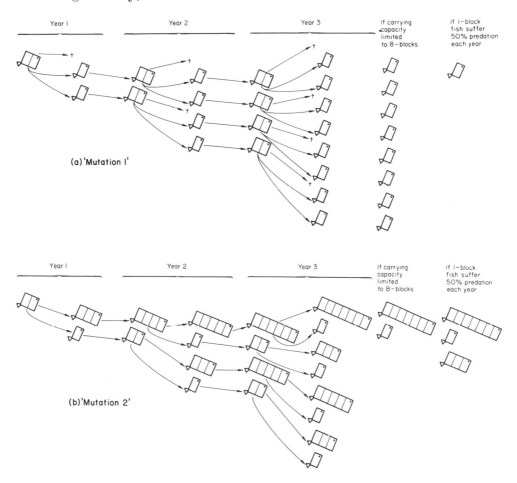

Fig. 3.4. The outcome of two different strategies for allocating resources in a hypothetical animal, the block-fish: (**a**) Mutation 1 in which summer productivity (2 blocks) is entirely allocated to reproduction, (**b**) Mutation 2 in which one block of summer productivity is allocated to reproduction and one block to growth and, hence, to adult survival.

The optimum allocation of resources between reproduction and maintenance, illustrated above, has been explored algebraically by several workers (e.g., Schaffer, 1974; Smith and Fretwell, 1974). Their conclusions are essentially those that are intuitive from the above: the optimum strategy depends on the different competitive advantages (including, of course, ecological efficiency) and survival probabilities of the various ages and sizes. Experimental evidence that high rates of reproduction early in life lead to lower survival and less subsequent reproduction has been obtained with the grass, *Poa annua* (Law, 1979).

The allocation of resources to territory holding may be viewed in the same way. In general it relates to condition (e) above: but sometimes, as in the lizard *Sceloparus jarrovi*, larger individuals hold larger territories (Simon, 1975). An analysis of the costs of territory defence in relation to food availability has been made in the sunbird *Nectarinia reichenowi* (Gill and Wolf, 1975a; Pyke, 1979) and of flock foraging in bats (Howell, 1979) and birds (Caraco, 1979a, b).

3.5   The K-strategy

The K-strategists have a stable habitat (τ/H is very small), and consequently they evolve towards maintaining their population at its equilibrium level, and towards increasing their interspecific competitive ability. Thus they will often be selected for large size, which will increase generation size and lower r. But this will not be a disadvantage: the evolutionary pressures are for the population to remain at or close to K, but not above it—this could lead to habitat degradation. High levels of fecundity are thus not essential if the reduction in births can be matched by increased survival. Plants and animals that are K-strategists make a significant investment in defence mechanisms. Parental care is facilitated by low fecundity (small litters or clutches but large size offspring), longevity and size. This reduction in mortality may be considered to lead to more efficient use of energy resources (Cody, 1966): indeed extreme K-strategists like the Andean condor, the albatross and large tropical butterflies (e.g., *Morpho*) are noteworthy for their low energy movement, namely gliding.

However, when K-strategists do suffer mortality and perturbations, their populations need to return quickly to equilibrium levels or competitors may seize the resources. This implies that b in eq. (3.3) should approximate to unity, and, because there is little mortality, this will tend to be accomplished through the birth rate. In other words the birth rate will be very sensitive to population density, and will rise rapidly if density falls (Southwood *et al.*, 1974). In many vertebrates this is achieved by increased litter and clutch size and by 'bringing' non-breeders into breeding, either by a shortening of the prereproductive period or by a modification of the social structure that had previously excluded them from territories (Wynne-Edwards, 1962; Southwood, 1970). The numbers of small trees that are found in forests, growing rapidly only when a clearing arises, provide another strategy for the same ends.

K-strategists can therefore be recognized by large size, longevity, low recruitment and mortality rates, high competitive ability and a large investment in each offspring (Pianka, 1970; Southwood *et al.*, 1974; Rabinovitch, 1974). Their population levels will stay close to the equilibrium level and their mating tactics will be geared to this density, e.g. the communal nuptual displays of birds such as the ruff, birds of paradise and the blackcock (Wynne-Edwards, 1962) and harems, whose size will maximize the longevity of the individual female (Elliott, 1975). In terms of molecular biology, Cavalier-Smith (1978) has suggested that K-strategists may be recognized by their high DNA content (high DNA C-value).

K-strategists are unlikely to be well adapted to recover from population densities significantly below their equilibrium level, and if depressed to such low levels they may become extinct. These organisms, rather than r-selected species, need the concern of the conservationist. The fossil record shows that many lines of animals increase continually in size until extinction: this has been called Cope's rule. These lines have become progressively more K-selected; more and more closely adapted to a specialized and hitherto stable habitat (Bretsky and Lorenz, 1970). Thus the extinction of the dinosaurs was probably due to their inability, because they were extreme K-strategists, to respond to the changes in climate at the end of the Cretaceous (Axelrod and Bailey, 1968; Southwood *et al.*, 1974).

3.6   The *r*-strategy

The r-strategists are continually colonizing habitats of a temporary nature (τ/H is not small), and they are exposed to selection at all population densities. Their strategy is basically opportunistic, 'boom and bust'. Migration will be a major component of their population process, and may even occur every generation (Southwood, 1962; Dingle, 1974; Southwood *et al.*, 1974; Kennedy, 1975).

Selection will favour a high r, arrived at by a large fecundity (large R_0) and short generation time (T_c); the potential of prokaryotes for extremely short generation times makes them the supreme r-strategists (Carlile, 1980). As the habitats they colonize are often virtual ecological vacuums, high competitive ability is not required, and they will typically be small in size. Mortality rate may be high; migration, which is an essential component of their fugitive existence, is invariably wasteful. Their main defences against predators, other than a high fecundity, are often a measure of synchrony (and hence temporary satiation of the predators) and their mobility—a hide and seek strategy. With plants this relative lack of defence is very neatly shown by Cates and Orians' (1975) studies on the palatability to slugs of plants from various successional stages (Table 3.1). Only those plants later in the succession make a significant investment in chemical or physical defence.

Table 3.1. Average palatability to slugs, *Arion* and *Ariolimax*, of herbs from various successional stages (after Cates and Orians, 1975).

Plant community	Number of plant species tested	Palatability index* for *Arion*	Palatability index* for *Ariolimax*
Early successional annuals and biennials	18	0·99	0·96
Early successional perennials	45	0·69	0·77
Later successional and climax plants	17	0·40	0·46

* Palatability index is defined as log (amount of test material eaten)/log (amount of control eaten).

Very high values of r lead to instability (Levins, 1968; May, 1974a, 1975a), but the extreme r-strategist is by definition not likely to be affected (Barclay, 1975). A population will not spend many generations, perhaps only one or two, in that particular habitat, and migration is such a capricious process that only a small fraction of mortality will be density dependent. Thus unrealistically high values of r would be necessary if they were, of themselves, to produce instability (Southwood, 1975). In other words, at the extreme end of the r-spectrum populations come into being in a new (or newly discovered) habitat with a few colonies; they pass through at the most a handful of generations when the models expounded in chapter 2 apply; and then as the habitat deteriorates (perhaps overcrowded with strong density dependence acting), migrants depart and establish new populations, freed of density restraints.

Because r-strategists occur at such varying densities, mate-finding tactics are likely to be efficient at low numbers and, although extinction will regularly be the fate of individual populations (as their habitats change), the species as a whole will be very resilient. Furthermore, their high mortalities, wide mobility and continuous exposure to new situations are likely to make them fertile sources of speciation.

3.7 Population dynamics and habitat characters

Given the differences in population parameters outlined above, the population growth curves of the extreme r- and K-strategists will be of very different forms (Fig. 3.5). The K-strategist will tend to have a stable equilibrium point (S in Fig. 3.5), to which the system returns after moderate disturbances; but if the population declines below some lower threshold (in Fig. 3.5) it cannot recover, and it decreases ineluctably to extinction. Conversely, the r-strategist's population grows

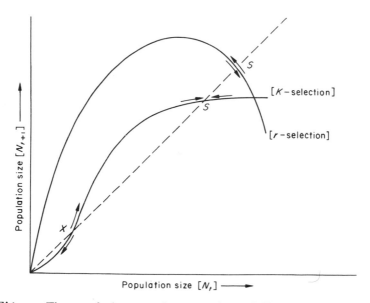

Fig. 3.5. The population growth curves of r- and K-strategists, showing stable (S) equilibrium points and an unstable point, the extinction point (x) (from Southwood *et al.*, 1974).

rapidly at low densities, has an equilibrium point (S in Fig. 3.5) about which it is liable to oscillate, and crashes down from high densities. Natural enemies are unlikely to be important for these extreme strategies: for the r-strategist because enemies will be unlikely to colonize in sufficient numbers sufficiently quickly; for the K-strategist because of its large size and high competitive ability. Most organisms are, of course, in an intermediate position and here natural enemies will have a significant role, and may provide at least one further equilibrium point part-way up the population growth curve. This concept is developed in chapter 5.

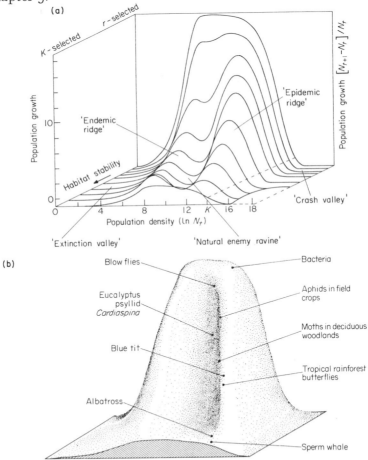

Fig. 3.6. The synoptic model of population dynamics showing **(a)** the general features of population growth in relation to population density and habitat duration and **(b)** the positions of some organisms on the landscape. **(a)** after Southwood and Comins, 1976.

On the basis of the accumulating evidence from natural populations, it is possible to construct a three-dimensional synoptic model of population dynamics (Fig. 3.6, after Southwood, 1975; Southwood & Comins, 1976). Besides illustrating the points made above as to the role of habitat, this model stresses that r- and K- are a convenient shorthand for the ends of a continuum and not the branches of a dichotomy.

In addition to the variations shown by the synoptic model (which is, of course, deterministic) other patterns will arise in species population dynamics due to spatial and temporal heterogeneities in the habitat. As already stressed, for an animal the extent of the spatial heterogeneity of a given area will be influenced by the size and mobility of the animal. In general, spatial heterogeneity tends to be stabilizing. Variations in many environmental factors are unlikely to occur synchronously in all the patches, and thus the species 'spreads the risk' through its occupancy of many small habitats (den Boer, 1968; May, 1974b). However, if the numbers in these populations drop to very low levels, so that population size can no longer be regarded as a continuous variable, then a new factor, the stochastic variation of small integer numbers, enters. This

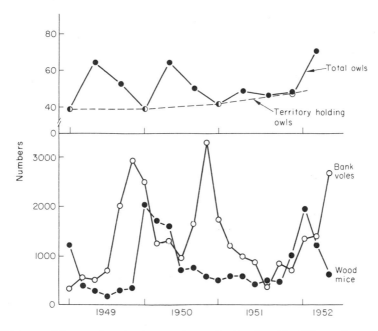

Fig. 3.7. Fluctuations in the numbers of animals with different longevities in the same habitat. (Tawny owls, Bank voles, and woodmice in Wytham Wood, 1948–52: data from Southern, 1970.)

increases instability and the chances of random extinction. The extinction of a Californian colony of the checkerspot butterfly, *Euphydryas editha* (Ehrlich *et al.*, 1975) and the many British colonies of the Large Blue Butterfly (*Maculinea arion*) (Muggleton and Benham, 1975) may be examples of this.

Temporal heterogeneity in the habitat will increase instability (Blackith, 1974; May, 1974b, 1976a), but again large, long-lived animals with low r-values and other strategies will average out the variations: species with short generation times will track the variation of the environment. The different population traces for tawny owls, voles and mice living in the same woodland illustrate the different scale and amplitude of fluctuations in different animals with different strategies; see Fig. 3.7.

3.8 Comparison of strategies between major taxa

As large organisms are basically K-selected and small ones r-selected, it is not unreasonable to consider vertebrates as K-selected and insects as r-selected (Pianka, 1970). It is interesting to note that the overall conditions prevailing at different periods of geological time probably encouraged the evolution of these different groups. The main bursts of evolution in the vertebrates occurred in the Jurassic, the lower Cretaceous, the Eocene and the Oligocene, periods of warm, moist and stable climates. The insects, in contrast, evolved rapidly in the Permian and Triassic when climatic conditions are thought to have been very variable.

All taxa however attempt to colonize the range of habitats available, although in this process they are, of course, encumbered with attributes due to their evolutionary history. Thus although one must go to the vertebrates and seed plants for the most extreme examples of K-selection, and to insects and bacteria for extreme r-strategies, within each group a spectrum has evolved. This is shown in the next section.

3.9 Comparisons of strategies within taxa

3.9.1 *Birds*

A measure of the reproductive effort of birds is easily obtained from the clutch-size, and there has already been considerable controversy over its significance (Wynne-Edwards, 1962; Lack, 1966, 1968; Mountford, 1973). Much of the disagreement seems to arise because of the different mechanisms operating in the K- and r-selected species (Cody, 1966; Southwood et al., 1974).

Examples of some of the most extreme r-selection in birds are provided by the zebra finch, *Taeniopygia castanotis* and the budgerigar, *Melopsittacus undulatus*, which are nomadic in Central Australia and have some of the shortest times to breeding. Another species with a short generation time (small T_c) is the Japanese quail, *Coturnix coturnix japonica*, a migrant. The largest clutch size (R_0) recorded by Lack (1968) for a passerine is that of the blue tit, *Parus caeruleus*. This bird may also be regarded as opportunistic: its breeding season food, moth larvae in deciduous woodlands, fluctuates greatly and its own population can show a two-fold change over a single season. All these birds are small, or small for their orders: the budgerigar is among the smallest of the parrots and the Japanese quail is tiny for a game bird.

At the other end of the spectrum are some excellent examples of K-selection. The large condors, *Gymnogyps* and *Vultur*, lay only a single egg every other season. The conservation of both is a matter of concern. *Vultur vultur*, with a wing span of over ten feet, is the largest of birds of prey, and this family (Cathartidae) contained *Teratornis*, the largest flying bird known from the fossil record. Less endangered, probably only because its breeding sites are relatively free from human disturbance, is the wandering albatross, *Diomedea exulans*. Again it only breeds in alternate years (when successful) and has a clutch size of one and a wing span of over ten feet; its period of immaturity (9 to 11 years) is the longest for any bird. One of its breeding populations, on Gough Island, is considered to have remained almost constant at 4,000 since 1889 (Eliott, 1971). All the Procellariiformes, the order to which the albatross belongs, and the related Sulidae are noteworthy for their low r values, arising from long pre-reproductive periods (large T_c), and low clutch size (small R_0). They select breeding sites free of predators and forage over great areas of ocean (Lack, 1968). This K-strategy can be accounted for in terms of individual Darwinian fitness (Goodman, 1974). Foraging for squid and fish may be considered to require experience, and therefore if young birds bred they would be unable to support their own offspring: indeed there is evidence that they often cannot support themselves (Jarvis, 1974). Recent studies on the South African gannet, *Sula capensis*, have shown that although birds may rear two young these are lighter than chicks from single clutches,

and their chances of survival through the post-fledgling period, the
time of heaviest mortality, are significantly reduced (Jarvis, 1974).
In other words the populations of these species are close to the effective
carrying capacity of the habitats around the breeding grounds; a pair
of birds cannot easily double the amount of food harvested. With a
small clutch size there will be strong selection against any activity
that lowers adult survival. The position is similar to strain 2 of the
'blockfish', with heavy mortality (> 50 per cent) of the juvenile 1-block-
fish. The population will depend on survival of the older block-fish to
reproduce in several years, and it will be seen that doubling their annual
reproduction would not be successful if it halved the survival of the
adults.

Goodman (1974) calculated from a model for *Sula sula* that fledgling
success needed to be increased at least 19 times more than the fractional
amount by which this increase was at the expense of decreased parental
survival: this model underestimated the function because it did not
allow for the high post-fledgling mortality, since shown by Jarvis (1974).
High fledgling loss is also a feature of the tawny owl, *Strix aluco*,
another *K*-strategist (Southern, 1970); see Fig. 3.7. Therefore any
tendency for parent birds to decrease their survival by foraging for too
large a clutch size, or by breeding away from the 'protected colonies',
will be selected against. There is thus no conflict between this *K*-strategy
and natural selection as it is normally understood. These birds are
exploiting habitats that, on the scale on which they forage, are very
stable; they cannot greatly increase the rate at which they harvest
resources. The long life span of adults makes the population very resistant to variations in recruitment, but recovery from abnormal adult
mortality is difficult. These are characteristics of *K*-strategies.

The more the habitat varies, the larger the clutch size; the species
can exploit the additional food by faster population increase. This
general tendency is illustrated in Table 3.2, which shows the relations
of clutch size to habitat for passerine birds.

Table 3.2. Clutch size in passerine birds in relation to habitat (from
Southwood *et al.*, 1974).

Habitat	Number of species	Average clutch size
Tropical forests	82	2·3
Tropical savanna, etc.	260	2·7
Tropical arid areas	21	3·9
Middle Europe	88	5·6

3.9.2 *Insects*

In common with a number of other invertebrate groups, insects have a complex metamorphosis which allows the possibility of any individual having more than one habitat during its life. However, the major part of the food harvesting normally occurs in the larval stage, and thus it is normally the character of this habitat that dominates the species strategy.

Lepidoptera

Various tropical butterflies have been noted for their longevity and/or the stability of their populations: *Morpho* (Young and Muyshondt, 1972); *Heliconius* (Ehrlich and Gilbert, 1973); *Charaxes* (Owen and Chanter, 1972); *Hamadryas* (Young, 1974). These butterflies are often large and territorial. Some *Morpho* may take up to 10 months to reach maturity. These characteristics place them towards the K-selected end of the spectrum. Adult survival is relatively high and, like the sea birds discussed above, the number of eggs laid at any one time is small (compared with most *Lepidoptera*), although the total for the whole adult life span is not dissimilar to that of related species. The wide distribution of eggs in space and time allows these species to compete successfully in habitats where there is heavy competition from predators and other herbivores, and where resources are scattered and limited. Again like the sea birds, their populations are resilient to fluctuations in recruitment.

Relatively migrant pest species exhibit r-strategies. The army worm, *Spodoptera exempta*, may lay up to 600 eggs: generation time is just over 3 weeks and large populations (outbreaks) occur on the young growth of graminaceous plants (Brown, 1962).

Parasitic Hymenoptera

As larvae, members of this group are mostly internal parasites of other insects. In the Ichneumonidae the females place the eggs in or on the host's body: if the host is inaccessible the ovipositor will be very long, as with *Rhyssa* that bores through timber to reach its host, wood-boring sawflies. Generally, the longer the ovipositor the more effort the female has to expend ovipositing each egg and, although less precisely correlated, the better protected the host (and hence the parasite larva) from predation. Within Ichneumonidae, Price (1972, 1973) has found a close relationship between a long ovipositor, lowered fecundity and

increased egg size. More specifically, he has shown that with the same
host, the sawfly *Neodiprion*, the fecundity (expressed as number of
ovarioles per ovary) falls the later in the host's development the attack
is made; see Fig. 3.8a. When this curve is compared with a survival
curve for the host (Dahlsten, 1967, see Fig. 3.8b), the correlations
among high host availability, low host life expectancy, and high parasite
fecundity are apparent.

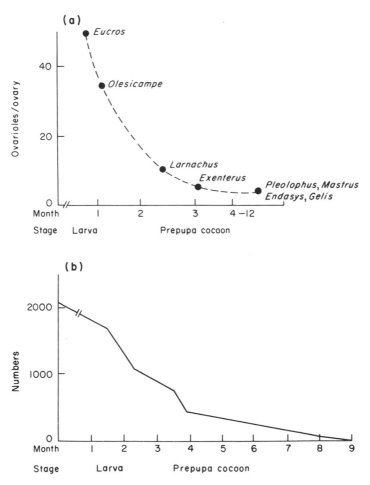

Fig. 3.8. The fecundity of various parasitoids in relation to the survival
of their host, the sawfly *Neodiprion*, at the time of parasitism: (**a**) the
number of ovarioles per ovary in various parasitoids against the
chronology and stage of attack (after Price, 1972); (**b**) survivorship curve
for a species of *Neodiprion* in California (after Dahlsten, 1967).

Sarcophagidae

This family of dipterous flies are commonly known as flesh flies and the viviparously produced larvae of most species are carrion feeders— strongly opportunistic and *r*-selected. However, *Blaseoxiphia fletcheri* larvae live in the fluid in the tubular leaves of pitcher plants. This is a much more stable habitat. Forsyth and Robertson (1975) found that the larval populations of this fly remained close to the carrying capacity of the habitat, the larvae were territorial, the fecundity was lower than normal (11 larvae per female compared with 50–170) and the individual first stage larvae much larger (6·9–7·0 mm compared with a range of 1·2–5·4 mm for other species). This provides an excellent example of how the stability of the habitat influences the range of characters, covered by the shorthand '*r–K* continuum', in a relatively narrow taxon.

3.9.3 *Plants*

One of the characters of a *r*-strategy is the high allocation of energy to reproduction, whilst *K*-strategists allocate more to mortality avoidance and competitive ability. Following the lead of Harper and Ogden (1970), several recent studies have assessed the resources plants allocate to reproduction. For example, the dandelion, *Taraxacum officinale*, exists in various biotypes; the more disturbed and transient the habitat, the larger the proportion of those biotypes that devote more of their biomass to reproduction (Gadgil and Solbrig, 1972).

Another plant with a range of habitats is *Polygonum cascadense*. Hickman (1975) found that the allocation of resources to reproduction was highest in those habitats where species diversity was lowest, vegetative cover was least and sap potential low. This is illustrated in Fig. 3.9.

A number of different species of goldenrod, *Solidago*, occur in eastern USA in sites ranging from woodlands to open dry disturbed sites, which are early successional stages. The biomass allocation to flowers and stems in these two extremes are compared in Fig. 3.10 (Abrahamson and Gadgil, 1973). This clearly shows the greater allocation to reproduction in the more unstable habitat, a distinction that is greater where two different species (*S. rugosa* and *S. memoralis*) are involved than between different forms of the same species (*S. speciosa*).

Reedmace or catstails (*Typha* spp.) are plants of damp habitats, and seeds will only colonize new habitats, because *Typha* litter inhibits germination (McNaughton, 1975). Thus an *r–K* selection continuum

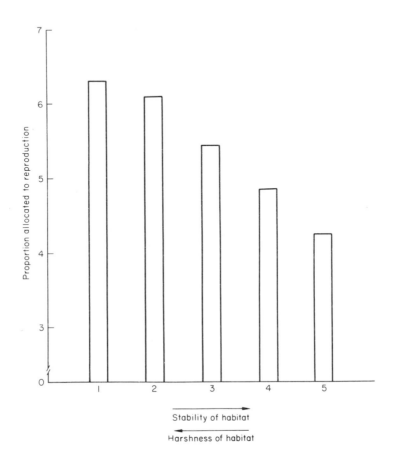

Fig. 3.9. The proportion of total weight allocated to reproductive organs in five populations of the plant, *Polygonum cascadense* in Oregon (data from Hickman, 1975).

might be expected in relation to the extent of colonizing opportunities, i.e., habitat duration (τ/H). McNaughton found strong evidence for such a continuum. In the most unstable habitats *Typha* species had genotypes with greater developmental speed (small τ), with higher fecundity (high R_o), but with less energy in each offspring (i.e., smaller size). These genotypes had high colonizing ability, but low competitive ability: the genotypes in more stable habitats showed a reversal of these traits.

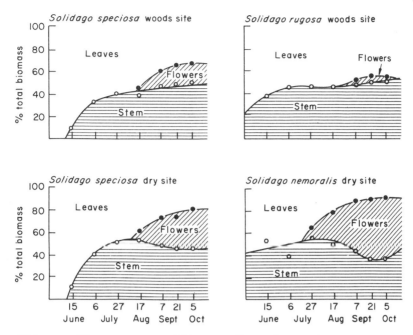

Fig. 3.10. The percentage of biomass allocated at various seasons, to stems and flowers in four goldenrod populations; the species (*Solidago speciosa*) that colonizes both types of habitat shows the differences, but to a less marked extent, that are found between the two species with more limited habitat range (from Abrahamson and Gadgil, 1973).

3.9.4 *Conclusion*

Thus one can see how the interrelated bionomic characters of an organism, such as size, longevity and fecundity, have evolved to give a pattern of population dynamics that is adapted to the features of its habitat. The organism's 'perception' of the scale of its habitat's heterogeneity in both space and time is itself influenced by these bionomic characters. All these variables interact to produce a broad continuum of strategies from small opportunists to large dominants, conveniently referred to as an *r–K* continuum. Within any taxa a portion of this continuum will be exposed; in addition to those referred to above it has been recognized in other groups from kangaroos (Richardson, 1975) to corals (Loya, 1976) and earthworms (Satchell, 1980). The temporal variability and adversity of the physical environment provide another type of selection, another axis, that of adversity (Whittaker, 1975a) or stress (Grime, 1974, 1977). The adaptations for that axis may confuse the consideration of a group purely in terms of r- and K-selection, but the two axes may be arranged to provide a 2×2 matrix with four main types of ecological strategy (Southwood, 1977) or three, if like Grime (1977) habitats of short duration stability and high adversity are excluded (Note: R-selection of Grime $\equiv r$-selection, C-selection of Grime $\equiv K$-selection and S-selection of Grime $\equiv$ adversity selection).

However, every species is to some extent a prisoner of its evolutionary history; the processes of mutation and selection govern the rate with which it can respond to changes in the features of its habitat. As the environment changes, often particularly rapidly under man's influence, the extent of adaptation may be incomplete. The interlocking features of bionomics, ecology and habitat as we see them today are, as it were, but a single frame in the film of the evolution of the biosphere.

4

The Concept of Population in Modular Organisms

JOHN L. HARPER

4.1 Introduction

Most of theoretical ecology is concerned with organisms that have unitary structure—the zygote develops as a determinate form through a life cycle that has a juvenile, a reproductive and a senescent phase. The individual is unitary, the number of its parts is determinate—the frog has four legs, two eyes; the wasp has six legs, four wings, two eyes and two antennae. Although some unitary animals are constructed by repetition of segments (*Taenia*, millipedes) the determinate character of the individual is still characterized by the specialization of segments and when injury occurs (as in planarians or lobsters) regeneration normally restores the integrity of the unit. The population biology of such unitary organisms is readily represented by counts of the number of units: the size of the population can be expressed by a symbol N where every member of the population represents an original zygote (or at least a single-celled stage) from which the unitary structure has been reproduced. A facetious way of expressing this very fundamental matter is to point out that one can determine the number of rabbits in a field by counting their ears and dividing by two or their legs and dividing by four—the result is the number of zygotes that have survived. There is no divisor that makes such a calculation possible from the leaves of a higher plant, the fronds of a fern, the zoids of an ascidian or a bryozoan.

Modular organisms, by contrast, develop from the zygote by the repeated sequential iteration of units of construction. The fundamental iterated units together usually form characteristically branched structures which may themselves be reiterated, as in the repeated (and regenerated) branch systems of a tree. The product of a zygote (genet) in modular organisms is usually branched, fixed and, except for a juvenile phase, immobile. My definition of module is deliberately wide: a repeated unit of multicellular structure, normally arranged in a branch system. It therefore includes a polyp, a hydroid or a zoid and in plants may be a leaf with axillary bud, a meristem, a shoot system, a ramet (or indeed a whole branch system, where such whole branch structures are themselves repeated in development from a zygote). The word 'genet' was coined deliberately to avoid the need continually to refer to the genetic individual—product of a zygote or single cell stage— when contrasting it with the terms 'ramet', 'shoot', 'tiller', 'hydroid', 'polyp' which describe the clearly recognizable units of construction that compose a genetic individual (Kays and Harper, 1974). Van Valen (1978) realized that such an approach could unify the ways in which we think of colonial animals and plants. Papers in a recent symposium on colonial invertebrates (particularly Rosen, 1979) show this view rapidly gaining hold.

Important properties follow from modular patterns of development and growth each of which has profound effects on the ways in which demographic and evolutionary theory is formulated.

4.2   The growth of populations of modules

There are two distinct levels of populational organization in modular organisms—the number of genets (products of zygotes or single cells) and the number of modules composing each genet. It is convenient to distinguish these two levels by symbols N and η respectively. We can then write two equations that describe populational events:

$$\eta_{t+1} = \eta_t + \text{module births} - \text{module deaths} \qquad (4.1)$$

which describes the *growth* of genets, and

$$N_{t+1} = N_t + \text{Births} - \text{Deaths} + \text{Immigrants} - \text{Emigrants} \qquad (4.2)$$

which describes the changes in *numbers* of genets.

A different population biology and demography is appropriate to the level of genet and module (N- and η-demography). The distinction is most clearly made by considering a zygote of the duckweed (*Lemna* spp.) or *Hydra* developing in a culture. After a period of growth each culture will consist of a number of *Lemna* fronds (or *Hydras*) which represent a genetic individual that has continually fallen apart as it has grown. Such an η-population grows at a rate determined by the births of new modules and deaths. The development of such an η-population is

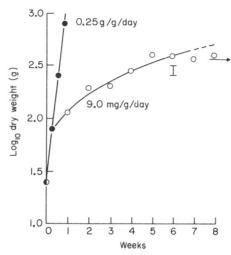

Fig. 4.1. The growth of a population of fronds of the duckweed *Lemna gibba* in nutrient culture. The 'population' in this case is made up of the parts of a single genet which continually separate from each other.
● growth under conditions in which self-crowding is prevented by repeated harvesting.
○ growth in self-crowding cultures.
→ yield per culture after 12 weeks.
(From Clatworthy and Harper, 1961.)

illustrated in Fig. 4.1 in which a culture of *Lemna gibba* passes through an exponential growth phase limited by the intrinsic growth rate, followed by a linear phase of resource limited growth (determined by the rate at which the culture as a whole intercepts limiting resources—in this case, light) to a stationary phase at which the birth and death rates of fronds are balanced. In such a sequence of phases we are seeing *the growth of a single plant* as a populational process. Of course, if the culture was started with several zygotes the population of fronds (or hydroids) developing in the culture would represent several genets, each in turn represented by a number of modules. The culture would then contain both levels of population structure (N = the number of genets present $\times \eta$ the number of modules per genet).

Most plants and modular animals do not fall to pieces as they grow: *Lemna* and *Hydra* are free-living aquatics and far from typical. Most modular organisms are fixed (attached or rooted) and the iteration of modules produces a more or less structured 'colony' spreading in 2-dimensions to form a mat or carpet or in 3-dimensions to form a canopy of modular units (gorgonian corals, herbs, trees and shrubs). Sometimes genets may retain their identity and be counted, particularly when all modules from a genet are held together on some common trunk or stem. However, in many laterally spreading herbaceous perennials, suckering trees and corals, parts of a genet or clone may die leaving physically and physiologically separate parts of one genet wandering (by growing) through a community of other plants or zoids. It may then be quite impracticable to recognize and count genets in the field. The difficulty becomes apparent the moment one attempts to count genets of white clover in a pasture, genets of aspens in a grove of suckering trees or of bryozoans on a kelp, though in all these cases the modules present may still be counted. N-demography is often much more difficult in the field than η-demography.

In some modular organisms physical fusion and physiological continuity may be established between genets, as when graft unions occur between the roots of neighbouring trees or between related clones of a modular animal such as the colonial ascidian, *Botryllus schlosseri* (Sabbadin, 1972). After fragmentation or physical union of genets there may be arguments about what constitutes a 'physiological individual' but the genet remains clearly defined as the product of a zygote or single celled stage—even if it may be difficult to recognize in the field.

In plants which depend on obtaining mineral resources and water from the substratum, a root system is usually present, either developing as an extension of the main axis of the genet or developed anew from successive modules or even serving as the source of new shoot modules (root budding or suckering). Root systems are again populations of modules and have populational properties often with a very rapid flux of births and deaths. To a physiologist, the presence of a root system on a part of a genet may give it some physiological individuality but it does not confer genetic status different from a single shoot on a tree or of a zoid in a colony. Natural selection will act on the products of zygotes and differentiate between them with respect to the number of descendant zygotes that they leave—defining their fitness in the process. In modular organisms the number of modules produced will directly affect the potential for leaving progeny and potential descendants. In some environments fitness may be maximized by the genet remaining whole and integrated and in others by breaking down into physiologically independent units. In some environments the fittest genotypes may be expressed as phenotypes with a single root system or attachment and in others by phenotypes in which individual modules form their own distinct root system. These contrasts in life cycle and morphology do not affect the fact that natural selection acts on the zygote-to-zygote transition (Williams, 1975). It is confusing and misleading to contrast sexual and vegetative reproduction. The relevant contrast is between reproduction and growth. Reproduction takes place from a single cell and in modular organisms most of the growth takes place by iteration (see Bonner, 1974).

The recognition of two distinct levels of populational process (N and η) in modular organisms avoids a great deal of semantic confusion from the use of the word 'individual' in discussing the ecology of such organisms. It is however a contribution to theory only in being 'A conception or mental scheme of something to be done, or of the method of doing it; a systematic statement of rules or principles to be followed' rather than 'A hypothesis proposed as an explanation' (*The Shorter Oxford English Dictionary*, O.U.P.)!

4.3 Form

The characteristics of modular organisms that are used by taxonomists are almost always features of the module rather than of the whole. Thus it is the shapes of flowers and leaves that are the major taxonomic characters used by botanists concerned with flowering plants (not the shape of the whole plant), and it is often the shape of a hydranth or a zoid (not the shape of the colony) that is taxonomically most useful for zoologists concerned with modular animals. The form of the whole genet is usually taxonomically unimportant—because it is highly plastic, whereas the form of the module is relatively highly canalized (and 'conservative' for the taxonomist). Nevertheless, the form of the whole genet may be quite distinctive and species specific. It is easy, for example, to distinguish the winter outline of an oak tree from that of a beech, though very much more difficult to find forms of words that express this difference.

The form or architecture that is produced by the repeated iteration of modules of construction is clearly of great importance in determining the fitness of a genet. A great many of the properties that in unitary organisms involve 'behaviour' and movement, such as the search for food and mates and escape from predators and competitors, are represented in spatially fixed modular organisms by aspects of form. The exposure of flowers to pollinators, the overtopping of neighbours in a canopy and the invasion of soil zones by roots are effected by the growth and placement of iterated modules. Although a shoot system may be said to 'search' for light and a root system to 'avoid' zones of drought, these activities involve the directions of growth—not chasing or running away.

The architecture of an array of modules is determined by a few characteristics, most notably branch angles, internode lengths and the probabilities that a bud will die, remain dormant or develop to produce daughter buds. The form of a tree is the accumulated record of the births and deaths of a population of buds arranged in space according to more or less precise rules. Recently Bell has developed models of the

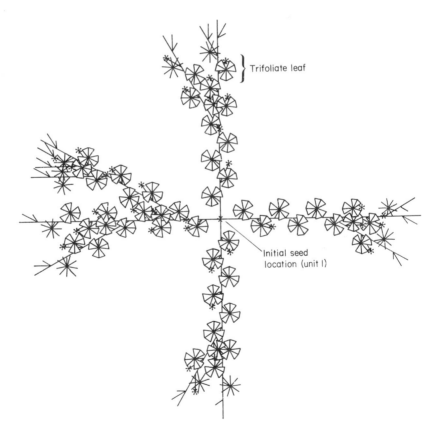

Trifoliate leaf

Initial seed
location (unit 1)

Fig. 4.2. Computer simulated growth form of *Oxalis corniculata*. The early branching of this plant produces 4 stolons at 90° to each other and later branching is at narrower angles. (Details and the computer programme are given in Bell, 1979.)

graphic computer simulation of growing plants (Fig. 4.2) including rhizomatous perennials and trees; the models are equally well suited to describe the growth of an animal colony such as that of *Obelia* (Bell, 1974; Bell, 1976; Bell, 1979; Harper and Bell, 1978). This procedure immediately makes the study of form in modular organisms both quantitative and analysable and may well represent a major growth point in population biology of modular organisms comparable to the role of behaviour studies in the population biology of unitary organisms.

The way in which a population of modules is packed can have major effects on the way that a genet exploits environmental resources and interacts with neighbours. Clegg (1978) has described 'phalanx' and 'guerilla' growth patterns in herbaceous plants. Phalanx species have short, horizontal internodes and the tillers or vegetative branches are packed in a clump (e.g. *Deschampsia caespitosa* and most other tussock grasses, *Mercurialis perennis*). Guerilla species bear longer internodes and develop growth forms in which the modules are spaced far apart, wandering and spreading. Clearly, phalanx and guerilla are two ends of a continuous spectrum of growth forms that can be described quantitatively (e.g. in Bell's models). A genet with phalanx form grows as an advancing front, monospecific and monozygotic and may present an impenetrably occupied zone of land and resources. At the same time, such a form exaggerates the mutual shading of parts of the same genet ('narcissistic' competition) and confines the extent of root proliferation: the effect is to exaggerate the likelihood that some parts will reduce the resources available to other parts of the same genet. Presumably some optimality model could be written that takes account of the balance between such conflicts of costs and benefits.

At the other end of the spectrum, a plant with guerilla growth form (e.g. *Ranunculus repens, Trifolium repens*) puts more resources into stolons, rhizomes, etc. that separate individual growth modules, which do not then enter significantly into each other's zones of resource depletion. Such growth forms allow a single genet to spread over an area quickly. In a closed community such a growth form maximizes the chance that a genet will make intraspecific and interspecific rather than intraclonal contacts.

A very similar spectrum of growth forms can be recognized among sessile marine invertebrates. Buss (1979) contrasts forms that develop as 'runners' with those that form plates or sheets. He suggests that runner-like forms should exhibit higher growth rates and recruitment rates but a lower capacity to interfere with neighbours than sheet-like

forms. In a quite different group of organisms, species of lichens growing
on rock surfaces, crustose (mat-forming) and foliose (aerially branching)
forms represent a similar spectrum of growth forms (Pentecost, 1980).
When crustose forms meet there may be trucial zones of contact or
overgrowth of one form by another. Truces occur most often between
mats of the same species. Foliose forms (leafy, lobate with unattached
lobe ends) have the higher growth rates and can usually slide over the
tightly adpressed mats of the slow growing (0.03–9 mm per annum)
crustose forms.

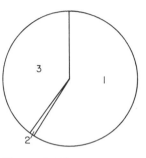

Thymus drucei
in dune grassland

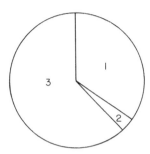

Lotus corniculatus
in neutral grassland

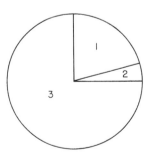

Ranunculus repens
in neutral grassland

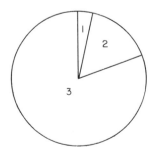

Hieracium pilosella
in dune grassland

Fig. 4.3. The frequency of leaf contacts made by plants of various
species in their natural habitats.
1 Intraclonal contacts.
2 Interclonal but intraspecific contacts.
3 Interspecific contacts.
(From Clegg, 1978.)

A fixed organism may have three types of neighbour and these can be defined in terms of their distance apart or the frequency with which they make physical contact. Figure 4.3 shows the leaf contacts that are made by four species of plant in their native habitats (Clegg, 1978). In each case the architecture of the plant is the major factor determining whether the neighbour of a leaf is likely to be on the same genet, on a conspecific or on a quite different species.

In the same way, the apically dominated growth form of some conifers (e.g. *Picea sitchensis, Abies* spp.) give the tree a conical structure in which dense cones of needles do much shading of each other but, high in the canopy where most of the photosynthetic action is taking place, rarely shade the needles of other trees. The contrasting architecture of an oak or a liane places bundles of canopy over and into the canopy of neighbours.

One odd consequence of fixed modular growth is that the various modules of a single genet may interface with or interlace among quite different neighbours so that quite different biotic selection is experienced by its various parts. The descendant zygotes left by a single genet will be the sum of the zygotes left by the different modules. Thus one side of an oak tree—abutting onto a birch—may have high fecundity while the other side, meeting the canopy of another oak, may be heavily suppressed. The two sides of the tree contribute quite differently to its fecundity and the selection pressures on the two sides may be quite different. Consequences of this to selection theory are not explored!

4.4 The age structure of a genet

Modular organisms have an internal age structure—parts are born at different times and deaths may occur throughout the life of a genet. Sabbadin (1972) says of the modular colonial ascidian *Botryllus* 'As a rule, there are three coexisting generations in the colony, adults and two successive generations of buds, and a gradual increase of the number of zoids, each of them giving rise to two or more buds.' Similarly a herb or a tree bears leaves or buds that have been developed (I see no harm in using the word 'born') at different times. An age structure for the population of leaves on a pine tree is shown in Fig. 4.4 and for shoots in a population of a sedge in Fig. 4.5. As a leaf or shoot ages its physiology changes and so does its contribution to the rest of the plant

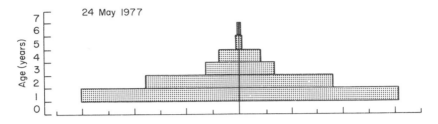

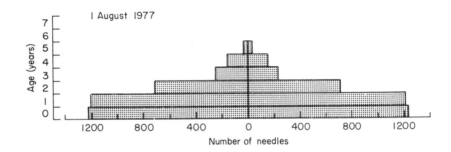

Fig. 4.4. The age structure of the leaf population on an open grown tree of *Pinus nigra*. (From Maillette, 1979.)

to which it is attached. Initially an importer of minerals and carbon assimilates, it becomes an active exporter of assimilates; this activity then declines and the leaf ends its life as an exporter of much of its mineral content (Hopkinson, 1964; Harper, 1977). A module has a life history and a changing physiological value. The growth of the whole genet is a product of the age structure and the age specific activity of its modules (Harper and Bell, 1979; Peters, 1980).

The age structure of a plant at the η-level can often be followed quite easily so that the effects of neighbours or of the physical environment can be measured in demographic terms. Figure 4.6 shows the effect of abundance or deficiency of mineral nutrients on the age structure of the leaf population of plants of *Linum usitatissimum*. A rather similar analysis has been made for the effects of the proximity of neighbours on the same species (Bazzaz and Harper, 1977). Such age structure is important not only to the plant for which its different parts have different value but also to a grazing animal for which young and old leaves may be nutritionally quite different.

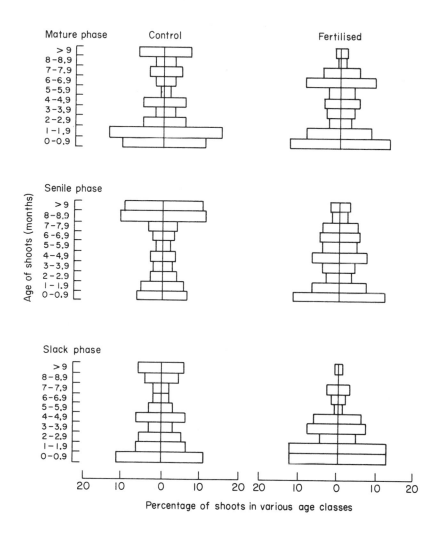

Fig. 4.5. The age structure of shoot (tiller) populations of *Carex arenaria* in three phases of a sand dune succession in July 1976 at Aberffraw, N. Wales. The fertilized areas had received NPK compound fertilizer in the previous year. In unfertilized plots the population in the juvenile phase of the succession contain a considerable number of old shoots, and this category dominates the senile plots. The effect of fertilizer application is to increase the proportion of juvenile shoots and greatly to reduce the density of old shoots. (From Noble *et al.*, 1979.)

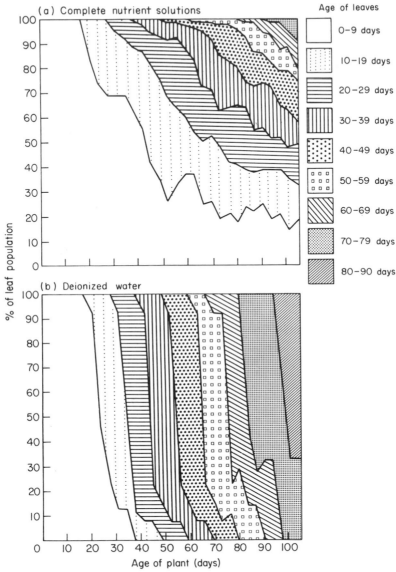

Fig. 4.6. The changing age structure of the leaf canopy of *Linum usitatissimum* during the growth cycle. (**a**) in full nutrient solution (Long Ashton). The plants accumulate leaves as they develop so that after 100 days a full age spectrum is represented in the canopy. (**b**) in deionized water and so dependent on the minimal resources present in the seed. Each cohort of leaves has only a short life and at any point in time only 2–3 age cohorts are present. (Original data of Harper, J.L. and Sellek, C.)

4.5   Polymorphism in modular organisms

The two levels of populational organization in plants and modular animals are represented by two levels of polymorphism. Conventional genetic polymorphism is represented at the N-level as differences *between* genets, e.g. the sexual polymorphism of dioecious plants (hemp, asparagus) and the colour of polymorphisms within a species and within a population of ascidians (Sabbadin, 1973). Such polymorphisms are between genets. Polymorphism at the η-level is seen in the division of labour between sexual and vegetative hydroids *within* a genet of *Obelia*, between long and short shoots on a tree, between male and female flowers on a monoecious hermaphrodite plant (maize or hazel) and between flowering and vegetative shoots occurring side-by-side on the same rhizome on a perennial herb, e.g. Iris.

4.6   The consequences of fixity of position

Most modular organisms develop spatially as fixed structures attached by rooting in soil or cemented to rocks, like kelps or as epiphytes. Many ecologically important consequences follow such fixity of position. Most particularly, individual genets develop in strict spatial relation-ship with specific neighbours (this is true of very few unitary organisms, though mussels and barnacles are rather conspicuous exceptions). In organisms with a fixed life style the concept of 'density' loses much of its value. Indeed, a botanist looking at the theoreticians' concern with 'density-dependence' and 'frequency-dependence' can see them as artificial concepts that have been forced on students of mobile organ-isms by the technical difficulty of discovering which individual interferes in the activity of another! The density of mobile animals is no more than a convenient but very crude statistical abstraction of the effects that individual organisms have upon each other. When such abstraction can be avoided, as in the analysis of the spatial organization of indi-viduals within a flock of wood pigeons (Murton *et al.*, 1966), it becomes clear that the concept of density disguises the real level at which organisms interact, rather than defining any profound quality of popu-lation behaviour. No such abstraction is needed in the case of fixed or rooted organisms where the depletion of resources (light, water, nutrient) by one individual is likely to affect the level of resources available only to its immediate neighbours. Zones of resource depletion may be created by the activities of any individual and the zone for most resources is narrow (e.g. it is unlikely that a root removing phosphate from the soil will affect the concentration more than 40–100 μm away from the surface of that root). The precision of neighbour relationships in a plant community is such that Mack and Harper (1977) were able to account for 69% of all variation in the fecundity of individual sand dune annuals from a knowledge of the size and species of the neighbours within 2 cm distance. Any generalization of such effects into terms of the density of plants per unit area is bound to blur or even entirely obscure the real level at which interaction between organisms is occurring. New developments in the theoretical ecology of plants and of other modular organisms need to take more account of the spatial positions of genets and modules than to spend time searching for yet more generalized density relationships. The development of theory that takes into account the behaviour of individual genets is particularly appropriate at a time when it is recognized that evolutionary forces need to be interpreted at the level of individual fitness rather than at the population or community level.

4.7   Generalized density phenomena

Although fixed and modular organisms react to the proximity of immediate neighbours, a great deal of energy has been given to establishing generalized density dependent relationships for higher plants. Most of the 'theoretical population biology' of higher plants has been of this kind. The reason for this emphasis is that a vast quantity of data exists in the literature of agriculture and forestry relating yield to sowing or planting density.

Two general principles emerge from analyses of these and other data: (a) *The 'law of constant final yield'*. As individual plants in a population grow, the yield of dry matter per unit area rapidly becomes independent of the number of genets present: yield (dry matter) reaches a plateau over wide ranges of sowing or planting density (Kira *et al.*, 1953; Shinozaki and Kira, 1956; Harper, 1977). This relationship is usually expressed as $W = K_i d^{-1}$ or $Y = Wd = K_i$ where W is mean plant weight, d is density, Y is yield per unit area and K_i is a constant. The form chosen for these equations emphasizes density, i.e. the number of genets (seeds sown or trees planted) per unit area. In line with the arguments in the previous section it may be more appropriate to emphasize the reciprocal of density, the space available per plant or the distance from neighbours, rather than the density and express the law of constant final yield as $W = K_i s$ where s is the space available for plant. This focusses attention more clearly on the space and resources available to individual plants and away from the holist concept of population density.

The law of constant final yield is an empirical law. It holds true partly because of the enormous plasticity in development of individual genets. Variations in the resources available to a plant (due for example to the proximity of competing neighbours) can result in gross differences in dry weight and fecundity. A single genet of the annual weed *Chenopodium album* may develop as an unbranched depauperate individual 5–10 cm high and bearing 1–4 seeds or, if isolated from neighbours and well provided with nutrients, form a much branched structure 100–150 cm high and bear 100,000 seeds. Such plasticity is largely a consequence of modular growth. Depauperate individuals are those in which few modules have been iterated and/or many have aborted or died. Vigorous growth (and high fecundity) reflects a high birth rate of modular parts and/or a low rate of modular abortion and death. At extremely high densities the death rate of modules exceeds the birth rate and whole genets are then lost (thinned) from the population.

The law of constant final yield breaks down only if the plants are so young or far apart that potentially limiting resources remain uncaptured or if the plants are so close together that gross disorders develop, such as an epidemic of a pathogen or the weakening and collapse of stems that results in the lodging of an over-dense crop of cereals.

(b) *The 3/2 thinning law*. This concerns the relationship between the numbers of plants that survive to maturity in a plant population and the size (mean weight) to which they grow: $W = Kd^{-3/2}$ or $Y = Wd = Kd^{-3/2} d = Kd^{-\frac{1}{2}}$ (Yoda *et al.*, 1963; White, 1980b), where W — mean weight per surviving plant, d = density, K is a constant and Y is yield per unit area. Again, the emphasis can usefully be switched from density to space per plant, emphasizing individual rather than population performance: $W = Ks^{-3/2}$ where s = area captured per plant. The relationship is illustrated in Fig. 4.7 in which each arrowed line represents the time course of a population starting from a particular sowing density. In populations that are started into growth at high densities (more properly, individuals that are sown close together) plants

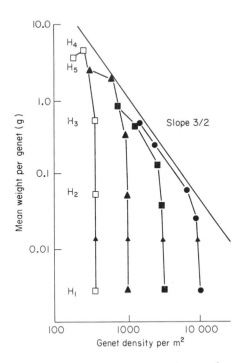

Fig. 4.7. The changing relationship between the number and weight of surviving plants in populations of the grass *Lolium perenne* sown at densities of 300, 1000, 3000 and 10,000 seeds per m². Each arrowed line represents the time course (H₁–H₅) of mortality and of the weight of survivors. A thinning line of slope −3/2 is drawn in to indicate the limiting condition to which all populations tend over time. (From Kays and Harper, 1974.)

quickly deplete the resources available to their neighbours and deaths soon occur. The populations quickly begin to track the 3/2 self-thinning line. At lower densities plants are further from neighbours and grow for longer before interfering with them. The approach of the population to the 3/2 thinning line is then delayed. At very low densities or in artificially thinned populations such as forestry plantations, the thinning line may never be reached. White (1980) and Gorham (1979) have listed more than 80 published sets of data for species ranging from mosses through herbaceous plants to forest trees for which there is remarkable conformity to the 3/2 thinning law (Fig. 4.8). Significant exceptions occur only when incident light intensity is very much reduced, which suggests that mutual shading may be a factor that controls this remarkable regularity in the behaviour of plant populations. The two processes, death of some individuals and continued growth of others, are closely interlinked. Neither can be specified as the variable that is dependent on the other. In a self-thinning population, if individuals

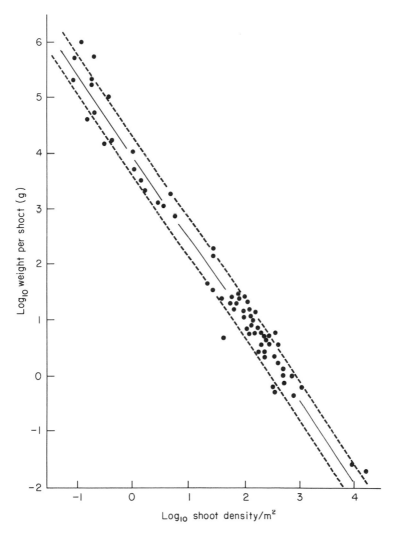

Fig. 4.8. The relationship between shoot dry weight and shoot density in 65 single species populations representing 29 different plant species ranging from shoots of annual and perennial non-woody species to saplings and full-sized trees. The dashed boundary lines represent plants half and twice as large as those represented by the equation $W = 9670d^{-1.49}$. (From Gorham, 1979.)

grow faster, more will die; if more individuals die, the survivors will grow faster.

It would be good to see some attempts to explain the generality of the 3/2 thinning law in terms of the physiology of growing and dying plants and it may be that this will come from studies of the way in which the birth and death rates of plant parts are affected by close neighbours. It is also important to ask how this generalization, which is made from studies of even-aged stands of single species, can be extended to mixtures of ages or species and also to populations of plants that are subject to predation. One experiment has been made (Bazzaz and Harper, 1976) to determine the relationship between growth and death in a mixed population of two plant species (*Lepidium sativum* and *Sinapis alba*). The relationship between the numbers and the mean weight of survivors (the mean of all plants of both species taken together) conformed closely to the 3/2 thinning law (Fig. 4.9) but nearly

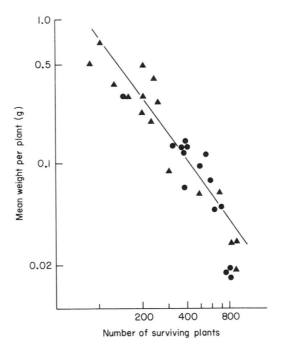

Fig. 4.9. The relationship between mean plant dry weight and the number of survivors in a population of *Sinapis alba* and *Lepidium sativum* sown in equiproportioned mixture. Grown at low ● and high ▲ fertility. N.B. The values shown are for the mean weight of plants of both species taken together. (From Bazzaz and Harper, 1976.)

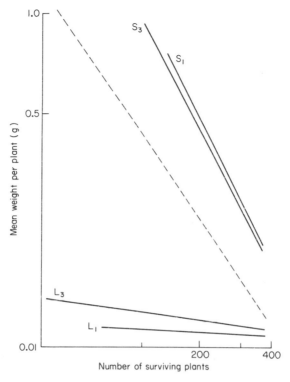

Fig. 4.10. Regression slopes of mean dry weight per plant on the number of surviving plants in a mixed population of *Sinapis alba* (*S*) and *Lepidium sativum* (*L*) (cf. Fig. 4.9) grown at low (S_1, L_1) and high (S_3, L_3) fertility. (From Bazzaz and Harper, 1976.)

all the mortality was suffered by *Lepidium* and nearly all the growth was made by *Sinapis* (Fig. 4.10). This suggests that processes of vegetational change might be analysed as multispecies participation in a mutual thinning process.

In a dense and naturally self-thinning population of plants the activities of predators might be expected to influence the way in which the thinning law operates. Again just one relevant experiment has been made. Dirzo (1979) sowed populations of pure stands of *Capsella bursa-pastoris* and of *Poa annua* at high densities at which natural self-thinning occurred. The populations were exposed to grazing by the slug *Agriolimax caruanae*. This slug tends to chew *Poa annua* at ground level and the damaged plants often die. The effect of predation was therefore significantly to reduce the density of survivors and these then grew larger than plants in populations with no slugs. In contrast, slugs rarely kill *Capsella*; they chew the lower rosette leaves so that grazed plants are smaller than ungrazed. These smaller plants had a greater chance of survival than those in the ungrazed populations where the survivors were bigger but fewer. It appears therefore that the 3/2 thinning law describes an upper limiting condition which may not be exceeded by any combination of surviving plant numbers and weights. If the growth rate of survivors is increased (e.g. by fertilizer application, Yoda *et al.*, 1963) fewer but bigger plants survive. If the growth rate of survivors is decreased (e.g. by leaves being grazed) more, but smaller, plants survive.

The generality of the 3/2 thinning law is not satisfactorily explained —it is derived from empirical data not from theory. Very simple models of growth and mortality in plant populations at different densities have been developed by Watkinson (1980) and Aikman and Watkinson (1980). They modify the logistic growth differential equation so that the increase in weight of an individual plant depends on the area available to it rather than on its weight. The effective area is reduced by an empirical function $f(si)$ with two terms, one of which expresses the constraints imposed upon the increasing total area of plants by the limited physical area of the plot in which they are growing. The other term allows for an increasing advantage or disadvantage for plants of varying sizes. Simulations produce growth and thinning behaviour that are highly realistic including close approximations to a slope of $-3/2$. The fact that we can now model a thinning system does not, of course, imply that we know how it works. No experiments have been made to study self-thinning under drought conditions. We do not know whether a similar or the same thinning law applies to populations of modular animals. It seems a reasonable guess that the law will apply to corals which, with their algal symbionts, are liable to suffer from mutual shading.

Not only is the slope of the thinning line for plants regularly close to $-3/2$ but the value of the proportion constant K lies in the same very narrow range for all species so far studied. Of published data from 65 plant stands, three-quarters of the values lie within the range $K = 4900-19500$ (Fig. 4.7). Presumably some aspect of packing theory underlies the generality but at present this can only be stated in the form of a truism; 'Only widely spaced plants, which can develop a large spatial area and so grow to great heights, are able to accumulate large amounts of biomass'. 'The remarkable fact is that a single simple rule relating shoot weight to the standing crop to density should apply over so many orders of magnitude and to plants as diverse in their architecture as mosses, ferns, gymnosperms, monocotyledons and dicotyledons' (Gorham, 1979). In the case of trees the law may reflect fundamental allometric relations between height, canopy area and basal area. It is not clear, however, that the same interpretation can be put on the thinning process among herbaceous plants with quite different forms such as grasses.

A further and very odd feature of the self-thinning law is that the relevant measure of the density of survivors is the number of genets not that of modules per unit area. Even in plants which grow clonally by the repeated iteration of aerial shoots or tillers it is not the surviving density of shoots but that of genets that fits the 3/2 thinning law (Kays and Harper, 1974). Clones of the rhizomatous *Mercurialis perennis* and other rhizomatous species do not develop to self-thinning densities (Hutchings, 1979). It is as if the internal control (correlative inhibition) of the placement of shoots in such a species, by regulating the lengths of rhizomes between shoots, holds the parts of a clone below a self-thinning density. Several new questions start here. What happens to the thinning process when two genets meet and the internal regulation within a genet has, added to it, interference from a neighbouring genet? What happens to the self-thinning process as it affects the birth and death of shoots within a canopy and at the junction between canopies? Does the regulation of module densities within a modular organism represent the beginning of a sociobiology of plants (all the modules of a genet share the same genotype, so all the conditions for the evolution of social structure are met with perfection!)?

The −3/2 thinning law is formulated as a relation between the number of genets in a population and their mean weight (occasionally mean volume in the case of trees). The mean weights are usually calculated by weighing a whole population and then dividing by the number of plants present; this procedure hides all the interesting variation between individual plants. Characteristic shifts occur in the frequency distribution of plant weights in a dense and self-thinning population. Initially normally distributed weights of seeds and seedlings become strongly skewed (and perhaps in some cases bimodal, Ford, 1975; Rabinowitz, 1979; Aikman and Watkinson, 1980). The survivors tend to be recruited from the larger and increasingly dominant members of the population (those composed of the most modules) and these will usually be the main contributors of progeny towards subsequent

generations. Although the study of means has revealed some fundamental properties of populations very relevant to understanding
community productivity it is, of course, the study of variance that is
most relevant to understanding evolution.

4.8   The predation of modular organisms

Organisms with modular organization are treated by predators in a
wholly different way to prey of unitary form. A predator that feeds on
a modular organism usually takes only parts and leaves a potentially
regenerating residue. I use the word 'predator' for herbivores in the
same way that Jackson (1977) used the word for organisms that feed on
colonial animals. It is extremely rare for a predator to eat a whole
plant—even a slug that eats a seedling will usually leave the below-
ground parts and, when germination is hypogeal, potentially regenerating buds are left below ground. More obviously, when a sheep grazes
grass or a giraffe browses on a tree it acts as a pruning instrument
leaving sites of regrowth, as does *Acanthaster* grazing on a coral. The
units of prey are not whole genets whose births and deaths can be
modelled as in a Lotka–Volterra equation for the interaction of predators and prey. Instead, a predator usually affects the growth rate of
its prey, reducing the assimilating area of hydranths, polyps, leaves or
roots. Predation may also affect architecture; when tent caterpillars
destroy the leader shoots on a tree lateral buds are released from
correlative inhibition and grow out in places where they would have
remained dormant. Thus repeated grazing or browsing modifies the
form of shrubby and herbaceous plants—sometimes preventing flowering and prolonging the vegetative phase of herbaceous perennials or
enforcing close tussocky form on woody shrubs such as heathers
(Barclay-Estrup, 1970). The effect of predation is then to change the
architecture of the whole genet and its competitive position within the
hierarchy of a population.

The relevant search image and food unit for a predator feeding on
prey with modular construction may often be the modular unit (e.g. the
leaf or shoot) not the whole genet. For this reason it is usually much
more useful for an entomologist working with forest pests to know the
number of leaves in a forest than to know the number of trees! Similarly
it is more useful for an agronomist to know the density of leaves in a
pasture than to know the number of genets that bear them. Many of the

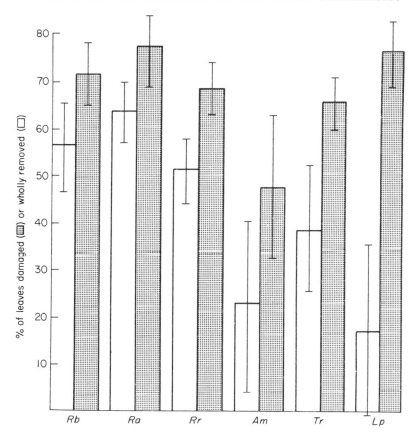

Fig. 4.11. The influence of predation on the leaves of six species of plant in an old permanent pasture. Leaves were marked or mapped as they emerged and their fate was determined at intervals from the type of damage done. Thus the individual leaf was used as the unit of prey in a study of the action of predators.

I = Standard Deviation; *Rb* = *Ranunculus bulbosus*; *Ra* = *R. acris*; *Rr* = *R. repens*; *Am* = *Achillea millefolium*; *Tr* = *Trifolium repens*; *Lp* = *Lolium perenne*.

Species with horizontal leaves (*Rb*, *Ra*, *Rr*) tended to suffer complete removal, whereas those with more vertical leaves (*Am,Lp*) often suffered the loss of only part of a bitten leaf. (From Peters, 1980.)

concerns of the theoretical ecologist take a different dimension when modular organisms are involved in a predator–prey or a parasite–host interaction. Thus when the predator is a leaf-eating caterpillar the effective density of prey and its grain size are determined by the way in which leaves are displayed rather than the way in which individual trees are spaced. The tight arrays of the needles of the spruce and the more widely spread clustered parcels of leaves in many broad-leaved trees, e.g. many species of Eucalyptus, present units of prey in quite different grain size to an insect predator.

A plant of white clover may provide quite different elements of diet to a series of herbivores. A leaf of white clover offers the food necessary for a full life cycle of pathogen (e.g. *Pseudopeziza trifolii*), a large fraction of the diet of a weevil and part of a day's food supply for a slug but only a fraction of one mouthful for sheep. The damage caused by the predator is also quite different: thus the pathogen may destroy a square millimetre of leaf surface, the weevil cut a hole of three or four times that area, the slug may cut out a sizeable segment from the leaflet or eat it entirely, whereas a sheep will most often remove a whole leaf and probably several neighbouring leaves from the same genet in one bite (Fig. 4.11). It is difficult to find comparable scale and grain effects in predator–prey relations when the prey are unitary organisms. Again, architecture becomes relevant. It is the architecture of the plant that determines the grain size of a leaf population that is on offer to a herbivore.

4.9　Reproduction

The fitness of a modular or a unitary organism will be expressed in the proportion of its descendants or descendant genes found in subsequent generations. The progeny, via zygotes (or other single celled stages) are usually produced in seasonal bursts, either as one big lethal bang or as repeated episodes in the life of the genet. The *growth* of the genet in modular organisms is a more or less continuous process (with a seasonal rhythm) of exponential increase in the number of modules present and can be modelled as a continuous process. Reproduction, however, is normally a step-wise process with time delays between episodes. The pace of modular growth is slowed down during phases of reproduction. In higher plants, meristems that become involved in reproduction are unavailable for further growth: a meristem may become a flower primordium, produce seed and die, *or* generate new meristems allowing further growth. Monocarpic (semelparous) plants are those in which *all* the meristems become involved in or die at reproduction; they are lost to further growth—reproduction is lethal to the genet. In perennial, polycarpic plants, not all available meristems are used up in producing flowering modules—a proportion remains to continue iteration in subsequent years. These proportions essentially determine the reproductive pattern of the life cycle. Thus most trees spend long periods in growth before reproductive episodes commence; the genet initially gains size by allocating all modules to growth at a sacrifice of reproductive precocity. Only when the plant is large are some modules allocated to reproduction. In many such perennials it is the very old stages in the life of a genet (extensive single clones of herbaceous plants or very large trees) that have the highest reproductive value. It is difficult to detect any phenomenon of senescence in such perennial modular organisms except when, as in the case of trees and large branched corals, they accumulate dead tissues which puts them increasingly at risk to physical forces such as storms (Harper, 1977).

The distinction between modular and unitary organisms made in this chapter is, of course, a generalization. There are plants that have unitary properties—a plant of *Welwitschia* has just two leaves. There are odd variants amongst animals (e.g. salps and some siphonophores) for which the concepts of N and η populations may need special pleading. In general, however, the thrust of this chapter is to disclaim any separate status for the population biology of plants and of animals and to foresee a new population *biology* appropriate for modular as opposed to unitary organisms.

5
Models for Two Interacting Populations

ROBERT M. MAY

5.1 Introduction

This chapter outlines some of the general dynamical features of models for two interacting populations.

Such pairwise interactions may have one or other of three basic forms: if the effects are such that the growth rate of one population is decreased, and of the other increased, we have the prey–predator situation (the ' − +' case); competition occurs if both growth rates are depressed by the co-occurrence of the two species (the ' − −' case); and reciprocal enhancement of growth rates corresponds to mutualism (the '+ +' case). These three cases are considered in sections 5.2, 5.3 and 5.4, respectively. Section 5.5 touches broadly on the many ways in which simple systems can exhibit two or more alternative equilibrium states, and section 5.6 looks ahead to multispecies systems.

The general discussion in chapter 5 sets the stage for the next three chapters, which describe some particular instances where these theoretical insights can be put more or less closely beside field and laboratory data.

5.2 Two populations: prey–predator

5.2.1 Continuous growth (differential equations)

Most elementary ecology texts contain an account of the simplest model for a prey–predator system, namely the classical Lotka–Volterra differential equations, which were first studied around 1920:

$$dN/dt = aN - \alpha NP \qquad (5.1)$$

$$dP/dt = -bP + \beta NP. \qquad (5.2)$$

Here the prey population, $N(t)$, has a propensity for unbounded exponential growth, aN, which is limited by predation: the effect of the predators upon the prey population is measured by the 'functional response' term, αNP. The predator population, $P(t)$, has an intrinsic death rate, $-bP$, and a growth rate or 'numerical response' which depends on the prey abundance as βNP.

This system has pathological dynamical properties, namely the neutral stability of the frictionless pendulum. The system oscillates with a period determined largely by the parameters of the model (the period is approximately $2\pi/\sqrt{ab}$, but does depend weakly on the amplitude of the oscillation), but with an amplitude determined solely, and forever, by the initial conditions. If the system is disturbed, it will then oscillate in some similar neutrally stable cycle, with a similar period but with a new amplitude determined by the disturbance. This pathological neutrally stable behaviour depends sensitively on the model possessing the exact structure of eqs. (5.1) and (5.2); the slightest alteration in the mathematical expression given to the various terms in these equations will tip the dynamics towards a stable point, or towards a stable limit cycle, as discussed below. Such a model is called 'structurally unstable'. Structurally unstable models have no place in biology (although they do have legitimacy in physics, where they derive from deep and special symmetries, such as translational invariance, in the physical world: for fuller discussion, see May, 1974b and 1975a, pp. 50–53).

These severe reservations having been expressed, it remains true that the classic Lotka–Volterra model does lay bare one of the general properties of prey–predator models, namely a propensity to oscillations. The mechanism is clear: high prey densities tend to produce high predator densities, which tend to depress prey densities, which makes for lower predator densities, which leads to higher prey densities, and so on. Whether or not this oscillatory tendency is damped depends on the details.

Relatively realistic models may be obtained by modifying the predator-free prey growth term (aN) to include density dependence, and by modifying the crude 'binary collision' functional and numerical responses (αNP and βNP) to allow for effects such as saturation in the predator's capacity to respond to increasing prey densities.

More specifically, the prey growth rate may be represented by the logistic eq. (2.3), or by one of the broadly equivalent forms referred to in section 2.2.2. A detailed discussion of various forms of predator functional response is given by Hassell in section 6.2. If the basic modification is the inclusion of saturation effects, we have a 'Holling type II' or 'invertebrate' functional response (see Fig. 6.1b); this has a destabilizing effect, because the predators are relatively less effective at high prey densities, just when they are most needed as a regulatory mechanism. More complicated modifications allow for predators to become increasingly efficient as prey numbers increase, until saturation eventually sets in (see Fig. 6.1c); such a 'Holling type III' or 'vertebrate' functional response is stabilizing at low prey densities, destabilizing at high densities. An important body of recent work, mainly on arthropod prey–predator systems, has shown that if the prey have a patchy or clumped distribution in space, and if the predators have an aggregative response (directing their attention preferentially to higher density patches of prey), the net effect is a stabilizing 'type III' functional response (see, e.g., Hassell, 1978; May, 1978a).

In short, a more realistic prey growth equation is

$$dN/dt = rN(1 - N/K) - PF(N,P). \tag{5.3}$$

Some forms for the function $F(N,P)$ are catalogued in Table 5.1, and illustrated in Fig. 6.1.

In a like fashion, the predator growth equation can be written more realistically as

$$dP/dt = P\,G(N,P). \tag{5.4}$$

In chapter 7, there is elaborated a distinction between 'laissez-faire' predators, where the per capita numerical response function G is itself independent of predator numbers, P,

$$dP/dt = P\,G(N), \tag{5.5}$$

and 'interferential' predators, where it is not. Some forms for the function $G(N,P)$ are summarized in Table 5.1. The functional and numerical responses are often linearly related, as in the general form (ix) of Table 5.1: since most of the forms (i) to (vii) for $F(N,P)$ are independent of P, this often leads to the 'laissez-faire' model (5.5).

Some of these modifications, leading to eqs. (5.3) and (5.4), tend to stabilize the system, and some to destabilize it, compared with the razor's edge of neutral stability manifested by the Lotka–Volterra eqs. (5.1) and (5.2).

A powerful mathematical theorem for 2-dimensional systems of differential equations, the Poincaré–Bendixson theorem, may be used to

Table 5.1. Explicit forms for the functional and numerical responses per predator, $F(N, P)$ and $G(N,P)$, in eqs. (5.3) and (5.4). (For references, and other forms, see May, 1975a, pp. 81–84.)

F or G?	Label	Formula	Remarks
F	(i)	αN	unsaturated (Lotka–Volterra)
F	(ii)	k	constant attack rate
F	(iii)	$kN/(N+D)$	Holling type II, 'invertebrate' (Holling)
F	(iv)	$k[1 - \exp{(-cN)}]$	Holling type II, 'invertebrate' (Ivlev)
F	(v)	$k[1 - \exp{(-cNP^{1-b})}]$	Holling type II, 'invertebrate' (Watt)
F	(vi)	$kN^2/(N^2+D^2)$	Holling type III, 'vertebrate'
F	(vii)	$k[1 - \exp{(-cN^2 P^{1-b})}]$	Holling type III, 'vertebrate' (Watt)
G	(viii)	$-b+\beta N$	Lotka–Volterra
G	(ix)	$-b+\beta F(N, P)$	F and G linearly related. [Caughley's 'laissez-faire' case if $F = F(N)$ only.]
G	(x)	$s[1 - \gamma P/N]$	logistic, with carrying capacity proportional to N.

show that essentially all such prey–predator models have *either* a stable point *or* a stable limit cycle (as defined on pages 8, 9). This result was obtained in general form by Kolmogorov (1936), and has recently been more specifically applied to most of the models in the ecological literature (May, 1972a; 1975a, ch. 4). Figure 5.1 illustrates one particular example, constructed from eqs. (5.3) and (5.4) with the forms (iii) and (x) of Table 5.1 for F and G respectively:

$$dN/dt = rN(1 - N/K) - kPN/(N+D), \qquad (5.6)$$

$$dP/dt = sP(1 - \gamma P/N), \qquad (5.7)$$

The parameters in Fig. 5.1 are so chosen that the system exhibits stable cycles. Other parameter choices could give stable equilibrium point solutions.

These prey–predator systems tend to be in tension between the stabilizing prey density dependence, and the often destabilizing predator functional and numerical responses. As one or other of the relevant parameters is tuned, the pair of 'eigenvalues' which character- ize the stability of the system can be see to move from the left half of the complex plane (corresponding to damped oscillations), to cross the imaginary axis into the right half plane (giving rise by a Hopf bifurcation to a stable limit cycle); see May (1975a, ch. 2 and appendix I). The neutral stability of the Lotka–Volterra model is reflected in the eigenvalues always lying exactly *on* the imaginary axis.

Some general correlations between a system's parameter values and its stability behaviour are worth noting, and comparing with observed properties of real populations.

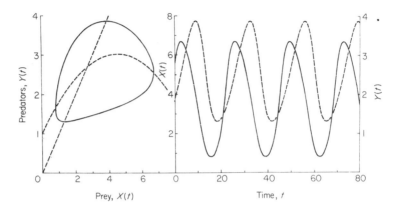

Fig. 5.1. A specific example of a prey–predator limit cycle. For the system obeying eqs. (5.3) and (5.4), with the functional and numerical responses F and G given by the forms (iii) and (x) of Table 5.1, respectively [see also eqs. (7.11) and (7.12)], we form dimensionless variables $X(t)$ and $Y(t)$ from the prey and predator populations: $X = N/D$, $Y = \gamma P/D$. To the left is shown the stable limit cycle endlessly traced out by these population numbers, for the parameter choice $r/s = 6$, $K/D = 10$, $k/\gamma r = 1$; the dashed lines are the isoclines of stationary prey ($dX/dt = 0$) and predator ($dY/dt = 0$) populations, and their intersection is the equilibrium point, here unstable. To the right we display the cyclic population numbers of prey (solid line) and predator (broken line) as functions of time, rt. If displaced from these stable cyclic trajectories, the system tends to return to them.

If the environmental carrying capacity for the prey, K, is much larger than the equilibrium prey density in the presence of predators, the stabilizing elements contributed to the dynamics by the prey density dependence will be relatively weak. This underlies the 'paradox of enrichment', whereby increasing K makes for lowered stability, and eventually for stable limit cycle behaviour. The 'paradox' was noted (and christened) by Rosenzweig (1971) on the basis of numerical studies and is possibly exemplified by natural populations such as the larch bud moth in Switzerland [whose population cycles in the optimal part of its range, between 1700 and 1900 m., but is relatively steady in marginal habitats (Baltenzweiler, 1971); but see ch. 14 for an alternative explanation of this particular cycle]. Laboratory studies, such as Luckenbill's (1973) on *Paramecium* and its predator *Didinium*, also tend to support this idea; there is much room for further such exploration of prey–predator dynamics under controlled conditions.

Stability studies of prey–predator models by May (1975a, ch. 4) and Tanner (1975) have further shown a propensity to stable cycles when the intrinsic growth rate of the prey population exceeds that of its predators. The typical situation is indicated by Fig. 5.2, which shows the stability properties of a model described by eqs. (5.6) and (5.7) (which is the same system as is illustrated in Fig. 5.1), as a function of the relevant parameter ratios K/D and r/s. The systems which tend to cycle are indeed those where the prey population has a relatively high growth rate (r/s

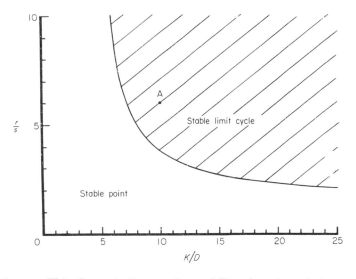

Fig. 5.2. This figure indicates the stability domains of the prey–predator system defined in Fig. 5.1. [see also the text, and eqs. (5.6) and (5.7)], in terms of the parameter ratios r/s (prey/predator intrinsic growth rates) and K/D (appropriately normalized value of the carrying capacity for the prey). In the unhatched region of this parameter space, the equilibrium point is stable; in the hatched region, there are stable limit cycles. The figure is drawn for $k/\gamma r = 1$, but qualitatively similar stability boundaries pertain to other values of this ratio. The point labelled A shows the parameter values corresponding to the explicit example illustrated in Fig. 5.1. After May (1975a, p. 192.)

large) in an environment with a relatively large carrying capacity (K/D large). The natural history data reviewed by Tanner is summarized schematically in Table 5.2, and it accords with the above notions. Roughly speaking (and this exercise is necessarily very approximate), the hare–lynx system is the only one of the eight whose population parameters lie in the hatched area of Fig. 5.2, combining relatively large K with an r/s that is not small, and it is also the only one which conspicuously exhibits stable cycles.

Plant–herbivore systems are, of course, special cases of the general prey–predator relationship, with the plants as 'prey' and the herbivores as 'predators'. Chapter 7 puts together theoretical insights and empirical observations for such systems. A more technical elaboration of this theme, applied to natural and managed grazing systems varying from intensive pastures to extensive ranges, is due to Noy-Meir (1975).

Table 5.2. Impressionistic summary of life history data for 8 natural prey-predator systems. (After Tanner, 1975).

Prey–predator	Geographical location	Is K relatively large?	Approximate* ratio r/s	Apparent dynamical behaviour
sparrow–hawk	Europe	no	2	equilibrium point
muskrat–mink	central North America	no	3	equilibrium point
hare–lynx	boreal North America	yes	1	cycles
mule deer–mountain lion	Rocky Mountains	yes	0·5	equilibrium point
white-tailed deer–wolf	Ontario	yes	0·6	equilibrium point
moose–wolf	Isle Royale	yes	0·4	equilibrium point
caribou–wolf	Alaska	yes	0·4	equilibrium point
white sheep–wolf	Alaska	yes	0·2	equilibrium point

* Estimated from Tanner's 'maximum survival and maximum fertility' column, as most appropriate for intrinsic r and s.

5.2.2 Prey–predator models and single species models

The relation between these 2-species models and the earlier single species models deserves mention.

Suppose we try to construct a single species equation for just the prey population in a prey–predator model: the population growth rate will depend on the predator population, which itself is derived (in a formal sense) as some integral over past prey populations. The upshot will be an equation involving only the prey population, but where the density dependent regulatory mechanism depends on past values of the prey population.

To be specific, suppose the growth of the predator population is described by an equation of the general form of eq. (5.5), with the numerical response per predator depending only on the prey density. This can be integrated to give

$$P(t) = P(o) \exp \int_0^t G[N(t')] \, dt'. \qquad (5.8)$$

The predator population depends on some explicit average over past prey populations. The growth rate of the prey population itself depends nonlinearly on both $N(t)$ and $P(t)$, as in the general eq. (5.3). Use of eq. (5.8) enables this to be reduced to a single equation involving only the prey population:

$$dN/dt = \text{function of} \left\{ N(t) \text{ and} \int_0^t G[N(t')] \, dt' \right\}. \qquad (5.9)$$

That is, we have an equation with a time-delayed regulatory mechanism, of the same general character as eq. (2.4). The time delay roughly corresponds to the time for the predator population to respond to prey changes (if our single species equation describes the prey population), or for the vegetation to recover (if our equation describes the herbivores). This point is discussed more fully by Caughley (1976).

As a variation on this theme, consider a plant–herbivore or substrate–plant system in which the 'predator' population, $N(t)$, obeys the logistic eq. (2.3), but where the carrying capacity $K(t)$ (representing the population of the plant or substrate 'prey') has a dynamical dependence on $N(t)$ via an equation of the general form

$$dK/dt = [K_o - K(t)]/\tau - bN(t). \qquad (5.10)$$

Here K_o is the saturation level of the resource (plant or substrate), and b the rate of depletion by consumers. The rate at which the resource recovers depends both on its intrinsic regeneration time, τ, and on the degree to which it is depressed below saturation, $K_o - K$. Equation (5.10) can be integrated

$$K(t) = K_o - b \int_{-\infty}^{t} N(t') \exp \left[-(t - t')/\tau \right] dt', \qquad (5.11)$$

thus again leading to a form of time-delayed logistic equation for the 'predatory' herbivore or plant population. Stability analysis of eq. (2.3) with eq. (5.11) for $K(t)$ shows that (like the ordinary logistic) the equilibrium point at $N^* = K_o/(1 + b\tau)$ is always stable, but that (unlike the ordinary logistic) perturbations die away in an oscillatory way once

$$b > (1 - r\tau)^2/(4r\tau^2). \qquad (5.12)$$

High depletion rates have a tendency to produce oscillations. For a detailed treatment, see McMurtrie (1978).

In short, the ideas developed in chapter 2 also, in a broad sense, include prey–predator situations.

5.2.3   Discrete growth (difference equations)

As discussed for single populations, there will be situations where generations do not overlap, and the prey–predator dynamics are described by difference equations. Arthropod prey–predator systems provide many examples, particularly in seasonal climates. Such systems are discussed in detail in chapter 6.

We content ourselves here with some general remarks about the dynamics of these systems. Although prey–predator systems with continuous growth (differential equations) usually possess either a stable point or a stable limit cycle, a similarly generic discussion of discrete growth models is difficult, and their dynamical behaviour is not typically so simple. Two studies by Beddington et al. (1975, 1976a), however, are illuminating.

These authors base their work on the classic Nicholson–Bailey model, generalized to include density dependence in the host population growth (for further details see section 6.7):

$$N_{t+1} = N_t \exp \left[r(1 - N_t/K) - aP_t \right], \qquad (5.13)$$

$$P_{t+1} = \alpha N_t [1 - \exp (-aP_t)]. \qquad (5.14)$$

Numerical studies show regimes of stable points, of stable cycles (which, unlike the single species models in section 2.3, need no longer have an integer period, and in general do not), and of chaos. The onset of chaotic behaviour, where the population trajectories are effectively indistinguishable from the sample function of a random process, takes place at lower r-values than for the corresponding single species model. The paper by Beddington et al. (1975) is rich in fascinating detail, illlustrated by photographs of oscilloscope trajectories. More recently, the analytic understanding which underpins section 2.3 has been generalized to multi-species situations by Guckenheimer et al. (1976). This elegant mathematical work paves the way for a generic discussion of the dynamic complexities of such models, and, as one special case, illuminates the numerical results mentioned above.

In the continuous prey–predator models, the stability properties tend to be global, in the sense that the system returns to its stable configuration regardless of the magnitude of the perturbations. A separate complication in discrete (difference equation) models is that such global stability behaviour is not typical: given the system possesses a stable point (or a stable cycle), the system will tend to return to the stable point following a small disturbance, but not following a large one. Large disturbances typically lead to the extinction of the predator population, if not of both populations. The term 'resilience' has been introduced (Holling, 1973; see also Orians, 1975 and May, 1975c) to characterize the magnitude of the population perturbations the system will tolerate before collapsing into some qualitatively different dynamical regime. Beddington et al. (1976a) have used a slight generalization of eqs. (5.13) and (5.14) to study general biological features of this resilience in arthropod prey–predator systems. To date, the concept of 'resilience' has been mainly a useful metaphor; the above work opens the door to experimental and theoretical studies of a quantitative kind, albeit on relatively simple systems.

5.3   Two populations: competition

Simple models for two competing populations, $N_1(t)$ and $N_2(t)$, were also studied in the 1920s by Lotka and Volterra. These models are the direct extension of the single species logistic equation:

$$dN_1/dt = r_1 N_1[1 - (N_1 + \alpha_{12} N_2)/K_1], \tag{5.15}$$

$$dN_2/dt = r_2 N_2[1 - (N_2 + \alpha_{21} N_1)/K_2]. \tag{5.16}$$

Here K_1 and K_2 are the carrying capacities of the environment, as seen through the eyes of species 1 and 2, respectively; r_1 and r_2 are the intrinsic growth rates; α_{12} is a competition coefficient which measures the extent to which species 2 presses upon the resources used by species 1; and α_{21} is the corresponding coefficient for the effect of species 1 on species 2. The multi-species generalization of these competition equations is discussed in chapter 8 [eq. (8.1)].

The stability character of these equations is well known. If intraspecific competition is stronger than interspecific (corresponding to $\alpha_{12}\,\alpha_{21} < 1$), there can be a stable equilibrium point with both species coexisting (provided also that the carrying capacity ratio obeys $1/\alpha_{21} > K_1/K_2 > \alpha_{12}$, which will easily be true if both α_{12} and α_{21} are small); if interspecific competition is stronger than intraspecific ($\alpha_{12}\,\alpha_{21} > 1$), no stable coexistence is possible. Particularly interesting is the special case when the two species use the resources in identical fashion (whence $\alpha_{12} = \alpha_{21} = 1$ and $K_1 = K_2$); again the species cannot persist together. Other models, which generalize other single species equations of the same generic character as the logistic, lead to similar conclusions (see, e.g., the early discussion by Gilpin and Justice, 1972, and the recent extensions and review by Nunney, 1981).

This early theoretical work led to the enunciation of the 'competitive exclusion principle', that is, the notion that species which make their livings in identical ways cannot stably coexist, which in turn stimulated many interesting experiments. This material is reviewed in chapter 8 (see also Hutchinson, 1975). The concept of 'niche' is used as a shorthand for the constellation of ecological factors which specify just how a species does make its living in the world; the competitive exclusion principle then says that no two species can occupy the same niche. Empirical and theoretical aspects of the niche are developed in detail in chapter 8, which emphasizes the difficulties involved in quantifying a species' niche once several ecological dimensions are relevant. A good definition of both the niche and the competitive exclusion principle have been given by Dr Seuss (Geisel, 1955), and put into the literature by Levin (1970):

'And NUH is the letter I use to spell Nutches
Who live in small caves, known as Nitches, for hutches.
These Nutches have troubles, the biggest of which is
The fact there are many more Nutches than Nitches.
Each Nutch in a Nitch knows that some other Nutch
Would like to move into his Nitch very much.
So each Nutch in a Nitch has to watch that small Nitch
Or Nutches who haven't got Nitches will snitch.'

Although these ideas provided much early impetus, they deserve close scrutiny. At a semantic level, one may ask how identical is 'identical'? On a more pragmatic plane, one may point in nature to many instances where organisms appear to co-occur with, at very least, substantial overlap between niches. MacArthur's guild of five very similar warblers, *Dendroica*, which are found together as insectivores in spruce trees in northern New England, provides one such apparent paradox, made famous by his resolution of it.

Following Hutchinson's (1959) 'Homage to Santa Rosalia, or why are there so many kinds of animals?', and the work of MacArthur and others, contemporary attention is focussed on the more substantial question of how similar can competing species be, yet stably persist together? What are the limits to niche overlap, the limits to similarity, among coexisting competitors?

Theoretical investigations of these questions begin with models for the detailed mechanism of competition, which may involve interspecific territoriality, or selection of items from some shared spectrum of resources, or whatever. The overall competition coefficient, α_{ij}, in some population equation such as eqs. (5.15) and (5.16) may then be calculated on this basis. In this way, the macroscopic population dynamics are related to the underlying microscopic mechanisms of competition, and one can ask how much niche overlap is compatible with the 2-species system having a stable equilibrium point.

In particular, consider the admittedly highly special case where the competing species differ *only* in their use of some 1-dimensional spectrum of resources. This situation is illustrated in Fig. 5.3. Here $K(x)$ represents the resource spectrum, which may be amount of food as a function of size or weight, or amount of vertical habitat as a function of height, or in general amount of a resource K as a function of some variable x. The function $f_i(x)$ describes the way the ith species utilizes this resource; each species has its preferred position in the spectrum (with the mean separation between the utilization functions of adjacent competitors being characterized by d), and a degree of variability about this preferred position (typified by a width w). The degree of niche separation clearly may be characterized by the ratio d/w, which will be large for well separated niches, small for highly overlapping ones. A more detailed and realistic version of Fig. 5.3 is given and discussed in chapter 8 (Fig. 8.1).

Following Levins (1968) and MacArthur (e.g., 1972, chs. 2 and 7; see also May, 1975a, pp. 142–147), the competition coefficients in eqs.

(5.15) and (5.16) and their multi-species version eq. (8.1) may then be regarded as the overlap integrals of the utilization functions for the species involved:

$$\alpha_{ij} = \int f_i(x) f_j(x) \, dx. \tag{5.17}$$

For 2-species competition, we may now choose a particular shape for the utilization functions (e.g., bell-shaped gaussian curves), and then study the macroscopic population dynamics of eqs. (5.15) and (5.16) as a function of the niche separation d/w, and of the parameters K_1 and K_2 which describe the shape of the resource spectrum as perceived by the individual species. We first examine the *statics* of the situation, to see if a 2-species equilibrium exists, and then study the *dynamics*, to see if the equilibrium is stable.

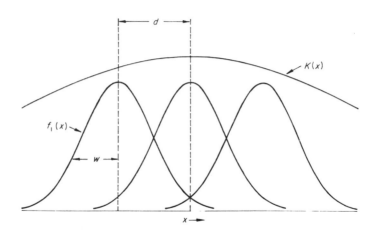

Fig. 5.3. The curve $K(x)$ represents some resource continuum (say amount of food, K, as a function of food size x), which sustains various species whose utilization functions $f(x)$ are, as illustrated, characterized by a width w and a separation d.

For the *statics*, we find the potential equilibrium populations N_1^* and N_2^* by putting $dN_1/dt = dN_2/dt = 0$ in eqs. (5.15) and (5.16). The results are shown in the 'horizontal plane', d/w versus K_2/K_1, in the schematically '3-dimensional' Fig. 5.4. For small niche separation, $d/w \lesssim 1$, a 2-species equilibrium point exists (i.e., both N_1^* and N_2^* are positive) only in a narrow band of K_2/K_1 values; otherwise the equilibrium configuration contains only species 1 or only species 2, depending on whether K_1 is significantly larger or smaller than K_2.

Conversely, for large niche separation, $d/w \gg 1$, a 2-species equilibrium point exists for essentially all resource spectrum shapes. Extensions to 3 or more competing species confirm the general tendency for a multi-species equilibrium point to be possible for a wide variety of resource spectrum shapes if d/w is large, but to impose very narrow constraints on the shape if d/w is small. Studies of this kind were begun by MacArthur and Levins (1967), and carried forward by Roughgarden (1974a), May (1974c, 1975a) and others; a critical review is by Abrams (1976a). Although not setting any limit to niche overlap in

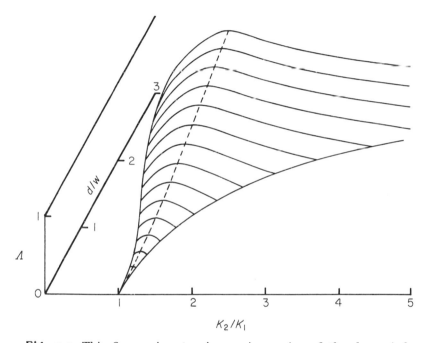

Fig. 5.4. This figure aims to give an impression of the dynamical behaviour of the model for two competing species which is outlined in the text. The quantity Λ represents the rate at which disturbances are damped (i.e., the reciprocal of the characteristic return time for the system); it is plotted as vertical height above the 'horizontal plane' of values of d/w (the degree of niche separation) and K_2/K_1 (the carrying capacity ratio). The value of Λ is shown along contours for fixed values of d/w from 0 to 3, spaced at intervals 0·2 apart. These contours are bounded by the two curved lines in the d/w versus K_2/K_1 plane, and outside these boundaries no 2-species equilibrium point exists. The dashed line indicates the ratio K_2/K_1 which gives the fastest damping rate, for given d/w. The features of this figure are as emphasized in the text.

principle, this line of attack suggests that most resource spectrum shapes will require $d/w \gtrsim 1$ if the competitors are to coexist.

The *dynamics* of the situation enters Fig. 5.4 via the vertical axis, Λ. This quality Λ characterizes the rate at which the system tends to return to equilibrium, following a disturbance. For simplicity, we have in Fig. 5.4 used $r_1 = r_2 = r$, and expressed Λ as a ratio to r; thus Λ involves the multi-species generalization of the characteristic return time T_R of section 2.2, and here the explicit normalization is $\Lambda = 1/(rT_R)$. In the absence of competition, $\alpha_{ij} \to 0$, the damping rate tends to r as for the ordinary logistic, and consequently $\Lambda \to 1$. We notice from Fig. 5.4 that *if* a 2-species equilibrium point exists (N_1^* and N_2^* both positive) it is necessarily stable, that is it necessarily has some finite positive damping rate, Λ. But this rate is very small (and the recovery time correspondingly very long) for small d/w, whereas the rate is in general high (and the recovery rapid) for large d/w.

This analysis of the *dynamics* may be pushed one stage further by noting that, in the real world, the environment is likely to exhibit elements of unpredictable variability. Thus the resource spectrum (as measured by quantities such as $K(x)$ or K_1 and K_2) is unlikely to be deterministic, but rather will, to a greater or lesser degree, contain randomly fluctuating components. In place of the deterministic differential eqs. (5.15) and 5.16), which lead to explicit dynamical trajectories for $N_1(t)$ and $N_2(t)$, we will now have stochastic differential equations for the populations, and their behaviour will be described in statistical terms, as fuzzed out probability clouds for the population values. Once such a spectrum of random environmental fluctuations is woven into the fabric of the model, the requirement that the damping rate merely be positive ($\Lambda > 0$) is no longer sufficient to ensure long-term stability, but rather the damping must be strong enough to prevent one or other of the populations from drifting to such low values as to court extinction. (Very crudely and qualitatively, we require $\Lambda > \sigma^2$, where σ^2 characterizes the variance in the environmental noise spectrum.) As is made plausible by the relation between Λ and d/w in Fig. 5.4, the analysis of competition between species in a randomly varying environment suggests the qualitative requirement that d/w should be greater than unity if the species are to persist together. Such studies were initiated by May and MacArthur (1972) and May (1973b). This initial work makes many mathematical approximations of one kind or another, some of which have been refined in subsequent studies by Ludwig (1975), Abrams (1976b), McMurtrie (1978), May (1974c, d, 1975a) and others.

In particular, if the noise spectrum for the random environmental fluctuations has a high degree of correlation (either between different points in time, or between different points on the resource axis), a high degree of niche overlap may be possible (Feldman and Roughgarden, 1975; Abrams, 1976; May, 1973b). In the limit of perfect correlations, an effectively deterministic environment is recovered, and we have the situation discussed above where there is in principle no limit to d/w, although exquisitely precise resource spectrum shapes are required if species are to coexist with d/w significantly less than unity. Other complications include sensitivity of the results to the way in which the competition coefficients α_{ij} are defined (one may argue that eq. (5.17) is not always the appropriate definition, and that other definitions can permit much greater niche overlap; see Abrams, 1975), and subtleties having to do with how the environmental stochasticity is introduced into the equations. A critical review of these several complications, which make any general conclusions difficult, is by Turelli (1978).

All the above work is based on the static or dynamic behaviour of models in which the utilization functions are immutably fixed. A different, and much more fundamental, approach has recently been developed by Roughgarden (1975b, 1979): in essence, he allows for evolution in the way the species use the resource spectrum. That is, the shape of the functions $f_i(x)$ in Fig. 5.3, particularly as reflected in their modal position along the resource axis (and thence in the parameter d), are subject to natural selection from generation to generation, and the system evolves to maximize the fitness of the constituent species. The result is, in general, that the model tends to evolve towards a configuration such that d/w is of order unity. This is path-breaking work, lying squarely at the interface between population genetics and population biology; between evolution and ecology (see also section 8.5 below).

No one of the above approaches is decisive. But it is suggestive that three quite distinct lines of attack—static and dynamic behaviour with fixed utilization functions, and models in which utilization functions evolve—give the same answer, namely that there tends to be a limit to niche overlap, such that the average difference *between* species exceeds the typical variability *within* a species. That is, in these special 1-dimensional models,

$$d \gtrsim w. \tag{5.18}$$

It is strongly to be emphasized that these estimates as to the limits to similarity among coexisting competitors are rough, order of magnitude, ones. The sign in the equations is most decidedly $\sim$ not $=$. Thus factor-of-2 differences between calculations based on differently shaped utilization functions, or on different criteria as to how small the population must be before it is highly likely to become extinct, are not really subject to experimental test (at least as matters stand now).

The reason why the different approaches lead to similar conclusions is basically simple. All the models tend to have the feature that the *total* population density, $\sum N_i(t)$, is density dependent in an essentially logistic fashion (with $T_R \sim 1/r$), independent of the value of d/w. The density dependence of the individual population is, however, sensitively influenced by the degree of niche overlap: for large d/w, each population is well regulated; but for $d/w \ll 1$, the individual populations have very weak density dependence, and they tend to drift up or down (keeping the total population relatively constant), until one or other fluctuates to extinction (see May, 1974b, particularly Figs. 5, 6). That is, individual populations tend to be robustly density dependent for $d/w \gg 1$, and to exhibit fragile density independent behaviour for $d/w \ll 1$. Hence the tendency to infer the limiting similarity described very roughly by eq. (5.18), from various superficially different approaches.

The bulk of this section's discussion of models for two competitors has been devoted to the light they shed on the issues of limiting similarity. This is because I regard it as the central contemporary problem in the area. Chapter 8 goes on from this point, to discuss pertinent observational data. Chapter 8 also discusses other aspects of competition theory, as well as many ways in which the theoretical enquiry into limiting similarity is profitably being extended, to include: direct interference between species (such as territorial squabbles); multiple age classes and phenomena associated with animals with indeterminate growth; competition in many dimensions; time delays in the effects of the interactions; and other things. As Pianka observes at the end of chapter 8, the overriding need is, however, for more experiments and observations rather than for more theory in this area.

5.4 Two populations: mutualism

Most contemporary ecology books have chapters, complete with simple mathematical models, on competition and on prey–predator.

Analogous chapters on mutualism are usually absent. I think the reasons for this are partly historical (Lotka and Volterra studied models for competition and prey–predator, but not for mutualism), and partly because mutualism is a relatively inconspicuous feature in temperate zone ecosystems. Be this as it may, mutualism is a conspicuous and ecologically important factor in most tropical communities, and I hope that the next generation of ecology texts will treat all three types of pairwise interaction between species on a roughly equal footing.

One initial difficulty is that the simple, quadratically nonlinear, Lotka–Volterra models that capture some of the essential dynamical features of prey–predator (namely, a propensity to oscillation) or competition (namely, a propensity to exclusion) are inadequate for even a first discussion of mutualism, as they tend to lead to silly solutions in which both populations undergo unbounded exponential growth, in an orgy of mutual benefaction.

Minimally realistic models for two mutualists must allow for saturation in the magnitude of at least one of the reciprocal benefits. This tends to produce a stable equilibrium point, with one (and usually both) of the two equilibrium populations being larger than it would be in the absence of the mutualistic interactions; this, of course, is what mutualism is all about. On the other hand, this equilibrium configuration tends to be less stable, in the sense that perturbations are typically damped more slowly than they are in the corresponding system without mutualistic interactions. Some explicit such models are discussed by Whittaker (1975a, pp. 39–41). Qualitative features of the stability properties associated with mutualism are discussed more fully in May (1973c), and a very general formulation of the dynamical character of these systems is in Hirsch and Smale (1974, ch. 12, problem 2).

To illustrate these remarks, consider the unrealistically simple case of two populations $N_1(t)$ and $N_2(t)$ which obey

$$dN_1/dt = rN_1[\mathrm{I} - N_1/(K_1 + \alpha N_2)], \qquad (5.19)$$

$$dN_2/dt = rN_2[\mathrm{I} - N_2/(K_2 + \beta N_1)]. \qquad (5.20)$$

Here each population obeys a logistic equation, eq. (2.3), with the modification that each species has its carrying capacity increased by the presence of the other, so that $K_1 \to K_1 + \alpha N_2$ and $K_2 \to K_2 + \beta N_1$. This overly simple model lacks any saturation effects, and we must limit the amount of mutualistic interaction by demanding $\alpha\beta < \mathrm{I}$, or else the system will 'run away', with both populations growing unboundedly large. As a further simplification, we take the populations to have equal intrinsic growth rates, r, in order not to lose the message of eq. (5.22) in a fog of distracting notation. In this model, the equilibrium populations are made larger by the mutualistic effects:

$$N_1^* = (K_1 + \alpha K_2)/(1 - \alpha\beta) > K_1 \qquad (5.21a)$$

$$N_2^* = (K_2 + \beta K_1)/(1 - \alpha\beta) > K_2. \qquad (5.21b)$$

On the other hand, the characteristic return time, T_R (see section 2.2), for the system to return to equilibrium following a disturbance is now

$$T_R = \frac{1}{r(1 - \sqrt{\alpha\beta})}, \qquad (5.22)$$

which is to be compared with that for each population in the absence of interactions, namely $T_R = 1/r$. The larger the degree of mutualism (the larger $\alpha\beta$), the more pronounced are both these effects.

Many mutualistic systems evolve to the point where at least one of the partners finds the other not merely helpful, but necessary for its existence. This is the limit of 'obligate' mutualism. In this event, models will need to include not only saturation effects, but also threshold effects.

Figure 5.5 attempts to capture some of the dynamical essentials of this situation. To fix ideas, we may think of the figure as pertaining to a plant–pollinator system, in which the plant population, $X(t)$, is an obligate outcrosser and cannot reproduce in the absence of its only pollinator, with population $Y(t)$, which in turn is sustained solely by the nectar of this particular plant; clearly the ideas extend to plants and seed-dispersers, ants and acacias, and other mutualistic systems. The pollinator population, Y, may be thought of as obeying a logistic equation, in which the carrying capacity is proportional to the plant population, X. Then the possible equilibrium values of Y are proportional to X, and lie along a straight line in the X–Y plane (the pollinator 'isocline', where $dY/dt = 0$, as labelled in Fig. 5.5). If the plant density is too low, pollinators may have difficulty finding more than one plant, and plant reproduction may fall below replacement levels; conversely, there will be an upper limit to plant population growth, set by physical limitations regardless of how abundant pollinators may be. The upshot is typically that the equilibrium values of X follow the shape labelled $dX/dt = 0$ (the plant 'isocline') in Fig. 5.5, with a threshold density of pollinators necessary before any equilibrium is possible, and an upward slope as plant population density approaches some saturation value at large Y. The directions in which plant and pollinator population trajectories move in the various domains of $X-Y$ space are as shown by the arrows in Fig. 5.5. The point A is a stable point, and attracts all trajectories originating outside the hatched region abutting the axes. The point B is unstable. All trajectories originating inside the hatched region are attracted to the origin, i.e., the populations become extinct. This model is discussed with more mathematical detail in May (1976a).

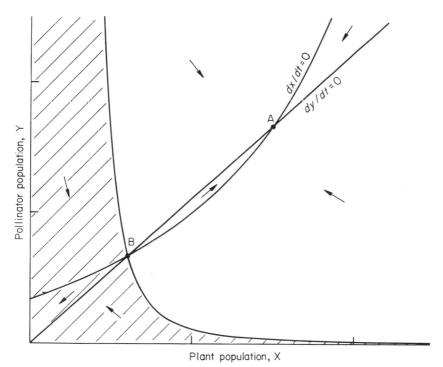

Fig. 5.5. This figure illustrates the dynamical behaviour of two mutualistically interacting populations, X and Y, as discussed in the text. The two labelled curves correspond to possible equilibrium values for X ($dX/dt=0$) and Y ($dY/dt=0$), respectively. Population trajectories originating outside the hatched region are attracted to the stable point A, while trajectories originating inside the hatched area are attracted to the origin (that is, extinction).

If the system whose dynamics is described by Fig. 5.5 is embedded in an environment which is subject to large amplitude perturbations, it is sooner or later liable to be swept into the hatched area in Fig. 5.5, whereupon extinction will probably follow. The long term persistence of mutualistic systems of this kind will be favoured by environments which are stable and predictable.

These dynamical considerations may go some way toward explaining why obligate mutualisms of this sort are less prominent in temperate and boreal ecosystems than in tropical ones. A good survey of·empirical evidence bearing on this point is given in Farnworth and Golley (1974, pp. 29–31), where it is observed that not only are there no obligate ant-plant mutualisms north of 24°, no nectarivorous or frugivorous bats north of 33°, no orchid bees north of 24° in America, but also within the tropics mutualistic interactions are more prevalent in the warm, wet evergreen forests than in the cooler and more seasonal habitats.

There are, of course, many instances of obligate mutualism that have gone so far as to blur the distinction between a single organism versus two mutualistic ones. Examples are the fungus–algae associations that constitute lichens; wood-digesting protozoans in the intestines of termites; nitrogen-fixing bacteria in the roots of leguminous plants; mycorrhizal fungal filaments on the roots of many higher land plants. Indeed such associations with various bacteria are essential for the functioning of all higher animals. However, the discussion in this section 5.4 has been conducted in the same spirit as the discussions of prey–predator and competition, in that we have in mind mutualistic interactions between two species each of which has, as it were, a life of its own.

Following my admonition in the opening paragraph of this section, I would have liked to include a chapter, along the lines of chapters 6, 7, 8, on mutualism. But I think that neither empirical nor theoretical aspects of the subject are yet developed to the point where such a synthetic overview is possible. There are, however, many insightful recent field studies, which shed increasing light on the interplay between the long sweep of evolution and the immediate dynamical effects of population interactions; it is possible that my laziness is the real reason why such a chapter does not appear in this second edition.

Work on plant–pollinator systems includes investigations of the interactions between plants and butterflies (e.g., Benson *et al.*, 1976; Gilbert, 1975, 1977), hummingbirds or sunbirds (e.g., Gill and Wolf, 1975b; Stiles, 1975; Colwell, 1973; Carpenter, 1979), bats (e.g., Howell, 1979; Howell and Hartl, 1980), bees (e.g., Janzen, 1971; Heinrich and Raven, 1972; Heinrich, 1975), lemurs and marsupials (e.g., Sussman and Raven, 1978), along with more general discussions (e.g., Kodric-Brown and Brown, 1979). The associations between fruit-producing plants and the animals that disperse their seeds provide further fascinating examples of mutualisms that are obligate (at least for the plants). In these systems, one design problem is producing seeds that can survive passage through the digestive tract of the fruit-eating animal; the result is often a seed that is unable to germinate unless its endocarp is abraded by such passage. Temple (1977) has shown the association between the large monoecious tree, *Calvaria major*, and the now-extinct Dodo on Mauritius to be an asociation of this kind; although the surviving *C. major* trees continue to produce seeds, none appear to have germinated for 300 years! The seeds of many species of African acacias apparently need similarly to pass through the digestive tract of elephants, and these trees are in trouble as elephant populations become smaller and more localized. Ants—seemingly the most abundant animals in the tropics—are involved in a variety of interesting mutualisms: in defending tropical plants bearing extrafloral nectaries (e.g., Bentley, 1976, 1977; Risch *et al.*, 1977); in ant–Homopteran associations (e.g., Hill and Blackmore, 1980), particularly ant-aphid associations (e.g., Addicott, 1978; Bristow, 1981); and in various more complicated systems (e.g., Tilman, 1978).

On the theoretical side, Levin has built on his empirical studies of the pollination ecology of plants, to explore simple models of plant–pollinator systems (King *et al.*, 1975). DeAngelis *et al.* (1979) have studied the persistence and stability of plants and seed-dispersers in patchy environments. More formal treatments of mutualistic interactions are by Goh (1979b), Vandermeer and Boucher (1978) and Travis and Post (1979). On a different tack, Briand and Yodzis (1980) have sought to document patterns in the phylogenetic distribution of obligate mutualisms. Furthermore, in many instances mutualistic and competitive interaction are inextricably woven together; this point is developed more fully in section 8.4.

In mutualistic models for such systems as plants and pollinators or plants and seed-dispersers, even more than for prey–predator or competition, it will often be necessary to introduce explicit time delays in the interaction processes, or else to work with difference equations. Whereas 2-dimensional systems of ordinary differential equations in general exhibit only stable points or stable limit cycles, these more general systems will display the rich range of dynamical behaviour alluded to earlier, from stable points, to stable cycles (with or without integer periods), to chaotic fluctuations. Such complications have been omitted from the discussion in this section.

5.5   Alternative stable states: thresholds and breakpoints

Figure 5.5 repeats a theme first sounded in Figs. 3.5 and 3.6, namely that simple models for the dynamical behaviour of interacting species can easily lead to situations in which there are two alternative equilibrium states. As this theme will arise again, in various forms, in chapters 6, 14, 7, 15, 16 and 11, it is useful to pause and discuss it very generally.

The existence of regimes with two alternative equilibrium points has arguably been documented recently in several contexts: for vegetation biomass (as a function of stock density in a managed pasture or rangeland; Noy-Meir, 1975); for fish and whale populations (as a function of harvesting intensity; Gulland, 1975; Clark and Mangel, 1979; Peterman *et al.*, 1979); for the Dungeness crab in California (as a function of egg-predation by a helminth; Wickham and Botsford, 1980); for near-shore marine communities (as a function of the densities

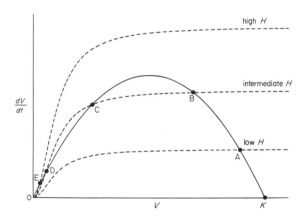

Fig. 5.6. The rate of change of vegetation biomass (or of fish population, or insect population), dV/dt, is shown as a function of V. The solid curve is the natural, ungrazed vegetation growth rate. The dashed curves are loss rates due to grazing (of type III pattern) at high, intermediate and low herbivore densities, H. Where the solid curve lies above the dashed one, the net growth rate is positive; where the solid curve lies below the dashed one, the net growth rate is negative; the points of intersection of the curves correspond to possible equilibrium points. For further discussion see the text. (From May, 1977a.)

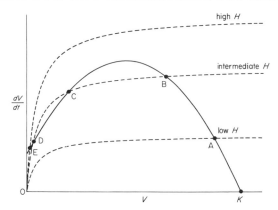

Fig. 5.7. As for Fig. 5.6, except here the natural growth rate is finite even for $V = 0$ (corresponding to immigration of seeds, or fish, or insects), and the loss rate due to grazing (or harvesting, or predation) is of 'type II'. (From May, 1977a.)

of sea otter populations; Simenstad *et al.*, 1978); for insect pest populations (as a function of predation levels; Ludwig *et al.*, 1978); and for various host–parasite systems.

Mechanisms common to all these examples are illustrated in Figs. 5.6 and 5.7. In Fig. 5.6 the solid line depicts the intrinsic biological growth rate of the population in question (vegetation, fish, insect pests) as a function of its density. To this logistic style of intrinsic growth is added mortality from predation (which may be herbivores on the vegetation, or fishers on the fish, or predators on the insect pests). The predation is 'type III', rising faster than linearly at low prey density but saturating to a constant per predator at high densities. The outcome is clearly a unique equilibrium point at high and at low predation levels, with an intermediate level of predation at which there are two possible stable equilibrium points (at B and D) separated by an unstable point (at C). Figure 5.7 illustrates an equivalent situation where the intrinsic biological growth curve for the population has a finite value at zero population density (corresponding to gains coming from immigration of seeds, or fish, or insect pests) and where the superimposed mortality from predation is of 'type II' form (rising linearly at low prey density). Again the upshot is a unique state at high and at low predation levels, with the possibility of two stable equilibrium states at intermediate levels of predation.

The net result is shown schematically in Fig. 5.8, which displays the possible equilibrium values of the vegetation biomass (or fish population, or insect pest population) as a function of the level of predation by

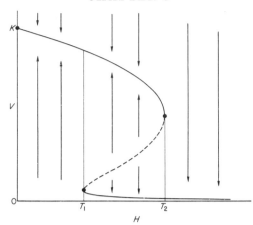

Fig. 5.8. The equilibrium values of the vegetation biomass (or fish population, or insect prey population), V, are shown as a function of the stocking density (or harvesting rate, or predator density), H. For a fixed value of H below the lower threshold at T_1, or above the upper threshold at T_2, there is a unique equilibrium value of V; any initial V value will move to this equilibrium, as indicated by the arrows. For H between T_1 and T_2 there are two alternative equilibria for V; as shown by the arrows, the system will move to the upper or lower equilibrium, depending on whether the initial value of V lies above or below the dashed 'breakpoint' curve. (From May, 1977a.)

grazing stock (or harvesting fisherman, or natural predators). At low predation levels there is a unique equilibrium state, corresponding roughly to that set by environmental resources. At some *threshold* level of predation, T_1, there discontinuously appears a second alternative stable state. Which of the two states the system settles towards depends on the initial conditions; the system will recover from small disturbances, but sufficiently large disturbances (crossing the dashed '*breakpoint*' line in Fig. 5.8) will carry the system from one stable state to the other. Finally, as the predation level increases beyond a second threshold level, at T_2, the original equilibrium state disappears discontinuously and the system again possesses a unique equilibrium state, now set predominantly by predation.

This qualitative scheme for the discontinuous appearance and disappearance of alternative stable states has been applied by the authors listed above to give an account of observed data on grazing ecosystems (mainly sheep), on various fisheries, on alternative near-shore communities in the Aleutian Islands, and on the behaviour of the spruce budworm in Canada. A quantitative explanation of the outbreaks of the eucalyptus psyllid *Cardiaspina albitextura* in Australia has been given in these terms (Southwood, 1977b), as has an explanation of the population history of the introduction of the European sawfly *Diprion hercyniae* in Canada (Southwood and Comins, 1976). An attempt to draw these threads together in a unifying review is by May (1977a).

Two additional comments are worth making.

First, notice that the dynamics in the neighbourhood of one or both of the alternative possible equilibrium points may be a stable point, or a stable cycle, or chaos, just as for the simpler models. This leads to complications that have not yet been fully resolved.

Second, some of the above remarks can be recast, *post hoc*, in the language of catastrophe theory. I doubt that much is gained thereby.

5.6   What next?

One logical next step would be to go on to consider *three* interacting populations. For the theoretician's usual case of continuous growth, leading to ordinary differential equations, the step up to three dimensions introduces not only a confusing proliferation of parameters, but also a qualitative change in the dynamical complexity. A 1- or 2-dimensional system of ordinary differential equations is rather tame, exhibiting only stable points or stable cycles (a fact which is intimately connected to the observation that here one can distinguish the inside from the outside of a closed curve); 3-dimensional such systems can display a full and rich dynamical complexity ('strange attractors'), akin to that we first met in section 2.3. Such messy behaviour is manifested even by the simple Lotka–Volterra equation for three competitors. This example is discussed as a sort of mathematical morality play by May and Leonard (1975; see also Gilpin, 1975a). One of its predictions, namely that there can be dynamically interesting situations of 'intransitive' competition where species A beats species B, which beats species C, which in turn triumphs over A, has indeed been observed among sessile marine organisms (Jackson, 1979; Buss and Jackson, 1979).

Partly to escape these complications, we count like the Australian Arunta tribe, 'one, two, many', and move on (after chapters 6, 7, 8,) directly to multispecies communities.

For some North American bird watchers, beatitude is to join the '600 club'; there are a little over 600 species of birds to be seen in the continent. These, and other broadly similar numbers, are shaped by the underlying limits to similarity among coexisting competitors. Although I believe competition usually to be the basic force, the numbers undoubtedly will be modified to a greater or lesser extent by prey–predator and mutualistic interactions among the multitude of organisms in the community. Other aspects of these general questions are explored in chapters 10 and 13, and in May (1975a, ch. 7). All this work represents only the first tentative steps, but I hope that as our understanding advances it will be possible to explain, in a fundamental way, why the number of bird species is 600 rather than 6 or 60,000, and maybe even why it is 600 rather than 60 or 6,000. The precision of the physical sciences will never be possible; it will never be possible to explain why the typical number is 600 rather than 500.

These are central questions, and ultimately as deep as the origin of species.

6
Arthropod Predator–Prey Systems

M. P. HASSELL

6.1 Introduction

The wealth of examples where animals show cryptic, mimetic or apo-
sematic colouration is itself testimony to the widespread importance of
predation as a factor reducing survival. Predation, like competition,
is therefore an important factor in the evolution of new species characteristics by natural selection, with selection favouring the efficient
predator and the elusive prey. This chapter is largely devoted to a
review of some theoretical studies on predation, focusing in particular
on the components of predator searching behaviour and how these
affect the interaction between predator and prey populations.

The predators we have in mind are arthropods—predominantly
insects—for two important reasons.

(1) There is considerable laboratory information available on many
of the components of search by arthropod predators. This has suggested
the functional form of many of the components included in the models
and also indicates realistic ranges of parameter values.

(2) There is amongst the arthropods one very large category of
insects whose life cycle has many of the features of the population
models to be discussed. These are the insect parasitoids (often loosely
called 'insect parasites'). They are very abundant, making up about
10 per cent of the one million or so known insect species, and largely
belong to the Diptera (flies) and Hymenoptera (ants, bees, wasps).
They differ from true parasites (such as liver flukes) in the strict
zoological sense, in that they almost always kill their host, which is
most often the egg, larval or pupal stage of some other insect. A
significant difference between predator and parasitoid life cycles is that
only the adult female parasitoid searches for hosts, and then primarily
to oviposit on, in or near the host rather than to consume it. The
parasitoid larvae may then feed from the outside or from within the

host, but in either case they cause little serious damage until approaching maturity when they feed on vital organs and the host is killed.

Several features of this life cycle represent important simplifications over that of other predators. In the first place, host mortality depends upon the searching ability of only a single stage in the parasitoid life cycle (the adult female), while males, females and often the immature stages of 'true' predators must all secure prey and will do so with differing searching efficiencies. The second simplification is that parasitoid reproduction is necessarily defined, in a simple, direct way, by the number of hosts parasitized. Such a close correspondence is lacking in predators, whose rate of increase depends upon the survival of each developmental stage and on the fecundity of the resulting adults. It is to insect parasitoids, therefore, that we should look for the most detailed correspondence between the simple population models discussed below and what has actually been observed. However, specific reference to parasitoids will only be made where some distinction from other predators need be made.

Predator–prey models have traditionally been couched in one of two mathematical formats—as differential or difference equations. Each of these is appropriate to quite different kinds of life history. Differential equations apply where generations overlap completely so that birth and death processes are continuous, while difference equations deal with population changes over discrete time intervals. They are, therefore, ideal for systems where generations are quite distinct. In limiting ourselves here to difference equation models for predator–prey interactions, we are restricting ourselves to the kinds of life cycle more typically found in temperate parasitoids and their hosts. Thus all the models to be discussed below will have the same basic form:

$$N_{t+1} = \lambda N_t f(N_t, P_t),\qquad (6.1)$$

$$P_{t+1} = cN_t\,[1 - f(N_t, P_t)],\qquad (6.2)$$

where N_t, N_{t+1} and P_t, P_{t+1} are respectively the prey and predator numbers in successive generations, λ is the finite nett rate of increase of the prey and c is the average number of predator progeny produced per prey attacked. Predation at time t remains at present an unspecified function of prey and predator population sizes. The use of this format does not necessarily mean that any conclusions to be drawn will not also apply to the many instances of slight overlap in age-classes and generations. This has recently been emphasized by Auslander et al. (1974) who have shown that difference equations can remain useful in describing some systems with complex age-class interactions which, by virtue of their internal dynamics, approach a state where the overlap in generations is reduced.

There are other details of predator biology which will shape the structure of population models. For instance, the degree of prey specificity is very important. Most predators are oligophagous or polyphagous, attacking several prey species; completely specific or monophagous species are few. Despite this, and for the obvious sake of simplicity, most theoretical studies have concentrated on single predator–single prey systems. [But see, for example, Roughgarden and Feldman (1975); Comins and Hassell (1976); and May and Hassell (1980).] These have the virtue of exposing some fundamental properties of predation, but we should bear in mind that they will be difficult to extrapolate directly to most natural systems.

In this chapter are distinguished two aspects of predation:

(1) The death rate of the prey (due to predation). In particular, we shall consider how predator efficiency is affected by the abundance of prey, the abundance of other predators and the relative prey and predator distributions.

(2) The rate of increase of the predator population. Here the influence of prey density on the survival and developmental rates of immature predators and on the fecundity of adults is to be considered.

This distinction [stressed by Beddington et al. (1976b) and Hassell et al. (1976)] parallels in some respects that proposed by Solomon (1949) and Holling (1959a) between functional and numerical responses.* It is, however, a somewhat broader framework since the functional response is restricted to relationships between the number of prey attacked per predator and *prey* density. The separate treatment of factors affecting the prey death rate and the predator rate of increase is much less important for parasitoids than for predators, each host parasitized often leading to one parasitoid progeny in the next generation. Indeed, this is exactly the assumption made in eq. (6.2) above (with $c=1$), and one that is also found in many of the so-called 'predator–prey' models in the literature (e.g., Lotka, 1925; Volterra, 1926).

It is convenient to commence with a brief discussion of the model of Nicholson (1933) and Nicholson and Bailey (1935). This provides an edifice upon which to build and explore the effects of more realistic

* A *functional response* defines the relationship between the numbers of prey attacked per predator at different prey densities, and the *numerical response* between predator numbers and prey density.

predator behaviour. Nicholson made three assumptions upon which all his models rest. He assumed: (1) that predators search randomly for their prey (i.e. their behaviour is not influenced by the density and distribution of prey and other predators); (2) that their appetite (or fecundity, if parasitoids) is unlimited; and (3) that the area effectively searched (i.e. in which all prey are found) is constant for a given predator population. Nicholson called this searching efficiency, the 'area of discovery' (a). Given these assumptions, the function in eq. (6.1) now becomes

$$f(N_t, P_t) = \exp(-aP_t), \tag{6.3}$$

leading to the population model

$$N_{t+1} = \lambda N_t \exp(-aP_t), \tag{6.4}$$

$$P_{t+1} = cN_t [1 - \exp(-aP_t)]. \tag{6.5}$$

Notice that the overall form of these equations is determined by the first term of the Poisson distribution, which serves to distribute the encounters between predators and prey randomly amongst the N_t prey available.

The properties of this model are well known. For each combination of λ and a there exists a unique equilibrium position, but one that is unstable, since the slightest disturbance leads inevitably to oscillations of increasing amplitude with the predator population lagging behind that of the prey. While such an unstable outcome has been observed in a few laboratory experiments [notably, Gause (1934) and Huffaker (1958)], it is not a feature of the real world where coupled predator–prey interactions have been seen to persist over long periods of time. This disparity in itself, however, does not condemn eqs. (6.4) and (6.5) as a description of predation, since the models may be made quite stable by simply allowing some resource limitation or other density dependence to act upon the prey population (Beddington et al., 1975). The effective rate of increase, λ, would now be a function of prey density as discussed in chapter 2. Having said this, it remains unlikely on a priori grounds that the persistence of all coupled predator–prey systems is dependent upon some external density dependent factors, rather than on the internal dynamics of the predator–prey system itself.

We now proceed to consider how eqs. (6.3)–(6.5) can be modified in the light of known predator responses to prey density, predator density and to the distribution of prey.

6.2   The response to prey density

It is implicit in eq. (6.3) that the number of encounters between a predator and a population of prey increases linearly with prey density as shown by the functional response in Fig. 6.1a. This, of course, is not feasible: no predator has an unlimited appetite, nor parasitoid an untold number of eggs. A further difficulty with eq. (6.3) is its implicit requirement that the time available for search (T_s) by a predator is constant at all prey densities. Without this, the area of discovery (a) could not remain a constant. It was Holling (1959b) who first argued forcibly that this too is impossible, and that searching time *must* depend upon prey density. Let us consider a to be the product of the proportion of prey discovered per unit of searching time (a') and searching time. Thus

$$a = a' \, T_s. \qquad (6.6)$$

Whenever a prey is encountered, a finite amount of time must be spent in quelling, killing and eating the prey, together with other related time-consuming activities, all of which Holling called the 'handling time' (T_h). There remains, therefore, progressively less time available for search (T_s) as more prey are eaten (parasitized), since

$$T_s = T - T_h \, N_a, \qquad (6.7)$$

where T is the initial, total time available to the predators for discovering prey and N_a is the number of prey attacked per predator.

This simple inclusion of handling time has the profound effect of changing the functional response to a negatively accelerating curve as shown by the example in Fig. 6.1b. Encouragingly, there are now numerous examples in the literature of such responses from arthropod predators and parasitoids, at least under laboratory conditions [see Hassell *et al.* (1976b) for a review]. It remains but a straightforward step explicitly to include handling time in eqs. (6.1) and (6.2). The precise form of the function, however, will differ between parasitoids and predators for the simple reason that hosts are not removed on parasitism and therefore remain available for rediscovery and further 'handling'. The different models that result are clearly derived and discussed in Rogers (1972a). Let us consider the case for parasitoids. The functional responses in Fig. 6.1 are described by the equation

$$N_a = a' \, T_s \, N_t, \qquad (6.8)$$

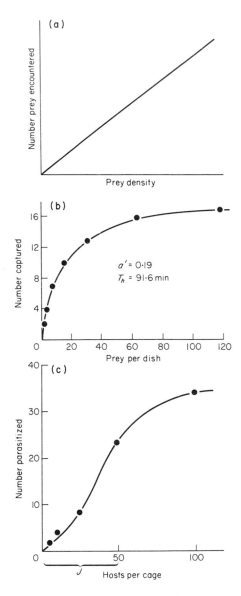

Fig. 6.1. Three types of functional response. (**a**) A linear response implicit in the Nicholson–Bailey model. (**b**) A concave response, obtained from second instar coccinellids (*Harmonia axyridis*) feeding on *Aphis craccivora* (Mogi, 1969). (**c**) A sigmoid response for the braconid parasitoid, *Aphidius uzbeckistanicus*, parasitizing the aphid, *Hylopteroides humilis* (Dransfield, 1975). *J* represents the range of prey densities over which the response is density dependent.

the form of the response depending on whether a' and T_s are constants, or vary with N_a. Substituting eq. (6.7) into eq. (6.8), we have

$$N_a = a' N_t (T - T_h N_a)$$

or

$$N_a = \frac{a' N_t T}{1 + a' T_h N_t}. \tag{6.9}$$

The function in eq. (6.1) now becomes

$$f(N_t, P_t) = \exp\left(-\frac{a' T P_t}{1 + a' T_h N_t}\right), \tag{6.10}$$

which clearly reverts to the Nicholson–Bailey model [eq. (6.3)] when $T_h = 0$.

The inclusion of the additional parameter T_h is inevitably destabilizing in a population model since predation is now proportionately greater at low rather than high prey densities. The extent of this additional instability depends, however, not on the absolute value of T_h but on the value relative to the time available (i.e. T_h/T). Table 6.1 gives estimated values for T_h and of T_h/T for a variety of parasitoids and predators. In all cases, handling is a very small fraction of total time and thus will have a very small destabilizing effect on population models. This is argued more formally in Hassell and May (1973).

A population model following from eq. (6.10) would be suitable as a first approximation to describe a parasitoid–host interaction where generations are discrete, search is random and there are no complications of parasitoid mutual interference (see below). It would be much less applicable to true predators, whose searching performance will vary during development. In other words, we may expect the parameters a' and T_h to scale markedly with both the age (or, equivalently, size) of the predator and, similarly, with the size of prey taken. The ways in which they may do so are illustrated in Fig. 6.2 from Thompson (1975). No longer will a single functional response serve as a description for predation as a whole. We now have a matrix of responses and must be able to predict the age structure of both predators and prey in order to model adequately the prey death rate.

Finally, we should note that while many functional responses *are* of the form shown in Fig. 6.1b, several arthropod predators exhibit

Table 6.1. Estimated values of handling time (T_h) for a selection of parasitoids and predators, using eq. (6.10) [see Rogers (1972a) for details of method] except for *Pleolophus* and *Nemeritis* where T_h was directly observed. The values of T_h/T are based on conservative estimates of longevity.

Parasite or predator species	Host or Prey	Handling time T_h (hrs)	$\dfrac{T_h}{T}$	Author(s)
Parasitoids				
Nemeritis canescens	*Ephestia cautella*	0·007	< 0·0001	Hassell & Rogers (1972)
Chelonus texanus	*Ephestia kühniella*	0·12	< 0·001	Ullyett (1949a)
Dahlbominus fuscipennis	*Neodiprion lecontei*	0·24	< 0·003	Burnett (1958)
Pleolophus basizonus	*Neodiprion sertifer*	0·72	< 0·02	Griffiths (1959)
Dahlbominus fuscipennis	*Neodiprion sertifer*	0·96	< 0·01	Burnett (1958)
Cryptus inornatus	*Loxostege sticticalis*	1·44	< 0·02	Ullyett (1949b)
Nasonia vitripennis	*Musca domestica*	12·00	< 0·1	DeBach & Smith (1941)
Predators				
Anthocoris confusus (5th Instar)	*Aulacorthum circumflexus*	0·38	< 0·001	Evans (1973)
Notonecta glauca (1st Instar)	*Daphnia magna*	0·76	< 0·005	B. H. McArdle (unpublished)
Ischnura elegans (12th Instar)	*Daphnia magna*	0·82	< 0·002	Thompson (1975)
Harmonia axyridis (2nd Instar)	*Aphis craccivora*	1·61	< 0·002	Mogi (1969)
Phytoseiulus persimilis (Adult ♀)	*Tetranychus urticae*	1·87	< 0·005	Pruszynski (1973)

rather different (e.g. humped, sigmoid) responses which are not adequately described by equations such as eq. (6.9). These are reviewed in Hassell *et al.* (1976b). Of particular interest are the sigmoid responses, since predation is now density dependent, and therefore may contribute to stability at least over a range of prey densities (J in Fig. 6.1c). The example in Fig. 6.1c results from more active search whenever prey are plentiful rather than scarce. In effect T or perhaps a' in eq. (6.9) is now itself a function of prey density. Such examples are probably more widespread amongst invertebrates than previously thought (Murdoch and Oaten, 1975; Hassell *et al.*, 1977).

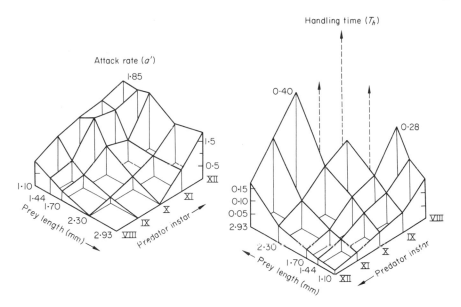

Fig. 6.2. Effect of both predator and prey size on the attack rate (a') and handling time (T_h), for the damselfly, *Ischnura elegans*, feeding on *Daphnia magna*. From Thompson (1975), courtesy of the British Ecological Society.

6.3   The response to predator density

Within the confines of a laboratory cage, predators and parasitoids often react markedly to the immediate presence of another individual of the same species. Thus, coccinellid larvae, predatory mites and the parasitoids, *Nemeritis canescens* and *Diaeretiella rapae*, have all been observed to respond to encounters by an increased tendency for local dispersal. The same tendency has been observed in *N. canescens* following the detection by the female that a host has already been parasitized (Rogers, 1972b). In addition, the considerable literature, particularly for egg parasitoids, on the ways that females mark their hosts and show aggressive behaviour to other females, suggests that some mutual interference occurs under natural conditions.

It is to be expected, therefore, that interference in a simple laboratory system will reduce the searching efficiency per predator as predator density is increased, the underlying cause being a reduction in searching time following each predator encounter. Clear examples of this, for both predators and parasitoids, are shown in Fig. 6.3. Some of the relationships are noticeably curvilinear (Fig. 6.3a, b), as indeed they must be over a sufficient range of predator density. Models that describe curvilinear relationships have been proposed by Royama (1971), Rogers and Hassell (1974) and Beddington (1975). The discussion is simplified, however, if a more empirical description of the data is adopted, where a linear relationship between log (searching efficiency) and log (predator density) is fitted as in Fig. 6.3c, d, e (Hassell and Varley, 1969). Thus,

$$\log a = \log Q - m \log P_t$$

or

$$a = Q P_t^{-m}, \tag{6.11}$$

where m is the slope and log Q the intercept of the linear regression.* The function for predation in eq. (6.1) now becomes

$$f(N_t, P_t) = \exp\left(-Q P_t^{1-m}\right). \tag{6.12}$$

Note that this reverts to the Nicholson–Bailey model, eq. (6.3), when $m = 0$.

We now have a simple population model that includes predator interference and whose stability properties may be readily analysed following the recipe in Hassell and May (1973). This gives the stability boundaries shown in Fig. 6.4 where stability hinges solely on the value of the interference constant m and the prey rate of increase λ. The third parameter Q only affects the equilibrium levels of the populations without affecting stability. In evaluating the significance of interference, it is a useful step to consider the values of m obtained from laboratory and field studies in the context of this stability diagram. The placing of points on the graph, however, is made difficult by having no real estimates of the prey rate of increase λ which, of course, is not merely the prey fecundity but is the rate of increase per prey after allowance for all prey mortalities other than predation (Hassell et al., 1976a). Known values of m are therefore given in Table 6.2, distinguishing between field and laboratory studies and also between predators and parasitoids.

The field values should be treated with caution since any errors in the estimates of the density of searching predators (which will usually be considerable) will tend to produce unrealistically high values for m

* Note that eq. (6.11) is dimensionally incorrect and should read as $a = Q(Q P_t)^{-m}$. It is retained, however, to be consistent with past usage.

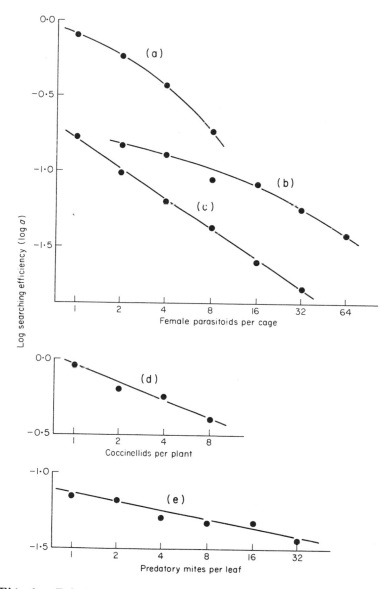

Fig. 6.3. Relationships between searching efficiency (log *a*) and log density of searching parasitoids or predators. (**a**) *Pseudeucoila bochei* (Bakker *et al.*, 1967); (**b**) *Encarsia formosa* (Burnett, 1958); (**c**) *Nemeritis canescens* (Hassell, 1971); (**d**) *Coccinella semptempunctata* (Michelakis, 1973); (**e**) *Phytoseiulus persimilis* (J. Fernando, unpublished).

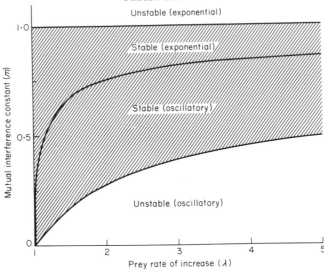

Fig. 6.4. Stability boundaries between the mutual interference constant
(m) and the prey rate of increase (λ) [see eqs. (6.1) and (6.12)]. The
shaded area denotes the conditions for stability and is divided into two
regions: where the equilibrium is approached monotonically and where
there are damped oscillations. The line between these regions indicates
the conditions for most rapid approach to the equilibria. From Hassell
and May (1973), courtesy of the British Ecological Society.

(Hassell and Varley, 1969). It is clear from this analysis that the degree
of interference found from laboratory studies would be a powerful
stabilizing mechanism if extended directly to natural population interactions. However, Free *et al.* (1977) conclude that significant amounts
of interference are in fact unlikely to occur at equilibrium population
densities. They argue that the value of m at equilibrium is largely
determined by the ratio of the time wasted by a predator following a
single interference encounter (T_w) to the duration of its searching lifetime (i.e. T_w/T), a ratio that must often be considerably less than the
m-values in Table 6.2. (This conclusion is along much the same lines as
that mentioned above for the ratio T_h/T.) Viewed in this way,
interference is only likely to be important when predator densities are
well above their equilibria.

So far we have only considered interference by randomly searching
predators (eq. 6.12), which makes it seem a maladapted behaviour
since the overall prey consumption per predator or hosts parasitized per
parasitoid is reduced. In the next section it becomes clear that the
adaptive significance of interference is better viewed in conjunction
with the predator's response to the spatial distribution of its prey.

Table 6.2. Some values for the interference constant m obtained from laboratory and field studies. (From Hassell, 1978.)

Species	m	Field (F) or Lab (L)	Parasitoid or Predator	Author(s)
Phytoseiulus persimilis $\begin{cases} \text{0·18*} \\ \text{0·44†} \end{cases}$		L	Predator	Fernando (1977)
Anagyrus pseudococci	0·18	L	Parasitoid	M. J. Berlinger (pers. comm.)
Dahlbominus fuscipennis	0·28	L	Parasitoid	Burnett (1956)
Aphytis coheni	0·33	L	Parasitoid	D. J. Rogers (pers. comm.)
Leptomastix flavus	0·33	L	Parasitoid	M. J. Berlinger (pers. comm.)
Aphidius uzbeckistanicus	0·35	L	Parasitoid	Dransfield (1975)
Coccinella septempunctata	0·38	L	Predator	Michelakis (1973)
Cryptus inornatus	0·38	L	Parasitoid	Ullyett (1949b)
Encarsia formosa	0·38	L	Parasitoid	Burnett (1958)
Alaptus fusculus	0·39	F	Parasitoid	Broadhead & Cheke (1975)
Bracon hebetor	0·44	L	Parasitoid	Benson (1973)
Bathyplectis anurus	0·47	L	Parasitoid	Latheef, Yeargen & Pass (1977)
Telenomus nakagawai	0·48	F	Parasitoid	Nakasuji, Hokyo & Kiritani (1966)
Cyzenis albicans	0·52	F	Parasitoid	Hassell & Varley (1969)
Chelonus texanus	0·54	L	Parasitoid	Ullyett (1949a)
Aptesis abdominator	0·60	L	Parasitoid	von B. Sechser (pers. comm.)
Diaeretiella rapae	0·65	L	Parasitoid	Chua (1975)
Nemeritis canescens	0·67	L	Parasitoid	Hassell (1971a)
Pseudeucoila bochei	0·68	L	Parasitoid	Bakker, Bagchee, van Zwet & Meelis (1967)
Cratichneumon culex	0·86	F	Parasitoid	Hassell & Varley (1969)
Olesicampe benefactor	0·91	F	Parasitoid	Ives (1976)
Apanteles fumiferanae	0·96	F	Parasitoid	Miller (1959)
Phygadeuon dumetorum	1·13	F	Parasitoid	Kowalski (1977)

* Adult females searching for deutonymph prey.

† Adult females searching for larval prey.

6.4   The response to prey distribution

Most prey populations under natural conditions will exhibit a clumped distribution between units of their habitat, whether they be trees, branches, leaves, etc. Within this framework, the random search assumed in eqs. (6.3), (6.10) and (6.12) implies that on average the same number of predators search for the same period of time per unit area (or patch). In this way each prey individual has an equal probability of being discovered. While this is a convenient assumption mathematically, it is not in accord with observed, or indeed expected, predator behaviour. Figure 6.5 shows some examples where arthropod predators

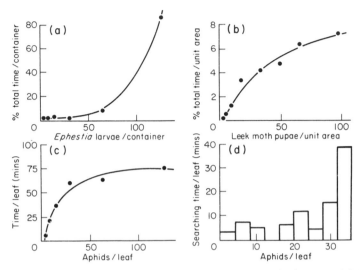

Fig. 6.5. Some examples of aggregative responses in insects. (a) the ichneumonid parasitoid, *Nemeritis canescens*, searching for different densities of flour moth larvae (*Ephestia cautella*) per container (Hassell, 1971); (b) the ichneumonid parasitoid, *Diadromus pulchellus*, searching for leek moth pupae (*Acrolepia assectella*) per unit area (Noyes, 1974); (c) the braconid parasitoid, *Diaeretiella rapae*, searching for aphids (*Brevicoryne brassicae*) per leaf (Akinlosotu, 1973); (d) the coccinellid, *Coccinella septempunctata*, searching for aphids (*B. brassicae*) per leaf (M. P. Hassell, unpublished). After Hassell and May (1974).

and parasitoids have clearly spent a disproportionate time in patches of high prey or host density. The possible means by which arthropod predators aggregate in this way are various. They may, for example, respond to a volatile substance emanating from the prey, as clearly shown for some predators or parasitoids of bark beetles (Wood *et al.*, 1968). Alternatively, or in addition, aggregation may be the end result of a tendency merely to remain for longer periods searching the region where a prey has already been encountered (Murdie and Hassell, 1973).

The importance of such behaviour lies in the way that aggregation can contribute to the stability of the interacting populations. This can be illustrated by a variety of models for predator aggregation. Let us commence with the Nicholson–Bailey model, eqs. (6.4) and (6.5), but now divide the prey and predator populations between n patches such that in the ith patch there is a fraction α_i of the prey and β_i of the predator population. The function for predation now becomes

$$f(N_t, P_t) = \sum_{i=1}^{n} [\alpha_i \exp(-a\beta_i P_t)], \qquad (6.13)$$

which serves to distribute in each generation P_t predators and N_t prey into the n patches in the proportion specified by α_i and β_i. This, of course, will reduce to the Nicholson–Bailey model in the event that the same fraction of the predator population searches in each ith patch.

Let us now consider a very simple model for generating the predator distribution per patch β_i in eq. (6.13), discussed by Hassell and May (1973):

$$\beta_i = c\alpha_i^{\mu}, \qquad (6.14)$$

where μ is an 'aggregation index' and c a normalization constant. The value of μ thus governs the degree of predator aggregation in patches of high prey density, with $\mu = 0$ corresponding to an even distribution of predators and $\mu = \infty$ to all predators being in the highest density patch, with all others being complete refuges from predation. Stability now hinges on three things:

(1) μ, the predator 'aggregation index'. Increasing values of μ tend to increase stability, provided the prey distribution is sufficiently uneven.
(2) The prey distribution. Stability is generally enhanced for a given value of μ if the prey distribution is made more clumped. Indeed, if there is too little contrast in the prey distributions per patch, no amount of predator aggregation will be sufficient for stability.
(3) λ, the prey rate of increase. Stability decreases as λ is made larger.

The underlying explanation for this stability rests on the differential exploitation of the prey in different patches. By aggregating in the highest prey density patches, the predators, at least initially, increase their encounter rate with prey above that possible from random search. The lower prey density patches thus become partial refuges from predation, and it is this refuge effect that contributes to stability. Interestingly, these effects can also be demonstrated from the relationship between searching efficiency (a) and predator density, as shown in Fig. 6.6. The curve A (where $\mu = 1$) shows that the searching efficiency at low predator densities is greater than that for random search (curve B, $\mu = 0$). This emphasizes the initial advantage to the predators that

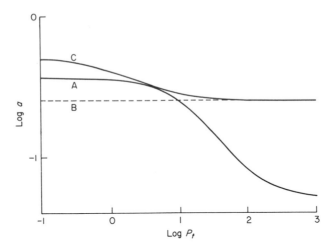

Fig. 6.6. The relationship between searching efficiency and predator density from three different models in which the predators forage over four patches (10, 6, 3, 1 prey). Curve A: a fixed aggregation strategy with $\mu = 1$ in eq. (6.14). Curve B: random search (i.e. $\mu = 0$). Curve C: optimal foraging. After Comins and Hassell (1979).

concentrate on high density patches. Increasing the predator density, however, leads to a sharp decline in searching efficiency, below that for random search. This arises because the predators have a fixed aggregation strategy defined by μ, and thus continue to search in the originally high prey density patches, even after almost all the prey there have been attacked. Clearly, real predators would have abandoned such patches to seek richer pastures elsewhere. Curve C in Fig. 6.6 stems from just such predators—in this case foraging optimally so as to maximize their overall rate of encounters with prey (Cook and Hubbard, 1977; Comins and Hassell, 1979; Hassell, 1979). The relationship arises in just the same way as for curve A; by the differential exploitation of patches of different prey density. The important difference between curves A and C is that curve C lies consistently above that for random search, however many predators are present. This emphasizes the selective advantage of having such a flexible foraging behaviour.

Curves A and C in Fig. 6.6 are reminiscent of the interference relationships in Fig. 6.3, but in this case they are entirely due to the non-random distribution of predators. For this reason Free *et al.* (1977) have dubbed such relationships as 'pseudo-interference'. In natural situations it will always be difficult to distinguish between this apparent interference due to non-random search and real interference resulting from mutual encounters between predators. In any event, the contribution to stability depends upon the overall relationship between a and P_t, irrespective of whether it is primarily due to mutual interference or is the result of non-random search.

So far we have considered mutual interference and predator aggregation as separate processes that happen to lead to similar relationships between searching efficiency and the extent of prey exploitation. In fact, they are likely to be further intertwined in that whenever predators aggregate in relatively restricted parts of their habitat, the likelihood of interference is enhanced. It is in this light that the adaptive role of interference should be viewed: as a mechanism ensuring a greater dispersal rate of predators from areas where prey are already, or likely to become, heavily exploited. The dispersing predators then have the chance of locating other less-exploited areas of prey. In this light, interference can now be viewed as a mechanism contributing towards a continually changing predator distribution per patch, such that each predator encounters prey at a higher rate than would occur if they searched at random, or had no means of responding to the distribution of available prey.

We saw in an earlier section that predators, rather than parasitoids, present the complication of differing functional responses depending upon predator and prey sizes. In a similar way, the level of interference or the degree of aggregation amongst predators is also likely to vary both throughout predator development and also for different prey sizes. Information on such inter-age class effects, however, is at present lacking.

6.5   The predator rate of increase

It has already been stressed that a parasitoid is a rather special type of predator whose rate of increase depends simply on the number of hosts parasitized by the adult female population. Such assumptions are quite inadequate for true predators, whose overall rate of increase depends crucially upon (1) the survival of the immature stages and (2) the fecundity of the surviving adults, both of which have a complex dependence on the number of prey eaten throughout the life cycle. The predator's rate of increase is, therefore, likely to be affected by any factors that have an important effect on searching efficiency; in particular, the predator's response to prey density, predator density and to the prey distribution. Unfortunately, adequate information is only available on one of these—the effect of prey density—which is very briefly examined in this section. A much fuller discussion is to be found in Beddington *et al.* (1976b).

The way in which the survival of immature predators to the adult stage can depend upon prey density is shown in Fig. 6.7a, b. Survival here is almost complete at high prey densities, but progressively declines

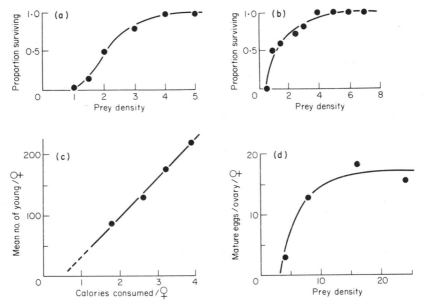

Fig. 6.7. (a) and (b) Relationships between mean prey density during particular instars and the proportion of individual predators surviving to the end of the instar. (a) First instar of the coccinellid, *Adalia bipunctata* (Wratten, 1973); (b) survival through first and second instar of the spider, *Linyphia triangularis* (Turnbull, 1962). (c) Reproductive rate of *Daphnia pulex* var. *pulicaria* as a function of feeding rate (Richman, 1958). (d) Fecundity of *Adalia decempunctata* as a function of aphid density available (Dixon, 1959). After Beddington *et al.* (1976b).

as less and less food is available. In addition to this all-or-none effect (starvation), low prey densities will reduce the predator's growth rate and hence prolong the developmental period. Apart from increasing the likelihood of mortality from other sources, this may have complex effects on the weight and fecundity of the ensuing adult.

Once in the adult stage, there remains an important dependence of fecundity per female on the number of prey eaten. Typically, in arthropods, this takes the form of a linear relationship above some critical level of prey consumption, as shown in Fig. 6.7c; below this, all the energy intake is allocated to maintenance of the predator and none to egg production. Following this, the relationship between fecundity and prey *density* rather than prey *eaten* (Fig. 6.7d), must reflect the form of the female predator's response to prey density (its functional response) except again for the displacement along the prey axis.

First steps in the mathematical description of these components have been taken by Beddington *et al.* (1976b). There now remains the task of exploring (analytically, if possible) the effects of such more realistic predator reproduction on the dynamics of predator–prey interactions. This is made somewhat more difficult by generation times no longer being absolutely fixed, but dependent on prey density itself. The effects on stability that will emerge are not easy to predict since there are now important time lags associated with larval survival and development rates affecting future adult fecundity.

6.6 Models and biological control

Biological control has emerged as a powerful pest control technique during the last one hundred years or so, although its origins go back much further. The normal procedure in 'classical' biological control programmes against exotic pests is to seek natural enemies, and especially parasitoids, in the general region where the pest originated. These are then imported, screened for harmful side effects and bred in large numbers prior to release. A successful programme is characterized by the parasitoids becoming established and causing a marked decline in the pest population, followed by both parasitoid and pest populations persisting at relatively low densities.

Many examples of such successes now exist and are reviewed in several recent texts on biological control (e.g. Huffaker, 1971; De Bach, 1974; Huffaker and Messenger, 1976). It is clear from these that biological control has been much more successful against pests of such standing crops as fruit and forest trees than against pests of annual crops (see Table 1 of Southwood, 1977b). The most plausible explanation for this is simply that a perennial standing crop permits a continuous interaction between host and parasitoid, without the ecological upheavals associated with the cultivation of an annual crop. We shall focus, therefore, on the perennial system since it relates much better to the structure of eqs. (6.1, 6.2).

A successful biological control programme requires two things: (1) that the average abundance of the host population (≡ equilibrium) be depressed below some economic level; and (2) that this equilibrium be sufficiently stable to prevent the host population from sporadically re-emerging as a pest. Two examples showing this are given in Fig. 6.8. These evocative pictures, which must broadly resemble those from many of the completely successful biological control projects (e.g. Huffaker and Kennett, 1966; DeBach *et al.*, 1971) prompt two obvious questions: (1) what determines the extent to which the parasitoids depress their host equilibrium? (2) what is responsible for the stability of this equilibrium? These will now be considered in turn.

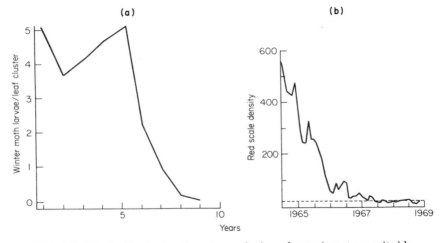

Fig. 6.8. The decline in two insect populations due to insect parasitoids. (a) Winter moth (*Operophtera brumata*) in Nova Scotia following the establishment of two parasitoid species, *Cyzenis albicans* and *Agrypon flaveolatum* (after Embree, 1965). (b) California red scale (*Aonidiella aurantii*) after a DDT-induced outbreak. The decline in density is due to the re-establishment of *Aphytis melinus* and *Comperiella bifasciata* (after DeBach *et al.*, 1971). From Hassell (1981).

6.6.1   *Equilibrium levels*

In simple population models such as that of eqs. (6.4, 6.5), the level of the host equilibrium (N^*) depends upon the host's rate of increase (λ) and the parasitoid's searching efficiency (a), with an increase in λ and a decrease in a both raising N^*. The problem in relating this simple statement to natural interactions is the neglect in the models of the many other hazards that normally affect host and parasitoid populations. It is quite unrealistic to assume that a host population suffers no mortalities other than parasitism, or that the parasitoid population is itself free from mortality. Such factors can be crucially important in determining equilibrium levels (Hassell and Moran, 1976; Hassell, 1977). It is, therefore, the balance between the *effective* host rate of increase after taking into account all factors other than parasitism, and the *overall* performance of the parasitoids after due allowance for all factors affecting the survival of the immature parasitoid progeny, that determine the observed equilibrium levels. A graphic example of how important these considerations can be is given in Hassell (1980) for the interaction of the winter moth (*Operophtera brumata*) and one of its parasitoids (*Cyzenis albicans*) in England and in Canada.

By comparing average host abundances before and after the introduction of parasitoids, several examples of how parasites can depress their host equilibrium can be extracted from the literature (Beddington *et al.*, 1978). A convenient means of quantifying this depression is from the expression

$$q = \frac{N^*}{K}, \qquad (6.15)$$

where q is the ratio of host equilibrium in the presence of the parasitoid

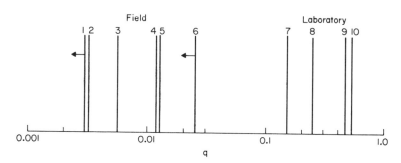

Fig. 6.9. Some values of $q(=N^*/K)$ from field (1 to 6) and laboratory (7 to 10) studies of parasitoid-host interactions. (1) *Aphytis melinus–Aonidiella aurantii*; (2) *Trioxys pallidus–Chromaphis juglandicola*; (3) *Aphytis maculicornis–Parlatoria oleae*; (4) *Aphytis melinus–Aonidiella aurantii*; (5) *Cyzenis albicans–Operophtera brumata*; (6) *Olesicampe benefactor–Pristiphora erichsonii*; (7) *Nasonia vitripennis–Musca domestica*; (8) *Neocatolaccus mamezophagus–Callosobruchus chinensis*; (9) *Nemeritis canescens–Anagasta kuhniella*; (10) *Heterospilus prosopidus–Callosobruchus chinensis*. From Beddington *et al.* (1978), in which source references are given.

(N^*) to that in its absence (the host's carrying capacity K) (Beddington et al., 1975, 1978; Hassell, 1978; May and Hassell, 1980). Using the parameter q has the merit that a variety of laboratory and field examples can be compared on a common scale, as done in Fig. 6.9 (Beddington et al., 1978). Clearly, the successful biological control examples (1–6) are characterized by very small values of q, much smaller than those observed from laboratory cultures (7–10). A plausible explanation for this difference emerges when we consider the likely cause of the stability of these equilibria.

6.6.2 Stability

A characteristic of completely successful biological control projects is the absence of periodic pest outbreaks, unless caused by human intervention, indicating a relatively stable parasitoid–host equilibrium over many generations. It is important, therefore, that we should seek to identify the factors that may be responsible for this stability.

One possibility is that the persistence of interactions such as shown in Fig. 6.8 is not due to the parasitoid at all, but to some density dependence acting independently on the host population. It is well known that an otherwise quite unstable parasitoid–host interaction can be rendered stable simply by introducing a host fecundity or mortality that is density dependent (Beddington et al., 1975, 1976). This, however, is unlikely to be the general solution that we seek, on two counts: (1) host resource limitation cannot be invoked at the very low population levels achieved in successful biological control projects; (2) one should be uneasy about any theory demanding that parasitoid–host interactions are persisting only by grace of some quite independent host mortality (see p. 108).

Two other possibilities are that mutual interference or sigmoid functional responses are the major causes of stability. However, these too must be discarded. We have already seen that mutual interference can be strongly stabilizing, but that this is unlikely to be the case at equilibrium levels (Free et al., 1977), especially at the very low levels following successful biological control. Similarly, sigmoid functional responses must be rejected, at least when considering the synchronized, specific parasitoids of eq. (6.1). The time delays in such a system make it almost impossible for the density dependence from a sigmoid functional response alone to stabilize the populations (Hassell and Comins, 1978).

A further possibility arises when we turn to parasitoids that, instead of searching randomly, tend to aggregate where hosts are most abundant. Like mutual interference, this can also contribute strongly to stability, but unlike interference, it can be just as effective at very low equilibrium levels as when hosts are more abundant. From the stance of biological control this non-random search is important because, for the first time, we have a plausible mechanism by which parasitoids can stabilize those interactions in Fig. 6.9 where q-values are very small. Figure 6.9 also shows the q-values from laboratory systems to be much higher (0·15 to 0·5) falling into a quite different category from the successful biological control results. Such smaller depressions in host equilibrium are more in accord with predicted values from a Nicholson–Bailey model in which the sole cause of stability comes from the addition of a density dependent host rate of increase [see eq. (6.16) below]. Interestingly, in at least three of the four laboratory systems, the hosts were probably not sufficiently uneven in their distribution for non-random search to be a significant stabilizing influence.

Two crucial requirements related to host-finding by parasitoids therefore emerge for a successful biological control agent: (1) a high effective searching efficiency relative to the host's rate of increase, to achieve a sufficient depression in the host equilibrium ($q \ll 1$), and (2) marked parasitoid aggregation in patches of high host density, to ensure that this equilibrium is stable.

While these two features can be independently varied within our model framework [e.g. a and μ in eqs. (6.13, 6.14)], they are unlikely to be independent in the real world, at least for more-or-less specific and synchronized parasitoids. Such species are more likely than polyphagous ones to respond to particular cues from a given host species [e.g. host pheromones (Mitchell and Mau, 1971; Sternlicht, 1973)]. There are two obvious consequences of such efficient host location.

(1) The searching efficiency (a) for a particular host species will be higher than in a polyphagous parasitoid searching for the same host.

(2) Patches of high host density will be more efficiently discovered leading to a higher value for the 'aggregation index', in eq. (6.14).

A long standing tenet amongst biological control workers has been that success is more likely using natural enemies that are fairly specific to the pest species in question. It was supposed that introduced polyphagous natural enemies would not show an adequate numerical relationship with any one of their prey species. While this may well be true, we can now see that the use of specific species has also probably increased the likelihood of optimal searching characteristics.

The situation in annual crop systems is quite different. Under these regularly disrupted conditions, the concept of equilibrium populations is less appropriate, at least on an annual time scale. Much more important is that the natural enemies should be efficient colonizers and capable of rapid rates of increase, attributes more likely to occur in polyphagous species. This point has been convincingly made by Force (1972), Ehler and van den Bosch (1974), Ehler (1977) and Dowell and Horn (1977).

6.7 Predation and competition

Predation is only one of several mortalities that affect prey populations under natural conditions. Some of these will depend on the physical environment; others, such as competition, are biological interactions. While the components of predation in these more complex systems will continue to have effects similar to those already described, the overall stability properties of the prey population now depend on the nett effect of a variety of stabilizing and destabilizing factors. This is best illustrated by combining, within a single model, both predation and intraspecific competition amongst the prey.

Single species competition models and their full range of stability behaviour have been reviewed in chapter 2. In general, they have a stable equilibrium, at least for low rates of population increase, and hence should continue to contribute to stability when included in predator–prey models. This has been well demonstrated by Beddington *et al.* (1975) using an extension of the Nicholson–Bailey model [eqs. (6.4) and (6.5) above] in which they included a version of the discrete logistic model (the form B of Table 2.2). The expression for N_{t+1} in eq. (6.4) now becomes

$$N_{t+1} = N_t \exp\left[r(1 - N_t/K) - aP_t\right]. \tag{6.16}$$

Here K is the 'carrying capacity' of the prey population and $r = \ln \lambda$. In contrast to the Nicholson–Bailey equations, there is now a wide range of conditions in this model in which the equilibrium is stable. This work is discussed more fully in section 5.2.3 [see eqs. (5.11) and (5.12)].

A stimulating view of the interaction of predation and competition has been presented by Southwood (1975) and Southwood and Comins

(1976), in which a framework is presented for interpreting several kinds of population fluctuations (see chapter 3 for a fuller discussion of this synoptic approach). The essential ingredient is that the *total* response of the predators (expressed in terms of the numbers of prey eaten by the predators per generation $N_a P_t$) should show a sigmoid relationship with prey density. This can be achieved in several ways, two of which are shown in Fig. 6.10:

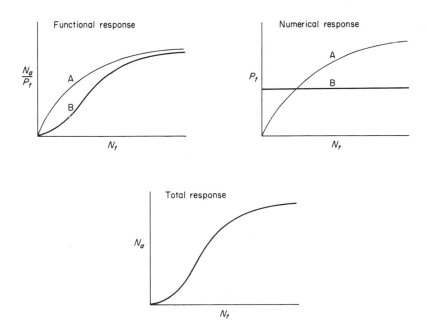

Fig. 6.10. Schematic figures showing alternative ways of obtaining a sigmoid total response between the prey eaten by P_t predators, N_a, and prey density N_t. The total response curve may be obtained by combining either functional response A or B with numerical response A, or from functional response B without any numerical response. From Hassell 1978.

(1) by the combination of a numerical response (A) with a convex or sigmoid functional response (A or B);
(2) by the combination of a sigmoid functional response (B) with no numerical response (B).

If intraspecific competition is now added, as in eq. (6.16), and we assume the second alternative above, we obtain the model:

$$N_{t+1} = N_t \exp\left[r(1 - N_t/K)\right] f(N_t, P_t), \qquad (6.17)$$

$$P_{t+1} = P_t = P^*, \qquad (6.18)$$

where

$$f(N_t, P_t) = \exp\left[-\frac{bN_tP_t}{1 + cN_t + bT_hN_t^2}\right], \qquad .(6.19)$$

which is a sigmoid analogue of the disc equation (6.10) with b and c both constants.

The properties of such a model can be conveniently illustrated by plotting the numbers of prey in successive generations as done in Fig. 6.11 (c.f. Fig. 3.5). An equilibrium, whether stable or unstable, will

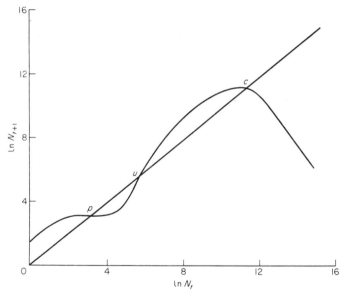

Fig. 6.11. Relationship between population densities in successive generations (ln N_t and ln N_{t+1}) from a predator–prey model with intra-specific competition acting upon the prey. Points p, c and u (where the curve intersects with the 45° line) represent stable equilibria due to predation (p) and competition (c) and an unstable equilibrium (u). After Southwood (1975).

occur whenever the curve crosses the 45° line. In this example, there is a lower stable equilibrium (p) maintained by the predator (due to the density dependent part of the predators' combined functional and numerical responses). Should prey density rise above point u, however, the prey population increases, unchecked by the predators which are now limited by handling time, satiation or their slower rate of increase, until finally the prey are regulated at a much higher stable equilibrium (c) by intraspecific competition. Whether the population remains at this level or invariably crashes, to be controlled again by the predator, depends upon the parameters of competition—in particular, on whether or not the prey 'scramble' and overexploit their resources. In this way, the model describes instances where the prey are normally regulated by predators, but occasionally 'escape' and increase to the limit of their resources. Furthermore, it is easily extended, by altering the properties of the predators, to include examples between the extremes of a globally stable predator–prey equilibrium to no such equilibrium at all.

In coupled predator–prey systems where time lags become very important, this elegant picture will be somewhat blurred since predation will now also hinge importantly on the way that the predator population is fluctuating. However, the general picture should remain, provided that predation tends to be more-or-less density dependent up to a critical level of prey abundance.

7

Plant–Herbivore Systems

GRAEME CAUGHLEY and JOHN H. LAWTON

7.1 Introduction

'Herbivore' is a catch-all word, encompassing not only cows, horses, sheep and their wild relatives, but also phytophagous insects, and beasts as different as giant tortoises, slugs and grouse; *Daphnia* and protozoa in ponds; limpets and urchins on the sea-shore and teeming zooplankton in the open ocean. They graze on food ranging from blue-green algae to beech leaves, and *Lemna* to *Laminaria*. Not surprisingly there is no such thing as a typical herbivore. At the risk of enraging vertebrate ecologists and range managers, who use 'grazing' in the strict sense of 'eating grass and herbs', we have used the word to mean 'eating any kind of plant or part of a plant, by any kind of animal'.

In order to make the chapter manageable, we have focussed on two main taxa: phytophagous insects because there are so many of them, and mammalian herbivores because what they lack in species they more than make up for in economic importance, impact on ecosystems, and aesthetics. The main insect groups feeding on plants are described by Southwood (1973): the total number of species involved is unknown, but there are roughly 120,000 different species of caterpillars in just the one order Lepidoptera for example, compared with a paltry 212 ungulates (Proboscidea, Perrissodactyla and Artiodactyla). Even the rodents muster less than 2000 species. Phytophagous insects and mammals aside, our examples are an eclectic selection from the ark, guided more by expediency than the gathering of a representative zoo.

The chapter opens (section 7.2) by providing a classification of grazing systems, to which examples can be assigned with varying degrees of confidence (section 7.3–7.4). Lotka–Volterra models provide the underlying theoretical frame, building directly on classical predator–prey interactions (chapter 5). This has been done deliberately, even though we know plants and their herbivores are different in many ways from predators and prey. Unfortunately the empirical data are lacking to build more realistic general models. Instead we have chosen to work from a familiar base, pointing to obvious flaws and elaborations as the chapter progresses. Eventually, we would hope to be able to abandon the base altogether. The most interesting elaborations and the most promising framework on which to construct the next generation of models are dealt with in section 7.5. Perhaps more than any other chapter we would expect this one to look totally different in 10 years time.

7.2   A classification of grazing systems

Table 7.1 is a classification of grazing systems in which the primary division is made according to Monro's (1967) dichotomy: between those systems in which the herbivores influence the amount of food available

Table 7.1. A classification of relationships between plants and herbivores.

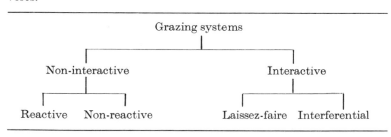

to themselves and subsequent generations, and those where they do not. The first is an 'interactive system', the second 'non-interactive'.

The essence of a *non-interactive system* lies in the herbivore's inability to influence the rate at which its resources are renewed. It subdivides uneasily into 'reactive systems' where the rate of increase of herbivores reacts to the rate at which plant material is renewed (there being no reciprocal reaction), and 'non-reactive systems' in which the herbivores increase at a rate largely independent of the parameters of plant growth.

In an *interactive system* the herbivores do influence the rate of renewal of the vegetation, which in turn influences the rate of increase of the animals: the two components interact. This class divides again into 'laissez-faire systems' in which the herbivores do not interfere with each other's feeding activities, and 'interferential systems' in which an animal may reduce the ability of another to obtain food that would otherwise be available to it. These two systems have different dynamics.

7.3 The dynamics of grazing systems: non-interactive systems

7.3.1 *Theory and examples of reactive systems*

This, our first and simplest example, is also, not unreasonably, the most unrealistic.

If the rate of renewal of the food of the herbivores, g, is independent of the standing crop of plants, V, food availability changes as

$$dV/dt = g. \tag{7.1}$$

When a herbivore population of size H feeds on these plants, with each herbivore eating at a constant rate c, the change in V might become

$$dV/dt = g - cH. \tag{7.2}$$

Alternatively, if each herbivore has a maximum rate of intake, c, which declines as the food supply is thinned out, then

$$dV/dt = g - cH(1 - \exp(-dV)). \tag{7.3}$$

Real field examples approximating to such a system are extremely rare. In the deserts of the south-western United States, the larvae of several species of *Drosophila* feed on cacti. Strictly they are not herbivores but eat yeasts which live in rot-pockets in the cactus (Fellows and Reed, 1972; Starmer *et al.*, 1976). The adults however share many of the problems of host–plant location and exploitation confronting real insect–herbivores, sufficient for us to make them honorary members and give some biological flesh to the bones of an otherwise abstract argument. To avoid misunderstanding, we are here only concerned with what might happen to the cactus–*Drosophila* interaction over a reasonably large area of the desert. Within each rot-pocket larvae and yeast are highly interactive.

Rot-pockets are generated at a rate g, quite independently of the rate at which the *Drosophila* exploit them; generally, the stem of a cactus will undergo some injury (e.g. freezing), rendering the damaged tissue vulnerable to infection. The *Drosophila*'s dynamics can be pictured as follows. If b is defined as the rate of intake sufficient to maintain a herbivore and allow its replacement in the next generation, then the proportion of available food used for maintenance is bH/g, leaving $(1 - bH/g)$ that can be channelled into generating an increase of herbivores. With rising H, the herbivores' rate of increase slows progressively until finally all the food is utilized for maintenance, and the population stabilizes at $H = g/b = K$. Hence the herbivore population grows as

$$dH/dt = rH(1 - bH/g). \qquad (7.4\text{a})$$

Here r is the herbivore's intrinsic rate of increase, the rate at which it would increase in that environment if no resource were limiting. Substituting K for g/b turns the equation into the common form describing logistic growth, eq. (2.3):

$$dH/dt = rH(1 - H/K). \qquad (7.4\text{b})$$

The assumptions underlying this model can therefore be summarized as the characteristics a population must possess if it is to grow logistically:

(a) A population's resources are renewed at a constant rate, independent of both the standing crop of the resource and the standing crop of the population, the population having no influence over the amount of resource available to the next generation.

(b) The members of the population have insatiable appetites.

(c) Rate of increase is a linear function of average food intake per head.

(d) There is no lag between cause and effect.

Of these four conditions for logistic growth only the first is critical.

How clearly the cactus-living *Drosophila* conform to these assumptions is unknown, but we would be surprised if their large-scale dynamics were not roughly logistic. Gause's (1934) classical laboratory populations of *Paramecium*, 'grazing' on the bacterium '*Bacillus pyocyaneus*' (= *Pseudomonas aeruginosa*) do conform to these major assumptions, and pleasingly they behave more or less as predicted (Fig. 8.4a). At the beginning of each experiment Gause placed twenty *Paramecium* in a tube and guaranteed conformity to eq. (7.1) by adding a constant quantity of *Bacillus* every day. Gause (1934), Kostitzin (1939) and Andrewartha and Birch (1954) all fit logistics to the data.

We had ulterior motives for developing the conditions for logistic growth in some detail, because in subsequent sections we use a logistic equation to model the growth in biomass of plant (rather than herbivore) populations. This does not mean we necessarily believe this equation describes how most plant populations behave, but rather because it provides a standard of ideal behaviour against which reality can be judged and measured (Harper, 1977). To the extent that plants cannot influence the rate of input of key resources, particularly sunlight and water, to the next generation, our use of the logistic is justified and has some empirical support (Brougham, 1955; Clatworthy and Harper, 1962; Davidson and Donald, 1958; Ikusima *et al.*, 1955), though even the simplest of systems do not always conform (Laws, 1980: see also 7.4.1). Unfortunately data on plant population dynamics are too sparse to justify a general analysis of plant–herbivore models in which the growth of the vegetation is markedly non-logistic. Providing plant biomass approaches an ungrazed asymptote along any roughly 'S' shaped curve, our models will not be seriously in error.

7.3.2 *Phytophagous insects: reactive and non-reactive examples*

Figure 7.1 summarizes population data for the herbivorous insects on two species of plants, namely bracken fern (*Pteridium aquilinum*) and the grass *Holcus mollis* (Lawton and McNeill, 1979). The insects on both plants share two features typical of many others (see Lawton and McNeill, 1979 for a review). First a small proportion of species may become temporarily very abundant and, when they do, inflict serious damage upon their hosts. Aphids (*Holcaphis*) and thrips (*Aptinothrips*) on *Holcus*, and broom moth, *Ceramica pisi*, on bracken are good examples. (Others are provided by Dixon, 1971a, b; Newbury, 1980a, b; and Tilden, 1951.) Such species clearly belong in the 'Interactive' category and will be returned to later (sections 7.4.2, 7.4.3 and 7.5.1). Second, the majority of species are never very abundant: most in fact are rare or very rare relative to the abundance of their food plants. Table 9.3 makes the same point in a different way. We assume most (though certainly not all) of these rare species have little or no impact upon the dynamics of their hosts (see Harris, 1973), and hence are non-interactive.

Three hypotheses may account for the rarity of most phytophagous insects most of the time:

(1) Contrary to superficial appearances, large parts of the plant are quite unsuitable as food, either because of a low nutritional status (e.g. Dixon, 1970; McNeill, 1973; McNeill and Southwood, 1978; White, 1969); or because they are chemically defended in some way (Feeny,

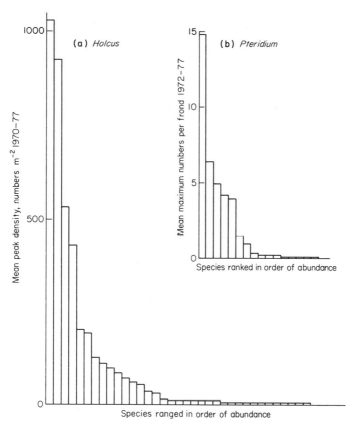

Fig. 7.1. Mean maximum densities of phytophagous insects on (**a**) the grass *Holcus mollis* and (**b**) the fern *Pteridium aquilinum* (Lawton and McNeill, 1979). Most species are rare most of the time.

1976; Rhoades and Cates, 1976); physically defended by the microscopic equivalent of thorns (Pillemer and Tingey, 1976; Rathcke and Poole, 1975); or simply too large, too small or too tough to exploit properly (Dixon and Logan, 1973; Whitham, 1978). Southwood (1973) reviews many of these problems. Populations in this category are, despite appearances, food-limited (van Emden and Way, 1973) and hence lie towards the reactive end of the non-interactive spectrum.

(2) Many species are kept rare most of the time through the vagaries and harshness of their environment. Safe sites are either highly localized and/or unpredictable in space, or conditions are rarely good enough for long enough for species to bump against the ceiling of food limitation (Andrewartha, 1970; Andrewartha and Birch, 1954; Thompson *et al.*,

1976; Whittaker *et al.*, 1979). By definition, such species are non-interactive over most of their habitat, most of the time. About 25 years ago, ecologists debated at great length whether populations could exist in the complete absence of density-dependent controls (which *in extremum* is what is implied by this hypothesis). Probably the best known example is the controversy surrounding the dynamics of *Thrips imaginis*, which feeds on sap and pollen (Andrewartha and Birch, 1954; Varley, 1963; Varley *et al.*, 1973). Irrespective of whether or not there are density-dependent controls in this population (there are) it provides a good example of a predominantly non-interactive herbivore.

(3) Natural enemies, particularly insect predators and parasitoids (chapters 5 and 6), or birds (Holmes *et al.*, 1979), true parasites and diseases (see chapter 14) play a key role in keeping many phytophagous insects rare relative to their food resources (and see Hairston *et al.*, 1960). Where natural enemies are very effective, food-limitation may again be negligible (Fig. 7.13). Successful biological control of former insect pests (Beddington *et al.*, 1978; Fig. 9, chapter 6; see also chapter 15) has transformed some highly interactive grazing systems into deceptively peaceful, non-interactive non-reactive states.

Species of insects from one habitat and food plant may between them span the full gambit of possible behaviours (Fig. 7.2). A summary of life-tables for phytophagous insects is given in Table 7.2. For many of these species, natural enemies are clearly important controlling agents; others appear to be controlled by food limitation, or by disease; and in still others it has not been possible to demonstrate any density dependence. As a sample of the world's phytophagous insects, Table 7.2 is pathetically small, but it shows clearly that anything can, and does, happen. Some species are interactive: most appear to be non-interactive, and within the latter category we expect to see both reactive (food limited) and non-reactive examples.

In parentheses, note that exactly the same phenomena influence the interaction of other herbivores with their food-plants. (1) Grazing mammals are highly selective in what they eat (Freeland and Janzen, 1974; Leuthold, 1977) and pond snail populations (*Limnaea*) respond vigorously to small quantities of high quality food (Eisenberg, 1970). (2) A regular annual dry season depresses ungulate and small mammal populations in the Serengeti well below their apparent food supply during the rest of the year (Sinclair, 1975); and shortage of shady sites, safe from overheating, is a severe problem for giant tortoises on Aldabra (Swingland and Coe, 1979). Finally, (3) predators may tem-

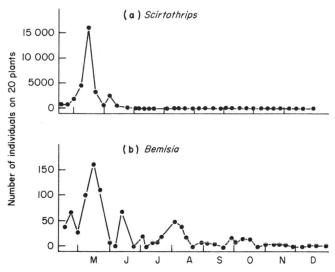

Fig. 7.2. The insects feeding on a small isolated plot of Cassava (*Manihot esculenta*) illustrate three alternative dynamical patterns (Samways, 1979). (a) The thrip *Scirtothrips manihoti* feeds on the upper leaves and buds. The population increased exponentially in the spring until the onset of lower temperatures, rain and heavy dew caused high density independent mortality and a rapid decline (hypothesis (2), p. 137). (b) The whitefly *Bemisia tuberculata* feeds on the lower and middle leaves. Predation and parasitism were negligible, yet the population never became very abundant 'apparently due to some lack of preference' (hypothesis (1), p. 136). (c) Not illustrated. The moth *Erinnys ello* also colonized the plant, but 83 per cent of eggs were parasitized by *Trichogramma*; the caterpillars were also parasitized by two hymenoptera and a tachinid, *Euphorocera*. As a result the moth remained extremely rare (hypothesis (3), p. 138).

porarily limit small mammal populations during the 'lows' of a population cycle (Pearson, 1966); or permanently in the case of sea otters and their sea urchin prey (Estes and Palmisano, 1974). Disease (rinderpest) can make terrible inroads into ungulate populations (Delany and Happold, 1979).

Obviously, the dichotomy between reactive and non-reactive species, and the larger one between interactive and non-interactive systems is one of convenience. Some food limitation is inevitable unless natural enemies or climate markedly depress the population, and some effects on the host plant must follow if the herbivores themselves experience food limitation. Thus, despite density-dependent pupal predation (Table 7.2), winter moth caterpillars become sufficiently abundant in

Table 7.2. Summary of main density-dependent controlling factors operating on populations of phytophagous insects, revealed by life-table studies (Varley *et al.*, 1973). Data are from Podoler and Rogers (1975), supplemented by Stubbs (1977), and other sources given in the table.

Species	Density dependent mortalities, discovered by original authors, or by subsequent analysis (see references for details).
	Parasitism, Predation and Disease
West Indian cane fly (Diptera) *Saccharosydne saccharivora*:	
'2 Dundas'	Egg parasitism by *Tetrastichus*
'D-Piece'	Parasitism of early nymphs by *Stenocranophilus*
Yew gall midge (Diptera) *Taxomyia taxi*	None identified: the life-cycle of this herbivore is very complex, and parasitism by *Mesopolobus* may be density-dependent in one-year life cycle galls (Redfern and Cameron, 1978).
Cabbage root fly (Diptera) *Erioischia brassicae*	Pupal parasitism by *Aleochara* and some predation on pupae.
Broom beetle (Coleoptera) *Phytodecta olivacea*	None identified: soil mortality due to predation up to emergence of adults in autumn may be density dependent.
Spring usher (Lepidoptera) *Erannis leucophaearia*	Parasitism and predation (Stubbs, 1977).
Pine looper (Lepidoptera) *Bupalus piniarius*	Larval mortality due mainly to parasitism by *Eucarcelia* and *Poecilostictus*. Infectious disease also important (Anderson. and May, 1980, 1981).
Mottled umber (Lepidoptera) *Erannis decemlineata*	Parasitism and predation (Stubbs, 1977).
Winter moth (Lepidoptera) *Operophtera brumata*: England	Pupal predation, by carabid and staphylinid beetles and small mammals. Infectious disease also important (Anderson and May, 1981).
Canada	None identified, but parasitism of caterpillars by *Cyzenis* (and *Agrypon*?) is strongly regulatory (Hassell, 1978: Beddington *et al.*, 1978).
Grey larch moth (Lepidoptera) *Zeiraphera diniana*	None identified: parasites or predators on eggs or pupae may be a 'major cause' (delayed density-dependence) of cycles (Varley and Gradwell, 1970). More likely, the cycles are driven by infectious disease (Anderson and May, 1980, 1981).
Black-headed budworm (Lepidoptera) *Acleris variana*	None identified: parasites or predators on eggs or pupae may be a 'major cause' (delayed density-dependence) of cycles (Varley and

Table 7.2 (continued)

	Gradwell, 1970). More likely, the cycles are driven by infectious disease (Anderson and May, 1980, 1981).
	Competition for Food
Grass mirid (Heteroptera) *Leptoterna dolobrata*	Competition for high nitrogen feeding sites by adults.
Colorado beetle (Coleoptera) *Leptinotarsa decemlineata*	Starvation of older larvae through food shortage. Parasitism of pupae by *Doryphorophaga* may also be density dependent (Stubbs, 1977).
Cinnabar moth (Lepidoptera) *Tyria jacobaeae*	Death of caterpillars from starvation, and delayed density-dependent reduction in adult fecundity (Dempster, 1975).
	Other
Large copper (Lepidoptera) *Lycaena dispar*	Decrease in adult fecundity (Stubbs, 1977).
Frit fly (Diptera) *Oscinella frit*	None identified, but adult mortality, seasonal migration and variation in fecundity appear to be density dependent in combination. Cause unknown.
Tea moth (Lepidoptera) *Andracea bipunctata*	None identified (Banerjee, 1979).
Olive scale (Homoptera)	
Parlotoria oleae: Hills valley	None identified.
Herndon	None identified.

some springs to defoliate the oak trees (*Quercus*) on which they feed. Defoliation impairs the growth of young oaks, and could greatly influence the outcome of competition between trees struggling for dominance (Gradwell, 1974), just as periodic outbreaks of tent caterpillars (*Malacosoma*) favour the main understory species (*Abies balsamea*) at the expense of the defoliated dominant, *Populus tremuloides* (Harper, 1977). This leads us naturally into a consideration of interactive systems.

7.4 The dynamics of grazing systems: interactive systems

One of the commonest grazing systems is that in which the rate of change of herbivores is a function of plant density, and the rate of change of plants is a function of herbivore density. The two components interact. Formally, the 'plant density' in this context is the quantity of vegetation per unit area which is available both for consumption by animals and for producing plant growth (Noy-Meir, 1975).

A model of this system must include several parameters: usually two for plant growth, two for grazing pressure and two or three for the growth of the herbivore population. Here is a representative set of parameters* for such a model:

r_1 = the intrinsic rate of increase of plants,

K = maximum ungrazed plant density, or biomass,

c_1 = maximum rate of food intake per herbivore,

d_1 = grazing (searching) efficiency of the herbivore when vegetation is sparse,

a = rate at which herbivores decline when the vegetation is burned out or grazed flat,

c_2 = rate at which this decline is ameliorated at high plant density,

d_2 = demographic efficiency of the herbivore; its ability to multiply when vegetation is sparse.

A number of choices are available to depict the relationship between plant density and rate of food intake per herbivore (the functional response), and between plant density and rate of increase of the herbivore (the numerical response): see Table 5.1. We have used simple functions that rise with increasing plant density, and saturate when food is abundant. The numerical response used here ignores the effect of 'underpopulation', that is, reduced fecundity at low density reflecting the difficulty of finding a mate when mates are scarce. This is a deliberate simplification. Underpopulation has not, to our knowledge, been identified for any herbivore, vertebrate or invertebrate, although it may sometimes occur.

7.4.1   *Laissez-faire systems*

In a laissez-faire interactive system the herbivores do not interfere with each other's search for food. Non-territorial ungulates provide a good example of this kind of grazing behaviour. Noy-Meir (1975) explored by graphic analysis the theoretical implications of holding ungulates at

---

* In particularizing from the general prey–predator system of Chapter 5, to the plant–herbivore system of this chapter, the symbol for the prey population, N, has already been replaced by V for vegetation, and the symbol for the predator population, P, by H for herbivore. Similarly, in what follows there is no attempt to keep the symbols for the various parameters precisely congruent with those used in Chapters 5 and 6 (e.g., those in Table 5.1).

an arbitrary constant density, as in farming, and Caughley (1976) looked at the properties of a system containing wild ungulates hunted by man. To keep it short and simple, the present discussion is limited almost entirely to systems in which wild ungulates range free of persecution.

May (1975a, and section 5.2) has summarized the different ways in which this system can be modelled, and he has emphasized that the qualitative behaviour of the model is insensitive to the details of its construction. Depending on the values of its parameters, the system may be characterized by a stable equilibrium point, or by a stable limit cycle, whose amplitude may be so severe as to produce extinction. Of the many available, one model will suffice to sketch in the outlines of this system:

$$dV/dt = r_1 V(1 - V/K) - c_1 H[1 - \exp(-d_1 V)], \qquad (7.5)$$

$$dH/dt = H\{-a + c_2[1 - \exp(-d_2 V)]\}. \qquad (7.6)$$

The symbols are as defined above.

The first equation expresses the rate of change of vegetation by two terms, the first depicting logistic growth and the second the rate of grazing. The functional response of diet to plant density is in Ivlev (1961) form; see form (iv) of Table 5.1. Equation (7.6) summarizes the rate of change of herbivores, H, in terms of their intrinsic ability to multiply, as modified by the availability of food. Herbivores can increase at a maximum rate of $\{-a + c_2[1 - \exp(-d_2 K)]\}$, which in most circumstances will equal their intrinsic rate of increase, $r_2 (r_2 = c_2 - a)$, because at high plant density the term inside the square brackets will tend to unity.

Before exploring the applicability of this model to real grazing systems, we will examine briefly the mathematical properties of a system conforming to these rules. Figure 7.3 shows the growth of a population of herbivores, and the resultant changes in plant density, as the two spiral towards their mutual equilibrium point. For this illustration, the parameters were set at

$$r_1 = 0.8 \qquad\qquad a = 1.1$$
$$K = 3000 \qquad\qquad c_2 = 1.5$$
$$c_1 = 1.2 \qquad\qquad d_2 = 0.001.$$
$$d_1 = 0.001$$

Although the example is imaginary it can be thought of, without contradicting current knowledge, as white-tailed deer colonizing a mosaic

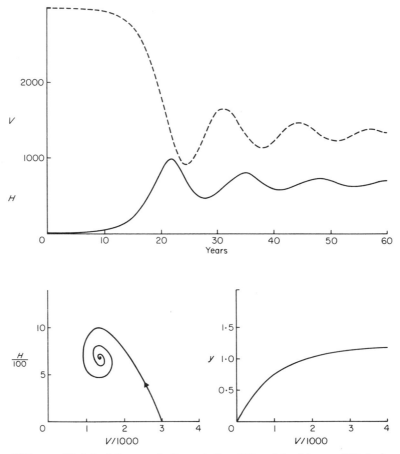

Fig. 7.3. Model of the trend of vegetation (V) and herbivores (H) during an ungulate eruption. Intake (y) of vegetation per herbivore is graphed against density of vegetation.

of grassland and forest. Wildlife managers will recognize the growth curve as a deer eruption. The model has an equilibrium at

$$V^* = (\text{I}/d_2) \ln [c_2/(c_2 - a)], \tag{7.7}$$

and

$$H^* = \frac{r_1\, V^*(\text{I} - V^*/K)}{c_1[\text{I} - \exp\,(-d_1 V^*)]}. \tag{7.8}$$

For the parameter values catalogued above, and illustrated in Fig. 7.3, the equilibria are $V^* = 1300$ and $H^* = 670$. Let $x = d_1 V^*$: then, depending on the specific parameter values, this equilibrium will be stable if

$$K < V^* \, (2\mathrm{e}^x - 2 - x)/(\mathrm{e}^x - \mathrm{I} - x) \tag{7.9}$$

or, if not, it will be the focus of a stable limit cycle. We can arrive at this conclusion by applying the graphic methods of Rosenzweig and MacArthur (1963). There will be one or more values of V at which $dV/dt = 0$ when H is held constant at some arbitrary value. The points may represent maxima or minima of cycles, or stable equilibrium. They can be plotted on a graph of H against V. If this procedure were followed for every possible value of H, the resultant points would fall along a curve, the zero 'isocline' of V. The form of this curve, expressing H as a function of V, clearly is given by eq. (7.8). Similarly, there is a set of points at which $dH/dt = 0$ when V is held constant. This isocline of herbivores is an inverted T, the crossbar coinciding with the uninteresting instance of $H = 0$, and the vertical rising from the V axis at the value V^* given by eq. (7.7). Figure 7.4 gives the zero isoclines of V and H for the numerical example above.

The salient feature is whether the position of the isocline of H is to the right of the highest point, indicated by a triangle, of the V isocline. That is a necessary and sufficient condition for point-stability of this model (Rosenzweig, 1971; Gilpin, 1972). The system's equilibrium

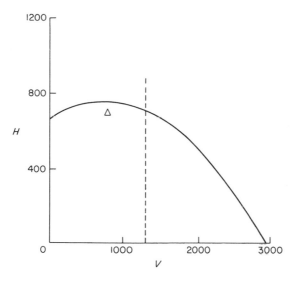

Fig. 7.4. Graph of the zero isocline for vegetation (the curve) and the zero isocline for herbivores (vertical line) for the 'white-tailed deer' model: the triangle indicates the peak of the V isocline.

point [the V^*, H^* of eqs. (7.7) and (7.8)] of course lies at the intersection of the two isoclines. If the equilibrium point lies to the left of
the hump, the system will cycle.

For relatively large values of d_1V^*, the stability condition eq.
(7.9) simplifies to

$$K < 2V^*. \tag{7.10}$$

This approximation is within 2 per cent for $d_1V^* > 5$. In this form the
biological implications are more apparent: a system of a vegetation and
a population of efficient herbivores has oscillatory behaviour if grazing
pressure tends to hold the standing crop of vegetation below half that
of its ungrazed state.

This simple conclusion requires two important qualifications. First,
it applies only to laissez-faire systems. Second, it describes what happens
in a spatially uniform world, and the food of most herbivores is very
far from being uniformly distributed; witness the spectacular migrations
of ungulates across the Serengeti (Delany and Happold, 1979), the
responses of the parsnip web worm *Depressaria pastinacella* to clumps
of wild parsnip *Pastinaca sativa* (Thompson, 1978), and the patchy
distribution of *Daphnia* and algae in lakes (Hebert, 1978). Just as the
dynamics of predator–prey interactions are markedly changed by the
inclusion of spatial heterogeneity (chapter 6), so too are herbivore–
plant interactions, although the problem has received much less attention. We return to the question of spatial heterogeneity in section 7.4.3.
A cautious, qualitative interpretation of eq. (7.10) therefore reduces to:
heavy grazing leads to oscillations, and no more.

The laissez-faire model is at its best as a summary of the interaction
of a population of predominantly non-migratory ungulates and their
food supply. For appropriate parameter values the trajectory of an
ungulate population increasing from minimal density while grazing
vegetation initially at density K follows an eruption to a peak, a crash,
and a 'stabilization' around an equilibrium level well below the peak
density (Caughley, 1970). Figure 7.5 shows the eruption of sheep introduced into western New South Wales. The parallel between this trajectory and that simulated in Fig. 7.3 is immediately apparent. Despite its
undoubted limitations the model captures the essence of the interaction
remarkably well.

Ungulate populations are also prone to erupt when the plant–
herbivore system is disturbed. Because the herbivores do not start
their upswing from minimal density, and because the vegetation is not

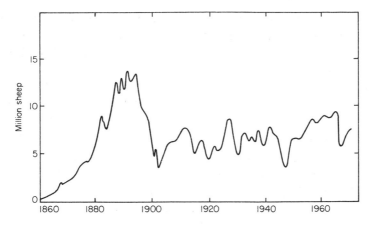

Fig. 7.5. An ungulate eruption: the trend of sheep numbers in the Western Divisions of New South Wales between 1860 and 1972. Data from Butlin (1962) and N.S.W. Yearbooks (1956–1972).

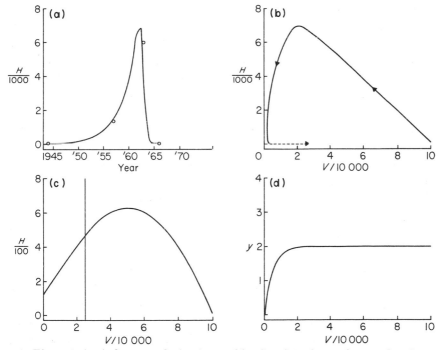

Fig. 7.6. A reindeer population in trouble. Graph (**a**) is a trajectory fitted by the laissez-faire model to Klein's (1968) estimates of numbers on St. Matthew Island. Graph (**b**) shows the same trajectory as a plot of herbivores (H) against vegetation (V); graph (**c**) shows the zero isoclines for herbivores (vertical line) and vegetation (curve), and graph (**d**) is intake (y) per head against density of vegetation.

at its K level, these 'sub-eruptions' are not as spectacular as that of Fig. 7.5. They are, however, much more common. Leopold *et al.* (1947) reported over a hundred for deer populations in the United States between 1900 and 1945.

Probably the most spectacular full-scale eruption on record is that of reindeer introduced onto St. Matthew Island in the Bering Sea. Figure 7.6 gives the available estimates of population size (Klein, 1968), to which a trajectory was fitted by eqs. (7.5) and (7.6); see Caughley (1976). The parameter values were guessed rather than estimated. As shown in Fig. 7.6c the intersection of zero isoclines is to the left of the hump, indicating that the system has no stable equilibrium point, and therefore that the reindeer numbers will oscillate cyclically (in this case violently, and probably to extinction).

Equations (7.5) and (7.6) also pass as reasonably good caricatures of coupled zooplankton–phytoplankton interactions. *Daphnia* are typical planktonic grazers, exploiting various species of algae in the upper waters (epilimnia) of lakes and ponds. In permanent water-bodies in temperate regions they overwinter as small numbers of adults, or ephipia (resistant eggs). Populations increase roughly exponentially in the spring, chasing a rapidly growing algal population, to reach a peak of 20–100 animals per litre. Then, during the latter phase of this increase, both brood size and the proportion of reproductive adults decline, as the animals begin to make significant inroads into their food supply (George and Edwards, 1974; Hebert, 1978). The population may collapse shortly thereafter, and undergo one or more similar cycles later in the year (Fig. 7.7a, c) or alternatively (recall eq. 7.9) stabilize (Fig. 7.7b) until winter switches things off. This qualitative sequence of events is exactly that described by eqs. 7.5 and 7.6 when the system starts with V and H both small, and the plants (V) are given a head start.

Inevitably there are complications. Sometimes, the *Daphnia* are dragged down by a decline in algal numbers, without contributing significantly to the crash through overgrazing. Instead, the algae exhaust essential nutrients (SiO_2, or P, for example), dramatically changing K (or r_1, depending upon your taste) in eq. 7.5 (e.g. Kilham, 1971, and references therein). Now, it is no longer true that plants cannot influence the supply of important resources available for the next generation (in the simple manner assumed on page 136), and plant growth must be modelled by choosing more complicated equations (e.g. Parker, 1968; Phillips, 1978).

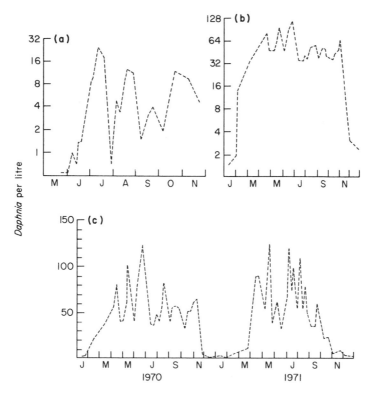

Fig. 7.7. *Daphnia* population dynamics in three different water-bodies. Growth in spring is approximately exponential until overgrazing limits further growth. The populations then fluctuate, or roughly stabilize around some average value until winter switches things off (George and Edwards, 1974; Hebert, 1978).

A second, planktonic example which certainly involves overgrazing is shown in Fig. 7.8.

7.4.2 *Interferential systems*

When herbivores interfere with each other's grazing, the previous model is no longer appropriate. In contrast to the laissez-faire systems of section 7.4.1, the per capita rate of increase of herbivores is a function of both plant density and animal density.

From among the many qualitatively similar possible models (see, e.g., Table 5.1), we choose one which has already been discussed in section 5.2 (see Figs. 5.1 and 5.2), and has had detailed attention in

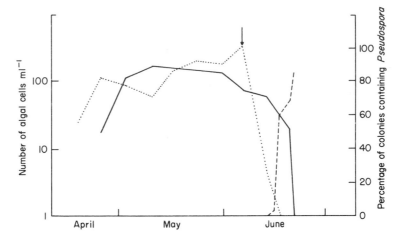

Fig. 7.8. Overgrazing of algal populations (*Gemellicystis imperfecta*) by protozoa tentatively ascribed to the genus *Pseudospora* (Canter and Lund, 1968). The protozoa are grazing herbivores, with an amoeboid feeding stage, ingesting algal cells whole. Algal populations in the north (———) and south (· · · · · ·) basin of Windermere decline rapidly as *Pseudospora* populations (– – – – –) build up in the north basin, and after the protozoa were first observed in the south basin (at the date indicated by an arrow).

the literature (May, 1975a; Tanner, 1975):

$$dV/dt = r_1 V(1 - V/K) - c_1 H[V/(V + D)] \qquad (7.11)$$

$$dH/dt = r_2 H(1 - JH/V). \qquad (7.12)$$

Here D is inversely proportional to grazing efficiency at low plant density (being the characteristic density of vegetation at which the herbivore functional response saturates), and J is a proportionality constant related to the number of plants needed to sustain a herbivore at equilibrium. The other symbols are as previously defined (note that r_1 and r_2 are to be identified with the earlier symbols r and s of chapter 5).

The zero isocline for the vegetation V, along which $dV/dt = 0$ for any given H, follows from eq. (7.11):

$$H = (r_1/c_1) (1 - V/K) (V + D). \qquad (7.13)$$

The zero isocline for H, along which $dH/dt = 0$, is similarly

$$H = V/J. \qquad (7.14)$$

This latter isocline slopes out from the origin, in contrast to the vertical isocline of eq. (7.7). A further contrast lies in the stability properties of this model, which allow the possibility of a stable equilibrium point to either the left or the right of the hump in the V isocline, at $V = (K - D)/2$. The details of the stability properties, as a function of parameters such as K/D and r_1/r_2, are discussed in Fig. 5.2 and the accompanying text; for a fully detailed exposition, see May (1975a, Appendix I).

Our example is provided by the moth *Cactoblastis cactorum*, which was introduced into Australia in 1925 to control the prickly-pear cacti *Opuntia inermis* and *O. stricta*. Extensive stands of dense pear averaging about 500 plants per acre were virtually wiped out within two years of their colonization by *Cactoblastis* (Dodd, 1940). Dodd's account of the initial spread is worth reading, as are the subsequent reverberations in the ecological literature (see for example, Annecke and Moran, 1978, and references therein). The post-crash equilibrium was first interpreted as a game of hide and seek between *Opuntia* and *Cactoblastis* (Nicholson, 1947; Andrewartha and Birch, 1954) and then as a stable equilibrium maintained by grazing, reinforced by larval interference (Monro, 1967, 1975; Birch, 1971).

The eggs of *Cactoblastis* are not laid at random. Their dispersion is doubly contagious in that they are laid in egg-sticks of around 80 eggs each and the egg-sticks are themselves clumped (Monro, 1967). Hence during the summer some cactus plants receive many more eggs than would be expected by random chance, and others escape infestation completely. The contagious distribution of *Cactoblastis* eggs greatly influences the outcome of grazing by the larvae. Since a loading of above about 1·5 sticks per cactus ensures destruction of the plant, much of the larvae's resource is wasted (Monro, 1967).

Dodd (1940) and Monro (1967) present enough data to allow a stab at the values of four of the six parameters of eqs. (7.11) and (7.12). A clue to r_1 is provided by Dodd's observation that *Opuntia* can increase from root stock to 250 tons per acre in two years. Assuming the root stock weighs one ton, the rate of increase is around $r = 2·7$. That will be an overestimate of r_1 because the growth is entirely vegetative. Our guess for a maximum rate of increase depending partly on sexual reproduction is $r_1 = 2$. The value of K is taken from Dodd's remark that 5000 plants per acre is a fair estimate. Two rough estimates of r_2 can be made. At the beginning of the experiment 2750 eggs were received from South America and hatched in the laboratory. The adults produced 100,605 eggs: $r_2 = 3.6$ per year [i.e. ln (100,605/2750)]. Dodd indicated that in the field *Cactoblastis* could erupt from 5000 larvae to 10,000,000 per acre in two years: $r_2 = 3.8$ per year. For different reasons both figures are liable to underestimate r_2 which is set at 4. J comes from Monro's measurements of the summer generation at his B1 and B2 sites: $J = 2.23$ in units of cactus plants per egg-stick. Parameters c_1 and D cannot be estimated. However, c_1 should be large to reflect plants damaged and killed by the larvae as a by-product of their feeding. This structural damage contributes most to c_1, vastly outweighing the contribution of ingested plant tissue. D should be small to reflect the uncanny ability with which female moths search out food plants on which to lay eggs.

Three outcomes are demanded of a simulation of this system: *Opuntia* must crash within two years to square with Dodd's observations, it must stabilize at around 11 plants per acre to reflect the summer density at Monro's B sites, and the equilibrium must be highly stable. These requirements are met by setting $c_1 = 6$ and $D = 4$. Figure 7.9 shows a simulation of *Cactoblastis* invading a stand of cactus. *Opuntia* density is expressed as plants per acre and *Cactoblastis* as egg-sticks per acre, the archaic 'acre' being retained to preserve uniformity with Dodd's round figures. *Opuntia* crashes in the required time to a highly stable equilibrium of 11 plants per acre carrying 5 egg-sticks.

This strategic model containing only six constants has produced a cactus crash looking much like the real thing. A tactical model would contain more detail, particularly on the survival of root stock after the crash which often allowed a temporary resurgence of the cactus; but it is unlikely to produce an outcome differing in kind from that provided by the simple interferential model. We can speculate cautiously that the model has described the major determining forces of the system, a conclusion reinforced by the result from the same data translated into laissez-faire parameters and fed through that model [eqs. (7.5), (7.6)]: without larval interference a spatially homogeneous system with these parameters is spectacularly unstable.

Interference amongst more advanced herbivores may involve much more than the highly contagious distribution of eggs, and the ensuing 'exploitation' competition seen in the *Opuntia–Cactoblastis* system, particularly when the herbivores have territories ('contest' competition). Now, each successful herbivore acquires for itself a large portion of the resources for its personal use (J in eq. (7.12) large). Something very

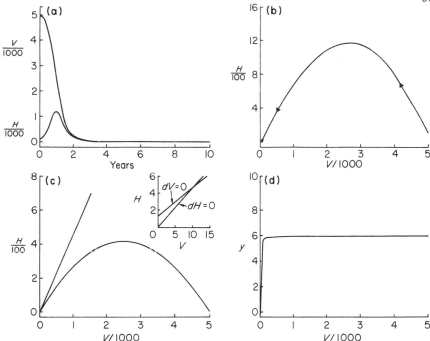

Fig. 7.9. The control of a plant by a herbivore. Graph (**a**) gives trajectories by the interferential model for a stand of cactus invaded by the moth *Cactoblastis*. Graph (**b**) shows the path to mutual equilibrium in phase space, graph (**c**) provides the modelled zero isoclines of cactus (curve) and *Cactoblastis* (line), and graph (**d**) is consumption per *Cactoblastis* egg-stick against cactus density.

like this apparently happens in an avian herbivore, the red grouse (*Lagopus lagopus scoticus*).

Shoots and flowers of heather (*Calluna vulgaris*, a small, woody, ericaceous shrub) make up at least 90 per cent of the diet of red grouse over most of the year, and heather is the dominant and sometimes virtually the only higher plant on the moors where grouse live. The population dynamics and nutrition of this unusual bird are described by Lance (1978), Miller and Watson (1978), Watson and Moss (1972) and references therein. The size of the breeding population is determined by the number of territories chosen by cocks in the previous autumn. Non-territorial birds are forced off the moor, suffering the brunt of the overwinter mortality, but will readily occupy territories if vacant lots are created by shooting. In other words, interference is very strong. As a result heather moors support rather a lot of heather and relatively few grouse, which eat somewhere between one and a half and two and a

half per cent of the total annual heather production (Savory, 1978).
Quite why grouse have such large territories in relation to their apparent food supply is not clear, although the heather's nutrient content
varies within and between areas, as well as with season, plant part, and
plant age: 'heather' in other words does not necessarily equal 'food'
(cf. hypothesis 1 on page 136). The grouse themselves are fastidious
feeders on the most nutritious plants and plant parts, with patchy
heather and heather of a high mean nitrogen content per shoot tip
favouring smaller territories, and a higher density of birds (reducing J
in eq. 7.12). Hence grouse populations respond to the availability and
rate of production of their food supply, caricatured by eqs. (7.11) and
(7.12.) The reciprocal impact of the grouse on the heather is much more
difficult to judge, because as a woody perennial, with mixed sexual and
vegetative reproduction, ill-defined individuals, and variable allocation
of resources to roots, shoots and flowers under different grazing regimes
(Harper, 1977), the plant is a paradigm of the problems confronting
plant population dynamicists. Thus, although the birds probably eat
less than three per cent of the total annual production of shoots and
flowers, as much as twenty per cent of the suitable heather in the most
heavily grazed age class might be consumed (Savory, 1978).

Obviously as a detailed model of the grouse–heather interaction,
eqs. (7.11) and (7.12) are rather poor. In particular, eq. (7.12) cheats because it specifies nothing about the behaviour of the birds; instead, it
simply defines its consequences in the term J/V. Models of exploiter-
victim systems with explicit social hierarchies in the exploiter populations are developed by Gurney and Nisbet (1979). Comfortingly, and
in keeping with the family of 'interference' models developed in chapter
6, their qualitative message is the same: behavioural interference is
strongly stabilizing.

7.4.3 *Interference or environmental heterogeneity?*

In most cases of clearly interactive systems we have no idea whether
the herbivores interfere significantly with one another or not. If we
move up a trophic level to chapter 6 we see strongly depressed host
populations stabilized *either* by real behavioural interference between
searching parasitoids *or* by spatial heterogeneity, the patchy distribution of hosts and the parasitoids' response to that distribution; and on
balance the latter seems much more important than the former (Beddington *et al.*, 1978; Hassell, 1978). For plants and herbivores this

problem appears to translate itself into the difference between Nicholson, Andrewartha and Birch's hide-and-seek explanation for the *Opuntia-Cactoblastis* interaction, and the interference model outlined in the previous section. In practice, the difference is more apparent than real.

The ratio $V^*/K = q$ provides a useful empirical measure of the impact of herbivores on plant populations, exactly analogous to q in predator–prey, or host–parasitoid interactions (Beddington *et al.*, 1978; chapter 6). Implicit in eq. (7.10) is the notion that strong, stable depression of the plant population ($q \ll 0.5$) can be accounted for by behavioural interference between the herbivores, or by spatial heterogeneity. Neither complication needs be invoked to explain the stability of the interaction if $q \gtrsim 0.5$.

The *Opuntia–Cactoblastis* interaction is but one of several examples where q can be estimated or 'guesstimated' in cases of successful control of weeds by phytophagous insects. The plant populations are often spectacularly reduced (Clausen, 1978; DeBach and Schlinger, 1964; Harris, 1973). Estimates of q are provided in Table 7.3; see also Fig 6.9. Arguing largely by analogy from Chapter 6, it is not difficult to believe that spatial heterogeneity is just as important in the persistence of these plant–herbivore interactions as it is for strongly depressed insect hosts and their parasitoids; either in the form of hide-and-seek between attacker and attacked, or a safe physical refuge for the victim in a part of its environment. A refuge from the ravages of the beetle *Chrysolina* exists in woodland for St. John's wort (*Hypericum*), and a reserve of plant biomass unavailable to grazing mammals provides a powerful stabilizing influence in Noy-Meir's general analysis of systems with fixed herbivore numbers (Noy-Meir, 1975).

Intriguingly, 'interference' in the *Opuntia–Cactoblastis* system was a direct consequence of the aggregated distribution of eggs by the female moths, leaving a reserve of plants undiscovered and unscathed. An identical clumped distribution of attacks by insect parasitoids also manifests itself as apparent interference (Comins and Hassell, 1979; Free *et al.*, 1977; May, 1978a) and for exactly the same reasons. Exploitation competition is most intense in regions of strong exploiter aggregation. Examples of the non-random distribution of plants, and of the herbivores' responses to these distributions are touched upon in section 7.4.1. Figure 7.10 shows the aggregative response of an insect herbivore, the cinnabar moth (*Tyria jacobaeae*) to clumps of its food plant, ragwort (*Senecio jacobaeae*). At high densities, exploitation competition between caterpillars is intense (Table 7.2). Even when plants are spread

Table 7.3. Estimates of the level of depression, q, of plant populations by insect herbivores in biological control programmes, and in an insect removal experiment (Cantlon, 1969). q is defined as the ratio of the average abundance of the plant in the presence of the herbivore (V^*) divided by its abundance (K) in the absence of control. Most of the estimates are very crude, and illustrate no more than $q < 0.5$ (see text, page 155).

Plant species	Main insect herbivore	Method of attack	$V^*/K = q$	Authority
Opuntia inermis & O. stricta	Cactoblastis cactorum	Larvae mine the pads	0·002	See text, page 152.
Hypericum perforatum	Chrysolina quadrigemina (& also C. hyperici)	Larvae attack winter basal growth, with complete loss of flowers.	<0·01 0·005	DeBach and Schlinger (1964); Harper (1969).
Senecio jacobaea	Tyria jacobaeae	Caterpillars eat leaves.	0·003 (Durham, Nova Scotia)	Harris, et al., (1978).
Carduus nutans	Rhinocyllus conicus	Larvae feed on thistle heads.	0·05	Kok and Surles (1975).
Chondrilla juncea	Cystiphora schmidti in conjunction with a rust Puccinia	Gall midge.	0·36 (Wagga) to 0·03 (Tamworth)	Cullen (1978).
Melampyrum lineare	Atlanticus testaceous	Eats seedlings, and later foliage of mature plants.	0·25	Cantlon (1969).

uniformly in experimental swards, damage by herbivores may be highly
non-random. Grazing by the slug *Agriolimax reticulatus*, for example,
damages some plants and leaves others entirely unharmed (Harper,
1977). In some cases, the differential susceptibility of individual plants
or clumps of plants is known to have a genetic basis (Cooper-Driver
and Swain, 1976; Edmunds and Alstad, 1978).

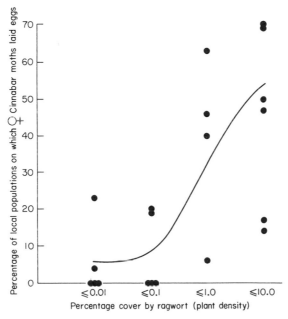

Fig. 7.10. Aggregative response by ovipositing female cinnabar moths,
Tyria jacobaea, to clumps of varying density of their larval food plant,
Senecio jacobaea (ragwort). The solid line joins the means of each
group of observations (van der Meijden, 1979). Cinnabar moths aggregate their attacks on the denser patches of ragwort (cf. Fig. 6.5).

In short, it is a matter of mathematical semantics whether the
Cactoblastis–Opuntia interaction, and similar plant–herbivore systems
where herbivore aggregation or a host refuge leads to the differential
exploitation of patches of hosts, are stabilized by 'interference' or by
'spatial heterogeneity'. Sophisticated behavioural interference between
herbivores of the type seen in red grouse may actually be rather rare,
just as cases of significant mutual interference between adult parasitoids
are also rare at equilibrium densities (chapter 6). The reasons for, and
consequences of, patchy host–plant use by herbivores would undoubtedly repay further study.

7.5 Extensions and elaborations

7.5.1 *Extensions to more than two species*

The notion that plant–herbivore interactions usually involve two species is obviously cheating, if not actually a downright lie; indeed if we are to peer deeper into the role of herbivory in plant communities, pretending that only two species are involved no longer works. Instead we must consider sets of plants and guilds of herbivores.

Guilds of phytophagous insects feeding in concert markedly influence the growth of individual hosts (Gradwell, 1974; Kulman, 1971; Morrow and La Marche, 1978), changing the demography and hence the dynamics of plant populations (Waloff and Richards, 1977; Fig. 7.11; see also Clements, Gibson, Henderson and Plumb, 1978; Connell, 1975; J. Foster, in Harper, 1977; Janzen, 1970; Perkins, 1978; Stephens, 1971). Other herbivores, fungal and viral pathogens often compound these effects. Whether the impact of herbivore guilds would be as dramatic in the absence of the majority of rare, apparently non-interactive species (section 7.3.2) is unknown.

Harper (1977) and Whittaker (1979) show convincingly how moderate levels of grazing have quite disproportionate effects on the population dynamics of plants by tipping the balance between competitors (see also Simmonds, 1933). In the simplest case (that of 'predator mediated coexistence') grazing reduces or eliminates competitive exclusion (Paine, 1966; see May, 1977c for a review); but depending upon such factors as the abundance of the exploiters, their feeding preferences, and on whether these are changed by encountering different frequencies of plants ('switching': see Rausher, 1978) the diversity of the vegetation may be enhanced or markedly reduced by grazing. Appropriate examples are provided by Connell (1975), Estes and Palmisano (1974), Harper (1977), Lubchenco (1978), McCauley and Briand (1979), Patrick (1975) and Porter (1973), for herbivores as varied as rabbits, voles, sea urchins, periwinkles, freshwater snails and zooplankton.

The most remarkable interpretation of the role of herbivory in altering the competitive balance between species of plants is that of Stanley (1973a), who invokes exploiter mediated coexistence to account for the major flowering of multicellular algae and most of the known phyla of multicellular animals near the beginning of the Cambrian.

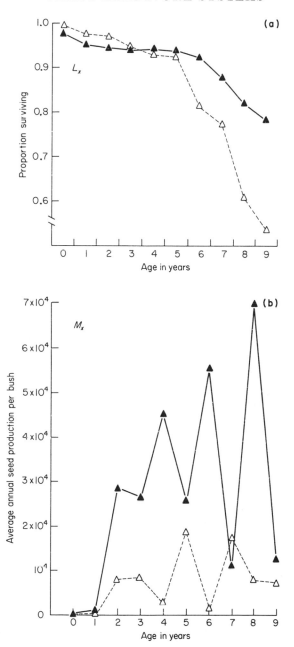

Fig. 7.11. The effect of removing herbivorous insects (by regular spraying with the insecticides demethoate and malathion) on survivorship (l_x) and fecundity (m_x) of broom bushes *Sarothamnus scoparius* (Waloff and Richards, 1977). Sprayed bushes (▲———▲) apparently survive better and have higher fecundities than unsprayed bushes (△– – – –△). (Only one experimental and one control plot were used in this work; hence differences between treatments may, in part, reflect differences unrelated to the insecticide application. Since the plots were only 30 m apart, such effects are likely to be small.)

Certainly the late Precambrian leap in diversity is as remarkable as it was slow to start. Stanley argues for a Precambrian world with a relatively impoverished number of producers, bacteria and single-celled algae, limited by inter-specific competition for resources. When grazers finally evolved, perhaps 2.5 billion years after life first appeared on earth, the ensuing reduction in competition amongst the producers made possible the coexistence of many more sorts of plants, and released an orgy of evolution in both exploited and exploiter trophic levels. Under Stanley's hypothesis grazing transformed the world.

Whether a particular herbivore has, or has not, a significant impact on its food plant also depends upon the trophic stack extending above it. We have already argued the case for many phytophagous insects being kept rare by natural enemies. Many promising herbivores imported for biological weed control failed because of the impact of indigenous predators (Goeden and Louda, 1976). Similarly, sea otters will sometimes so deplete sea urchin populations as to change the whole appearance of the coast (Estes and Palmisano, 1974). On the western Aleutian Islands, the urchin–otter interaction involves four levels. On islands where man (level four) has eliminated otters (level three), sea urchins (at level two, and a favourite food of otters), thrive and kelp beds (level one) are eliminated by urchin-grazing. Without man, otters and kelp thrive, and urchins play 'pig in the middle'. The general implications of this game are intriguing (Fretwell, 1977) but largely ignored. Other things being equal, systems with an even number of trophic levels (2 and 4) are more likely to be interactive at the plant–herbivore level than are systems with an odd number of steps in the food chain.

7.5.2 *Some evolutionary considerations*

The plethora of spines, irritating hairs, poisons and other protective devices evolved by plants bear strong witness to the selection pressures imposed by grazing. In consequence, most species of herbivores are unable or unwilling to eat most species of plants: herbivory is a trade for specialists, with adapted herbivores forced to attack only particular plants and parts of plants, often during a restricted season; or to feed and develop slowly and with high mortality on suboptimal diets (Lawton and McNeill, 1979). Such effects are sufficient on their own (hypothesis (1), page 136), or in combination with adverse climatic conditions and natural enemies (Lawton and McNeill, 1979) to keep most species of herbivorous insect (and presumably many other herbivores as well) rare, most of the time. This deceptively peaceful stalemate reveals its true colours when man unwittingly breeds vulnerable varieties of crops (Day, 1972; van Emden and Way, 1973; Pathak, 1975) and when drought, other stresses, or even unusually favourable growing conditions so alter a plant's defensive chemistry as to create a major herbivore explosion (Newsome, 1969; White, 1969, 1974, 1976).

The evolutionary response of plants to grazing need not simply be in permanent defences. Animals when attacked by pathogens or parasites display a bewildering battery of induced defences (chapter 14); are there equivalent 'immune responses' to grazing in plants? For an individual algal cell swallowed whole by a *Daphnia* the question is obviously meaningless. But for trees and caterpillars, grass and voles it is not. Its importance rests, amongst other things, upon our ability to model plant–herbivore interactions successfully, because an 'immune response' turned on and off as grazing pressures rise and fall (an example of which was discussed in section 2.4.4) transforms the constants c_1, c_2 and d_2 in eqs. (7.5) and (7.6) into variables; and if there are costs to the plant, alters r_1 (and conceivably even K) as well. As Haukioja and Hakala (1975) point out, such a reciprocal interaction between plant and herbivore gives an additional twist to, or provides a basic mechanism driving, herbivore cycles (section 2.4.4) and periodic outbreaks (see also Lomnicki, 1977, and references therein). Evidence for changes in plant-chemistry following grazing is provided by Haukioja and his co-workers (Haukioja and Hakala, 1975; Haukioja and Niemela, 1979); see also the discussion in section 2.4.4. Some, like the widespread production of wound induced proteinase inhibitors after localized tissue damage (Ryan and Green, 1974) play an enigmatic role in defence. The advantages of others, for example a marked reduction in the performance of caterpillars fed foliage from previously defoliated trees (Haukioja and Niemela, 1979; Thielges, 1968; Wallner and Walton, 1979; Fig. 7.12), are immediately apparent. We urgently need to know how quickly such responses decay, and whether any are permanent.

Plant defences, permanent or ephemeral, attest the obvious: being eaten is usually a bad thing. However, and unlike predation, grazing does not always or even often equal death, tempting both Harris (1973) and Owen and Wiegert (1976) to speculate that some grazing by insects may actually benefit the plant; see also the discussion in chapter 15. The chain of arguments in Owen and Wiegert's hypothesis is long, involving for example, secretion of copious quantities of honeydew

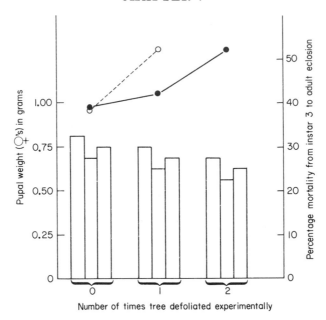

Fig. 7.12. Effects of experimentally defoliating grey birch trees, *Betula populifolia*, on the subsequent growth and survival of Gypsy moth, *Lymantria dispar*, with unrestricted access to leaves from control or treated trees (Wallner and Walton, 1979). Pupal weights (histograms) were significantly reduced in three populations reared on leaves from defoliated trees: the effect is more marked after two defoliations. Since fecundity in many Lepidoptera is proportional to female size, a herbivore outbreak leading to defoliation may reduce the rate of increase of subsequent generations of moths. Larval survival (solid and dotted graphs) is also worse on previously defoliated trees.

(excess sugars from the phloem sap) by aphids, with the honeydew then stimulating nitrogen fixation in the soil below the plant. Whatever benefits may in theory accrue to the plant through increased nitrogen availability (see Stenseth, 1978), they are not yet known to work in practice, because aphids impair rather than stimulate, the growth of trees (Dixon, 1971a, b; see also 7.3.2 and 7.5.1); at the moment, Owen and Wiegert's hypothesis has no experimental support.

More conventionally, compensatory growth upon being grazed is well established as a major adaptive component in the repertoire of many plants. McNaughton (1979a) provides a general review, and examines in detail the effects of heavy grazing by ungulates in the Serengeti. Here, moderate grazing stimulates productivity up to twice the levels in ungrazed control plots. Equation (7.5) implies plant productivity will be maximized when $V/K = 0.5$, but McNaughton's review hints at far more than a passive response as plant biomass is depressed below K. With ungulate grazing r_1 can no longer be viewed as a constant, but as a variable which will be increased, amongst other things, by: (1) increased photosynthetic rates in residual tissue; (2) reallocation of substrates from elsewhere in the plant; (3) mechanical removal of old, senescent tissue; (4) nutrient recycling from dung and urine; and last but not least (5) direct effects from growth-promoting substances in ruminant saliva (see also Dyer, 1980).

Finally, grazing may also drastically alter the partitioning of plant resources between roots, stems, flowers and leaves, and between sexual and vegetative reproduction (Harper, 1977; see also chapters 3 and 15); presumably many of these shifts in the resource budget are again adaptive. Thus defoliation may not only alter the growth rates and survival of trees (page 141) but also reduce their seed production (Rockwood, 1973). The herb *Solidago canadensis* responds to gall-forming insects by increasing stem growth and decreasing rhizome and seed production (Hartnett and Abrahamson, 1979); and the leek, *Allium porrum*, has a graded series of responses to attack by caterpillars of its own species of moth, *Acrolepiopsis assectella* (Boscher, 1979). Attack on one leaf stimulates axillary buds to flower; moderate levels of attack increase seed production; whilst heavy damage reduces seed production but increases bulblet formation (vegetative reproduction). The long-term consequences of these changes in resource allocation for the population dynamics of plants are largely unknown. But they are obviously extremely difficult to accommodate within the existing framework of Lotka–Volterra models.

7.5.3 *Models*

By now, it should be obvious that the art of modelling grazing systems, and hence our theoretical understanding, lags a long way behind the predator–prey and host–parasitoid interactions dealt with in chapter 6. The reasons are both sociological and biological: sociological because botanists have been peculiarly reluctant to study plant population dynamics, botanists and zoologists equally reluctant to abandon their training and straddle the first two trophic levels; and biological because the problems are not easy. Unlike 'prey', individual plants may be difficult, if not impossible to define; grazing rarely results in death, but can so alter the demography of the plant as to make it virtually a different beast; and vegetative reproduction or buried seed banks all add to the confusion. In brief, realism resides in a jungle of parameters, most of them as yet unstudied and unmeasured except in the most rudimentary way.

For many systems, particularly when the herbivores are small and short lived relative to the vegetation, the most promising way forward appears to be to model plants as a 'population of parts' (see chapter 3), coupling the herbivores' dynamics to the bits which they eat. The idea has been most clearly developed, and the best data gathered by Gutierrez and his co-workers, particularly for cotton (see Gutierrez, Wang and Jones, 1979; Gutierrez, Wang and Regev, 1979; and Wang, Gutierrez, Oster and Daxl, 1977 and references therein). Their model views the plant as separate, but interacting, populations of leaves stems, roots and fruits, with submodels describing the dynamics of each. These submodels are integrated via a 'carbohydrate pool' which distributes available photosynthetic material, and provides the essential link between, for example, the amount of leaf tissue and subsequent fruit production.

Population models for the herbivores (e.g. pink bollworm, *Pectinophora gossypiella*, and Lygus bugs, *Lygus hesperus*) are coupled to the plants by making their birth, death, immigration and emigration rates functions of food availability and their own population densities. The herbivores alter the dynamics of plant parts in one or more of several different ways. For example, they may change the abundance and age structure of the leaves by eating them; if they are bugs, they directly reduce the carbohydrate pool; and if they attack fruits, they reduce seed production. Each of these effects, in turn, has consequences for the 'carbohydrate pool' of the plant and hence the dynamics of other plant parts.

The models are complex and have to be solved by simulation, but they can be applied to a wide range of plants and insects (Gutierrez, Wang and Jones, 1979). To date, the crop models describe very well what happens during a single growing season; they have not been used to simulate the dynamics of vegetation and herbivores over many generations. In all probability, such long-term simulations would require so many parameters as to be virtually incomprehensible without simplification. A similar problem has confronted, and been solved by, workers on spruce budworm *Choristoneura fumiferana* in the coniferous (*Abies* and *Picea*) forests of Canada (see Holling *et al.*, 1976; Peterman *et al.*, 1979 and references therein).

Budworms can increase rapidly in a matter of months: the trees in contrast take a decade to recover after an outbreak and have a life span of seventy years. Holling and his co-workers used their very complex simulation model to find the equilibrium points for insects and trees under various forest conditions. The resulting 'manifold' (Fig. 7.13b) distills the essence from an enormous amount of intricate biological detail. An immature forest supports only very low numbers of budworm, but as the forest matures, there comes a point where the budworm population is bound to explode. After a few years of outbreak densities,

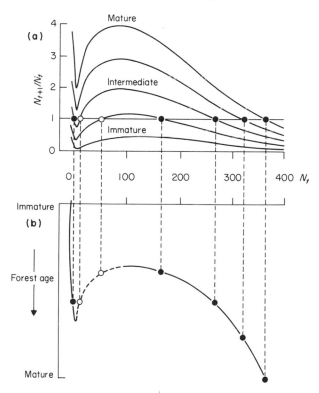

Fig. 7.13 (a). A family of budworm 'recruitment curves' for different levels of forest maturity. Plotted are points for each forest condition, where N_{t+1} and N_t are budworm numbers in two successive generations. (b) is a 'manifold' of equilibrium points, (where $N_{t+1} = N_t$) and should be viewed sideways with forest maturity as the abscissa. Stable equilibria are represented by solid dots, unstable ones by open circles (Peterman *et al.*, 1979). The lower equilibria in young forests are apparently maintained by predation (hypothesis (3), p. 138): as the forest matures, predators become less effective and eventually the budworm 'outbreak' to the upper equilibria set by food limitation.

the old forest then deteriorates and is replaced, once more, by young trees. This example is discussed further in chapter 15.

Most plant–herbivore systems in which the herbivores are insects, and probably many vertebrate herbivore–plant interactions as well, share the 'nested hierarchy' of time scales characterized by cotton, spruce, and the beasts which eat them. Obviously, the births and deaths of caterpillars and leaves occur on time-scales orders of magnitude shorter than the births and deaths of trees. Such a nested hierarchy of time scales is identical to that seen in the dynamics of parasites and diseases of man and other vertebrates with indirect life cycles (May and Anderson, 1979; see also chapter 14). To model these interactions analytically, processes with the shortest time-scales are solved first, before being embedded in models of long-term events. Coupled with a detailed understanding of how the loss of one plant part affects the dynamics of other parts, we now have an exciting new framework within which to study plant–herbivore dynamics. At the moment such an approach to the dynamics of natural vegetation is no more than a glimmer in our collective mind's eye: we urgently need more data.

+ then the whole plant association

8
Competition and Niche Theory

ERIC R. PIANKA

8.1 Introduction: Definitions and Theoretical Background

8.1.1 *Competition*

By definition, competition occurs when two or more organisms, or other organismic units such as populations, interfere with or inhibit one another. The organisms concerned typically use some common resource which is in short supply. Moreover, the presence of each organismic unit reduces the fitness and/or equilibrium population size of the other. Competition is sometimes quite direct, as in the case of interspecific territoriality, and is then termed *interference competition*. More indirect competition also occurs, such as that arising through the joint use of the same limited resources, which is termed *exploitation competition*. Because it is always advantageous for either party in a competitive interaction to avoid the other whenever possible, competition presumably promotes the use of different resources and hence generates ecological diversity. The mechanisms by which members of a community of organisms partition resources among themselves and reduce interspecific competition shapes community structure, and may often influence species diversity profoundly (see Schoener, 1974, for a review).

Ecologists are, however, divided in their attitudes concerning the probable importance of competition in structuring natural communities. Some, myself included, either tacitly or explicitly assume that self-replication in a finite environment must eventually lead to some competition. Other ecologists, particularly those that study small organisms and/or organisms at lower trophic levels, tend to be much more skeptical about the impact of competition upon organisms in nature. Still others have begrudgingly elevated competition to the status of a dogma—this school in turn emphasizes the importance of predation in structuring natural communities. While the persistence of this dispute over the strength of competition in natural communities could conceivably reflect a natural dichotomy, it might well be more realistic and more profitable not to view competition as an all-or-none phenomenon. An emerging conceptual framework envisions a gradient in the intensity of competition, varying continuously between the end-points of a complete competitive vacuum (no competition) to a fully saturated environment with demand equal to supply ('over-saturated' environments are also possible, with demand temporarily exceeding supply).

Competition lends itself readily to mathematical models, and an extensive body of theory exists, most of which assumes saturated communities at equilibrium with their resources, sometimes referred to as 'competitive' communities. Much of this theory is built upon the overworked Lotka–Volterra competition equations:

$$\frac{dN_i}{dt} = r_i N_i \left(\frac{K_i - N_i - \sum\limits_{j \neq i}^{n} \alpha_{ij} N_j}{K_i} \right) \qquad (8.1)$$

Here n is number of species (subscripted by i and j), r_i is the intrinsic rate of increase of species i, K_i is its 'carrying capacity,' N_i is its population density, and α_{ij} is the 'competition coefficient' which measures the inhibitory effects of an individual of species j upon species i. At equilibrium, all dN_i/dt must be equal to zero, giving the equilibrium population densities:

$$N_i^* = K_i - \sum\limits_{j \neq i}^{n} \alpha_{ij} N_j^*, \qquad (8.2)$$

for all i from 1 to n.

The Lotka–Volterra competition equations greatly oversimplify the process of interspecific competition (for examples, see Hairston *et al.*, 1968; Wilbur, 1972; and Neill, 1974). Indeed, the alphas in these equations could be illusory and may often obscure the real *mechanisms* of competitive interactions. Nevertheless, whatever flaws the Lotka–Volterra equations may have, they have clearly contributed much to current ecological thinking. Not only do they provide a conceptual framework, but they have helped to give rise to many exceedingly useful ecological concepts in addition to competition coefficients, including equilibrial population densities, the community matrix, diffuse competition, r and K selection, as well as non-linear isoclines (see, for example, Ayala *et al.*, 1973a; Schoener, 1974; Gilpin and Ayala, 1973). Some important papers, books and/or major reviews in the voluminous literature on competition include Crombie (1947), Birch (1957), Milne (1961), Milthorpe (1961), DeBach (1966), Miller (1967), MacArthur (1972), Grant (1972a), Stern and Roche (1974), Schoener (1974, 1976a, 1977) and Connell (1975).

8.1.2 *Niche theory*

Among the first to use the term niche was Grinnell (1917, 1924, 1928), who viewed it as the ultimate distributional unit, thus stressing a spatial concept of the niche. Elton (1927) emphasized more ethological aspects, and defined the ecological niche as the functional role and position of the organism in its community, stressing especially its trophic relationships with other species. Although the term niche has been used in a wide variety of ways by subsequent workers, the idea of a niche gradually became linked with competition. Empirical studies of Gause (1934) and others (see below) showed that ecologically similar species were seldom able to coexist in simple laboratory systems; hence species living together must each have their own unique niche. The one species per niche concept became accepted as ecological dogma, although a few dissenters urged otherwise (Ross, 1957, 1958). Because the term 'niche' has been used in a wide variety of different contexts and is rather vaguely defined, some ecologists prefer not to use the word [see, for example, commentaries on niche in Williamson (1972) and Emlen (1973)]. The ecological niche has, however, become increasingly identified with resource utilization spectra (Fig. 8.1), through both theoretical and empirical work of a growing school of population biologists (Levins, 1968; MacArthur, 1968, 1970, 1972; Schoener and Gorman, 1968; Pianka, 1969, 1973, 1974, 1975; Colwell and Futuyma, 1971; Inger and Colwell, 1977; Roughgarden, 1972, 1976, 1979; Vandermeer, 1972; Pielou, 1972; May and MacArthur, 1972; May, 1974d, 1975d; Cody, 1974; Schoener, 1968, 1975a, 1975b, 1977). Such an emphasis upon resource use is operationally tractable, although it largely neglects considerations of reproductive success (some earlier treatments of the niche such as the n-dimensional hypervolume concept of Hutchinson (1957) used fitness to define niche boundaries).

Although niche theory will ultimately have to include aspects of reproductive success as well as resource utilization phenomena, emphasis

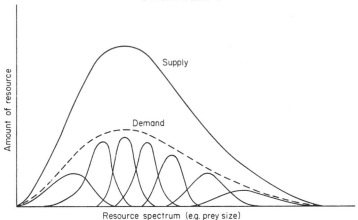

Fig. 8.1. Niche relationships among potentially competing species are often visualized and modelled with bell-shaped resource utilization curves. The uppermost bell-shaped curve represents the supply of resources along a resource continuum, such as prey size. The vertical axis measures the amount of resource available or used over some time interval. The lower small curves represent seven hypothetical species in a community, with those species that exploit the 'tails' of the resource spectrum using a broader range of resources (that is, they have broader niches) because their resources are less abundant. The sum of the component species' utilization curves (dashed line) reflects the total use or the overall demand along the resource gradient. Pressures leading to the avoidance of interspecific competition should result in a relatively constant ratio of demand/supply along the resource continuum, as shown. Discrete, rather than continuous, resource states can be handled in analogous ways.

must be on the latter here. Possibilities abound for significant further work, both theoretical and empirical, on the constraints and interactions between optimal foraging and optimal reproductive tactics (Pianka, 1976).

The concept of niche 'breadth' (niche 'width' or 'size' are frequent synonyms) has proven to be extremely useful. Niche breadth is simply the total variety of different resources exploited by an organismic unit. In the absence of any competitors or other enemies, the entire set of resources used is referred to as the 'fundamental' niche (Hutchinson, 1957) or the 'pre-interactive,' 'pre-competitive,' or 'virtual' niche. Any real organismic unit presumably does not exploit its entire fundamental niche since its activities are somewhat curtailed by its competitors as well as by its predators; hence its 'realized,' 'post-interactive,' or 'post-competitive' niche is usually a subset of the fundamental niche (Hutchinson, 1957; Vandermeer, 1972). The difference between the fundamental and the realized niche, or the niche change due to competitors, thus reflects the effects of interspecific competition (as well as predation).

Consideration of the variety of factors influencing niche breadth leads into the problem of specialization versus generalization. A fairly substantial body of theory on optimal foraging predicts that niche breadth should generally increase as resource availability decreases (Emlen, 1966, 1968; MacArthur and Pianka, 1966; Schoener, 1971; MacArthur, 1972; Charnov, 1973, 1975). In an environment with a scant food supply, a consumer cannot afford to bypass many inferior prey items because mean search time per item encountered is long and expectation of prey encounter is low. In such an environment, a broad niche maximizes returns per unit expenditure, promoting generalization. In a food-rich environment, however, search time per item is low since a foraging animal encounters numerous potential prey items; under such circumstances substandard prey items can be bypassed because expectation of finding a superior item in the near future is high. Hence rich food supplies are expected to lead to selective foraging and narrow food niche breadths. A competitor can act either to compress or to expand the realized niche of another species, depending upon whether or not it reduces resource levels uniformly (which leads to niche expansion) or in a patchy manner (which should often result in a niche contraction, especially in the microhabitats used).

Two fundamental aspects of niche breadth have been distinguished— the so-called 'between-phenotype' versus 'within-phenotype' components (Van Valen, 1965; Roughgarden, 1972, 1974b, 1974c). A population with a niche breadth determined entirely by the between-phenotype component would be composed of specialized individuals with no overlap in resources used; in contrast, a population composed of pure generalists with each member exploiting the entire range of resources used by the total population would have a between-phenotype component of niche breadth of zero and a maximal within-phenotype component. Real populations clearly lie somewhere between these two extremes, with various mixtures of the two components of niche breadth.

Another central aspect of niche theory concerns the amount of resource sharing, or niche overlap. Ecologists have long been intrigued with the notion that there should be an upper limit on how similar the ecologies of two species can be and still allow coexistence. Concepts that have emerged from such thinking include the so-called 'principle' of competitive exclusion (below), character displacement, limiting similarity, species packing, and maximal tolerable niche overlap (for examples, see Hutchinson, 1959; Schoener, 1965, MacArthur and Levins, 1967; MacArthur, 1969, 1970; May and MacArthur, 1972; Pianka, 1972; Grant, 1972a). A number of models of niche overlap in competitive communities, typically built upon eq. (8.1), have generated several testable predictions (MacArthur and Levins, 1967; MacArthur, 1970, 1972; May and MacArthur, 1972; May, 1974d; Gilpin, 1974; Rappoldt and Hogeweg, 1980; Roughgarden, 1974a, 1975b, 1976). Some of these models suggest that maximal tolerable niche overlap should decrease as the number of competing species increases, with such decreases in overlap approximating a decaying exponential (cf. Fig. 8.12). Indeed, MacArthur (1972) coined the term 'diffuse competition' to describe the total competitive effects of a number of interspecific competitors, implying that a little competitive inhibition per species when summed over many other species can be equivalent to strong competitive inhibition by fewer competing species. One model, that of May and MacArthur (1972), predicts that maximal tolerable overlap should be relatively insensitive to environmental variability.

Most existing theory on niche overlap is framed in terms of a single niche dimension (but see May, 1974d, and Yoshiyama and Roughgarden, 1977). Real plants and animals, however, differ in their use of just one resource only infrequently (for such an example, see Fig. 8.9). Rather, pairs of species frequently show moderate niche overlap along two or more niche dimensions (Fig. 8.2). Ideally a multidimensional analysis of resource utilization and niche separation along more than a single niche dimension should proceed through estimation of proportional simultaneous utilization of all resources along each separate niche dimension (Pianka, 1974, 1975; May, 1975d). In practice however, it is extremely difficult to obtain such multidimensional

---

Fig. 8.2. (a) Diagrammatic representation of resource utilization of two hypothetical species along two niche dimensions, such as prey size and foraging height above the ground. Note that, although the shadows of the three dimensional peaks on each separate niche dimension overlap, true overlap in both dimensions is very slight. Adapted after Clapham (1973).

(b). A similar plot for seven species, showing that pairs with substantial or complete overlap along one dimension can avoid or reduce competition by niche separation along another dimension.

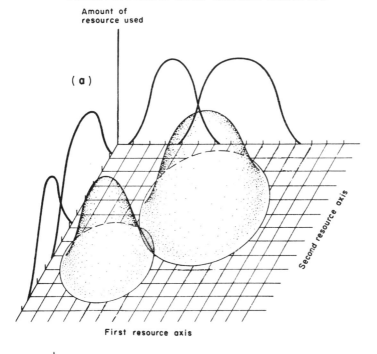

(a)

Amount of
resource used

First resource axis

Second resource axis

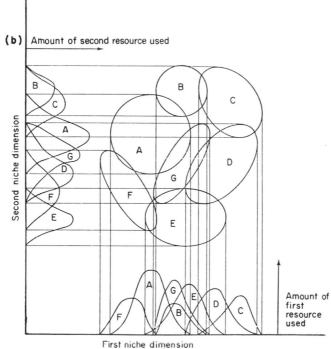

(b) Amount of second resource used

Second niche dimension

First niche dimension

Amount of
first
resource
used

utilization data because animals move and integrate over both space and time. An individual animal would have to be followed, and its use of all resources recorded continually, in order to obtain accurate estimates of its true utilization of a multidimensional niche space. Since this is often very tedious or even impossible, one usually attempts to approximate from separate unidimensional utilization distributions (Fig. 8.3).

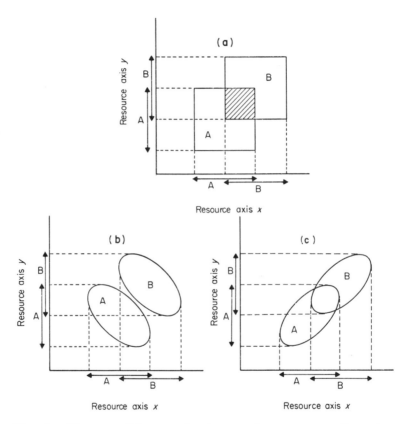

Fig. 8.3. Three possible cases for the use of two resource dimensions (assumed to be constant within boxes or ellipses) for two hypothetical species, A and B. Although unidimensional projections are identical in all three cases, true multidimensional overlap is zero in case (**b**). Case (**a**) illustrates truly independent niche dimensions, with any point along resource axis *x* being equally likely along the entire length of resource axis *y*; under such circumstances, niche dimensions are orthogonal and unidimensional projections accurately reflect multidimensional conditions. However, when niche dimensions are partially dependent upon one another (cases (**b**) and (**c**)), unidimensional projections can be severely misleading.

Provided that niche dimensions are truly independent, with for example any prey item being equally likely to be captured in any place, overall multidimensional utilization is simply the product of the separate unidimensional utilization functions (May, 1975d). Estimates of various niche parameters along component niche dimensions can then simply be multiplied to obtain multidimensional estimates. However, should niche dimensions be partially dependent upon one another (Fig. 8.3), there is no substitute for knowledge of the true multidimensional utilization functions. In the extreme case of complete dependence (with, for example, any given prey item being found only at a particular height above ground), appearances to the contrary, there is actually only a single niche dimension, and a simple average provides the best estimate of true utilization. Moreover the arithmetic average of estimates of unidimensional niche overlap obtained from two or more separate unidimensional patterns of resource use actually constitutes an *upper bound* on the true multidimensional overlap (May, 1975d). Certain pitfalls in the analysis of niche relationships along more than one dimension have been noted by May (1975d) and Pianka (1974, 1975).

An important aspect of niche dimensionality concerns the notion of the number of immediately adjacent species in niche space (MacArthur, 1972). An increased number of niche dimensions results in a greater potential for adjacent species in niche space (see Fig. 8.2) and hence may intensify diffuse competition (MacArthur, 1972; Pianka, 1974, 1975).

8.1.3 *Niche overlap and competition*

Because competition coefficients are exceedingly difficult to estimate directly except by population removal experiments (below), measures of niche overlap have often been used as estimates of the alphas in eqs. (8.1) and (8.2) (Pico *et al.*, 1965; Schoener, 1968; Levins, 1968; Orians and Horn, 1969; Pianka, 1969; Culver, 1970; Brown and Lieberman, 1973; May, 1975d). However, tempting though it may be, equating overlap with competition can be dubious and misleading (Colwell and Futuyma, 1971). Although niche overlap is clearly a prerequisite to exploitative competition, overlap need not necessarily lead to competition unless resources are in short supply. (In a competitive vacuum with a surplus of resources available, niches could presumably overlap completely without detriment to the organisms concerned.) Moreover, interference competition is unlikely to evolve unless there is a potential for overlap in use of limited resources (i.e., exploitation competition must be *potentially* possible). Avoidance of competition can lead to entirely non-overlapping patterns of resource utilization (disjunct niches), as for example occurs in interspecific territoriality. In the parlance of the mathematician, niche overlap in itself is neither a necessary nor a sufficient condition for interference competition; moreover, overlap is only a necessary but not sufficient condition for exploitation competition.

Indeed, the preceeding arguments suggest that there may often be an inverse relationship between competition and niche overlap. If so, extensive overlap might actually be correlated with reduced competition. Such reasoning led me to propose the 'niche overlap hypothesis', which asserts that maximal tolerable overlap should be lower in intensely competitive situations than in environments with lower demand/supply ratios (Pianka, 1972; see also Roughgarden and Feldman, 1975). The impact of predators on competition is poorly understood, but they may often reduce its intensity and hence facilitate coexistence (Paine, 1966; Connell, 1975).

8.1.4 Temporal variability and fugitive species

Most existing theory on both interspecific competition and niche relationships assumes that component populations have reached an equilibrium with their resources and differ enough in their use of these resources to coexist. Hutchinson (1951) formulated a somewhat more dynamic and perhaps more realistic concept of a 'fugitive' species, which he envisioned as predictably inferior, always excluded locally under interspecific competition, but which persists in newly disturbed regions merely by virtue of its high dispersal ability. Hutchinson's mechanism for persistence of a colonizing species by means of high dispersal ability in a continually disturbed and changing patchy environment, in spite of pressures from competitively superior species, has been modelled analytically (Skellam, 1951; Levins and Culver, 1971; Horn and MacArthur, 1972; Slatkin, 1974; Levin, 1974). This view of competitive pressures and processes varying in time and space is appealing and doubtlessly realistic, although perhaps operationally intractable in many instances. In a later attempt to explain the apparent 'paradox' of the plankton (that is, the coexistence of members of a diverse community in a relatively homogeneous physical environment with few possibilities for niche separation), Hutchinson (1961) suggested that temporally-changing environments may promote diversity by periodically altering the relative competitive abilities of component species, hence allowing their coexistence (Weins, 1977, recently revived this idea).

8.2 Empirical studies

Competitive effects can be measured in two ways: (1) by perturbation experiments in which rates of population growth of competitors are monitored as they approach equilibrium, and (2) by addition and/or removal experiments, in which the equilibrium population density of a species, N_i^* in eq. (8.2), is monitored at equilibrium both in the presence and in the absence of its competitor, in an otherwise unchanged environment. Either slowed rates of population growth, or an observed reduction in population density, or a niche shift in response to the presence of a competitor constitute direct evidence of interspecific competition.

Most experimental studies of competition have been performed under relatively homogeneous and constant laboratory conditions. Such studies, briefly considered below, led to the so-called 'principle' of competitive exclusion of Gause (1934) (see also Cole, 1960; Hardin, 1960; Patten, 1961; DeBach, 1966), which asserts that some ecological difference must exist between coexisting species.

Although competition is the conceptual backbone of much current ecological thought, it has proven exceedingly difficult to study in natural communities, probably partially because reduction or avoidance of competition is always advantageous when possible. Also, the great spatial and temporal variability characteristic of most natural communities demands a dynamic approach to the investigation of competitive interactions. Existing evidence of competition in nature is largely circumstantial (see below) and unequivocal removal experiments hold considerable promise (Connell, 1975). Indeed, because competition lies at the heart of so many ecological processes but has been studied so inadequately, carefully designed and well executed empirical investigations into the precise mechanisms and results of competitive interactions seem virtually certain to be of central importance to the future of ecology.

8.2.1 *Laboratory experiments*

Gause (1934) performed some of the earliest competition experiments with several species of *Paramecium*, using laboratory culture media, renewed at regular intervals. Population growth and population densities were monitored both in single-species cultures and in mixtures of two competing species grown together (Fig. 8.4). One experiment clearly demonstrated competitive exclusion of *P. caudatum* by *P. aurelia*. In another experiment with *P. caudatum* and *P. bursaria*, this pair of *Paramecium* species coexisted in a mixed culture, although at lower population densities than when grown in pure cultures of a single species (Fig. 8.4). Competition coefficients, reflecting the intensity of the competitive interaction, are readily calculated from such data.

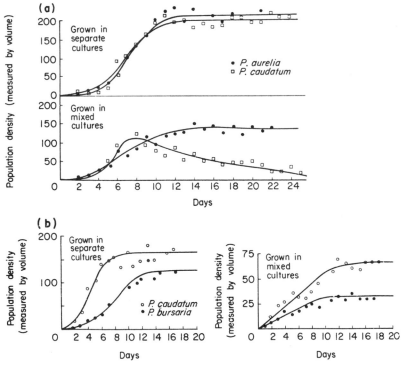

Fig. 8.4. (a) When *Paramecium aurelia* and *P. caudatum* are grown together in mixed laboratory cultures, *P. aurelia* excludes *P. caudatum*. Modified from Gause (1934) after Clapham (1973). (b) Laboratory competition experiments with two more dissimilar *Paramecium* species, *P. caudatum* and *P. bursaria*, result in coexistence at lower population densities than in pure cultures. From data of Gause, modified from Clapham (1973).

By far the most exhaustive laboratory studies of competition are those of Park (1948, 1954, 1962) and his associates, who worked with flour beetles, especially of the genus *Tribolium*. While Park's studies are much too extensive to review here, they also convincingly demonstrated competitive exclusion. In addition, a series of experiments showed that the outcome of interspecific competition depends upon (a) initial population densities (Neyman *et al.*, 1956), (b) environmental conditions of temperature and humidity (Park, 1954), and (c) on the genetic constitution of the strains of competing species (Park *et al.*, 1964).

Among the numerous other laboratory experiments on competition that have been undertaken since Gause's and Park's pioneering efforts, perhaps one of the more informative is the elegant work of Neill (1972, 1974, 1975). Using aquatic laboratory microcosms containing communities of four species of micro-crustaceans along with associated bacteria and algae under equilibrium conditions, Neill performed a series of replicated removal experiments and allowed the resulting systems to return to equilibrium. Each species of microcrustacean, as well as each possible pair of species, were removed and estimates of population densities of the various species were made under nearly all possible competitive regimes. Competition coefficients estimated from these equilibrium population densities varied with community composition. In both Neill's microcosms and in an amphibian community studied by Wilbur (1972), the joint effects of two species upon a third in a 3-species system cannot always be predicted from the separate interactions in the three component 2-species systems. Such results indicate that, if eq. (8.1) is to reflect reality, it must somehow be expanded to include 'interactive' competition coefficients reflecting the joint effects of two species upon a third, as suggested by Hairston *et al.* (1968) and Wilbur (1972).

8.2.2 *Field observations*

Removal experiments under field conditions are usually next to impossible, and have seldom been attempted (but see below). Instead, field studies of competition tend to rely heavily on 'natural' experiments, in which aspects of the ecology of a species are compared between areas where it occurs alone (allopatry) with other areas where it occurs with another competing species (sympatry). Provided that the two areas are otherwise basically similar, niche shifts observed in sympatry should reflect the response to interspecific competition. However, as pointed out by Grant (1972) and Connell (1975) among others, such observations often lack a suitable 'control', since other factors probably differ between allopatry and sympatry.

Niche shifts and character displacement

Such a situation occurs among two species of flatworms along temperature gradients in streams (Beauchamp and Ullyett, 1932). Figure 8.5 depicts the distributions of these two species of *Planaria* in streams

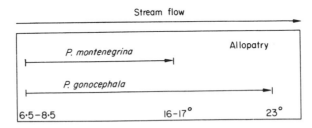

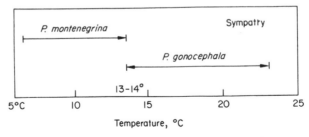

Fig. 8.5. Distributions of *Planaria montenegrina* and *P. gonocephala* along temperature gradients in streams when they occur separately (above) and together (below). Each species is restricted to a smaller range of thermal conditions when in competition with the other. From Beauchamp and Ullyett (1932) after Miller (1967).

where each occurs separately and where both exist together. Neither species occupies as broad a range in temperature when the two occur together as it does when it is the only species in the stream (Fig. 8.5). Similar observations of niche shifts in 'incomplete' faunas include studies on salamanders (Hairston, 1951) and birds (Crowell, 1962).

A related phenomenon, termed 'character displacement' by Brown and Wilson (1956), sometimes occurs when two wide-ranging species have partially overlapping geographic distributions, with a zone of sympatry and two zones where each species occurs alone in allopatry. Such species pairs are often very similar to one another where they occur in allopatry, but they typically diverge when they coexist (Fig. 8.6).

Hutchinson (1959) first commented on the apparent constancy in the magnitude of morphological character displacement, reporting ratios of

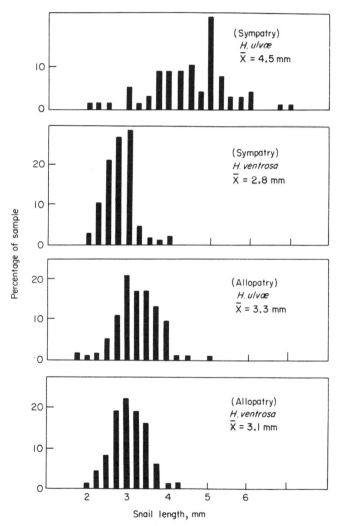

Fig. 8.6. Length-frequency distributions of the shells of two species of snails, *Hydrobia ulvae* and *H. ventrosa*, at a locality where they coexist (top two boxes) and at two localities where they occur in allopatry (bottom two boxes). From Fenchel (1975).

mouthpart sizes among coexisting congeneric species of insects, birds and mammals ranging only from about 1·2 to 1·4. Schoener (1965), Grant (1968) and Diamond (1973) found similar ratios of bill lengths among coexisting pairs of bird species, while Fenchel (1975) obtained comparable but slightly larger character displacement ratios in body sizes of two species of small marine snails (Fig. 8.6). Pulliam (1975) reports very constant, although slightly smaller (about 1·1), bill length ratios among seed-eating sparrows. However, sympatric *Anolis* lizards in the Caribbean often differ by considerably greater ratios, particularly on small islands where the ratio of sizes approaches 2 (Schoener, 1970). Indeed, Schoener (1977) suggests that niche variance along the food-size dimension and the between-phenotype component of niche breadth may be generally greater among lizards than in birds due both to lack of parental care in lizards and their slower indeterminate growth, which results in greater variation in size among individuals within a population. For an irreverent review of the Hutchinsonian 1·3 ratio, showing that it turns up in many contexts (including kitchen implements, children's bicycles, and ensembles of musical instruments), see Horn and May (1977) and Maiorana (1978).

Under circumstances in which there is a potential for intense exploitation competition, niche overlap may be reduced by interference competition leading to interspecific territoriality (Orians and Wilson, 1964), which could conceivably select for an actual convergence in phenotypic characteristics involved in recognition and territorial defence (Cody, 1969, 1974). In a critical review, Grant (1972a) evaluated existing evidence for and against both character convergence and divergent character displacement; he concludes that not only is the evidence for the ecological bases of character displacement (differential resource use) quite weak, but also that many putative examples of divergent character displacement (including the 'classic' *Sitta* nuthatch case!) could easily represent merely gradual clinal variation associated with various environmental gradients.

Two reasonably strong cases for ecological character displacement have appeared since Grant's review. In one study by Fenchel (1975), mentioned above, two small species of deposit-feeding marine snails, *Hydrobia ulvae* and *H. ventrosa*, have nearly identical size frequency distributions where each occurs in allopatry under different conditions of salinity and hydrography, However, in relatively narrow (and probably quite recent) zones of sympatry along certain salinity gradients, the two species appear to coexist in a stable equilibrium with the more widespread *H. ulvae* becoming conspicuously larger than in allopatry while the other species gets noticeably smaller (Fig. 8.6). Fenchel (1975 and unpublished) has expanded these studies and shown particle size selection by snails of different sizes, providing an ecological basis for differential resource utilization arising from the size difference.

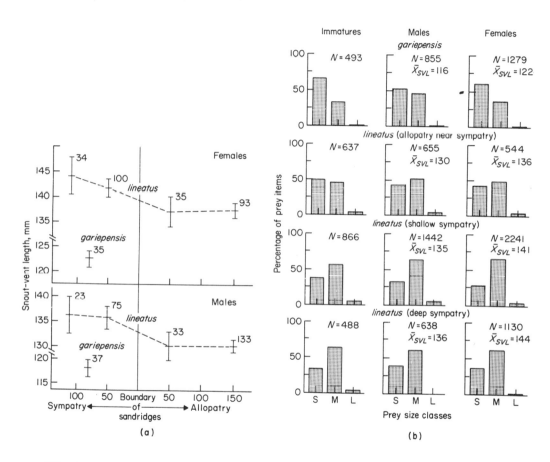

Fig. 8.7. (a) Step clines in mean snout-vent lengths of subterranean legless lizards (Scincidae: *Typhlosaurus lineatus*) associated with the presence of a smaller congeneric species *T. gariepensis*. Head proportions also change with proportionately larger heads occurring in sympatry.

(b) Distributions of prey sizes eaten by *Typhlosaurus* immatures, adult males, and adult females under various conditions of allopatry and sympatry. Sympatric *T. lineatus* eat more larger prey items than do allopatric *T. lineatus*. Both figures from Huey *et al.* (1974).

The second study of character displacement involves two species of subterranean skinks (genus *Typhlosaurus*), which occur in sympatry throughout the sandridge regions of the southern part of the Kalahari semidesert (Huey *et al.*, 1974; Huey and Pianka, 1974). Both lizard species eat almost nothing but termites, largely the same few species. Although the smaller of the two species (*T. gariepensis*) is known only from sympatry, the larger species (*T. lineatus*) also occurs in allopatry on adjacent flat sandveld areas. Snout-vent lengths of *lineatus* increase abrubtly at the boundary of the sandridges (Fig. 8.7a), suggesting divergent ecological character displacement in sympatry (although the possibility that sandridge habitats differ fundamentally from sandveld ones cannot be entirely discounted—see Pianka, Huey and Lawlor, 1979). Moreover, heads become *proportionately* larger in sympatry. Correlated with this increase in body size and head proportions is a dietary shift to larger castes and species of termites (Fig. 8.7b), which reduces overlap and probably competition with *T. gariepensis*. Ratios

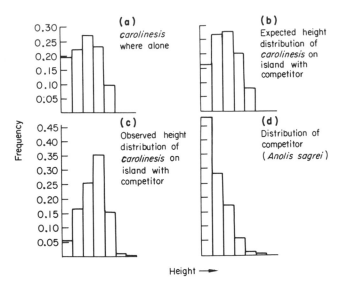

Fig. 8.8. Expected and observed frequency distributions of perch heights of *Anolis* lizards. (**a**) Observed height distribution of *A. carolinensis* where it occurs alone without competitors. (**b**) Expected distribution of perch heights of *A. carolinensis* on another island with different availabilities of various perch heights, assuming no niche shift. (**c**) Observed distribution of *Anolis carolinensis* on the second island with a competitor (compare with **b**). (**d**) Height distribution of the competing species, *A. sagrei*. From Schoener (1975a).

of snout-vent lengths of the two species in sympatry are about 1·17, while head length ratios are around 1·5.

Interpreting such niche shifts between allopatry and sympatry is often difficult due to geographical variation in various environmental aspects, such as resource availability (Schoener, 1969; Grant, 1972a; Connell, 1975). Schoener (1975a, 1975b) devised a technique to correct for such changes in resource availability and used it to demonstrate niche shifts in response to competition among *Anolis* lizards (Fig. 8.8).

Overdispersion of niches

Several different approaches to the theory of limiting similarity predict an upper limit on tolerable niche overlap (MacArthur and Levins, 1967; May and MacArthur, 1972; May, 1975a; Lawlor and Maynard Smith, 1976; Fenchel and Christiansen, 1976). Schoener (1974) suggests that such limits on niche similarity of coexisting species, coupled with interspecific competition, should result in regular (as opposed to random) spacing of species in niche space (but see next section

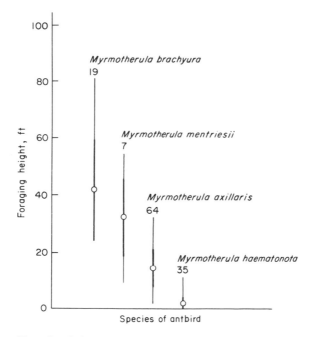

Fig. 8.9. Foraging heights of four species of sympatric antbirds (genus *Myrmotherula*). Means shown with dots and standard deviations by thickened bars. From MacArthur (1972) after data of Terborgh.

for an alternative). A number of sets of species that differ primarily along a single niche dimension do indeed appear to be separated by rather constant amounts (see, for example, Orians and Horn, 1969; MacArthur, 1972). One celebrated example is Terborgh's observation that various species of antbirds forage at different heights above the ground (Fig. 8.9).

However, as indicated above, ecologies of most potential competitors probably differ along several niche dimensions simultaneously. Thus, Schoener (1968, 1974) demonstrated that pairs of species of *Anolis* lizards with high dietary overlap tend to separate out in their use of microhabitats, while pairs using similar microhabitats overlap relatively little in prey sizes eaten (Fig. 8.10). Comparable inverse relationships

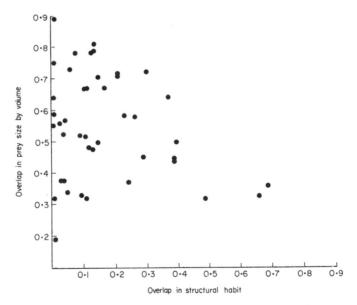

Fig. 8.10. Overlap in prey size plotted against overlap in structural microhabitat among various species of *Anolis* lizards on the island of Bimini. Pairs with high dietary overlap tend to exploit different structural microhabitats; conversely, those with high spatial overlap overlap relatively little in prey sizes eaten. From Schoener (1968).

between dietary and microhabitat overlap occur among many pairs of species of nocturnal gekkonid lizards in Australian deserts (Pianka and Pianka, 1976). In a similar vein, Cody (1968) found that grassland birds partition resources along at least three distinct niche dimensions, with the relative importance of various dimensions in separating species

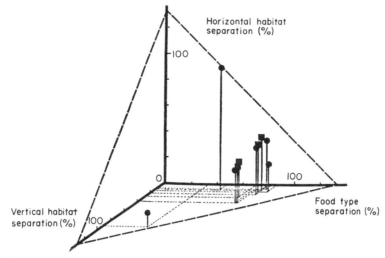

Fig. 8.11. Average niche separation along three dimensions in ten grassland bird communities, suggesting a relatively constant overall amount of separation. From Cody (1968).

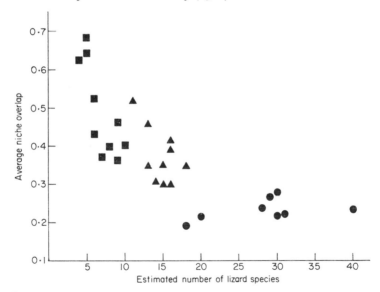

Fig. 8.12. Mean niche overlap plotted against the number of lizard species for 28 study areas on three continents. Squares represent North American sites (Great Basin, Mojave, and Sonoran deserts), triangles are areas in Kalahari semidesert of southern Africa, while dots indicate sites in the Great Victoria desert of Western Australia. This inverse correlation is highly significant statistically ($r = -0.73$, $p < 0.001$). Several estimates of maximal niche overlap are also inversely correlated with the number of lizard species. From Pianka (1974).

differing in various communities (Fig. 8.11). An intriguing degree of constancy in the overall niche separation along all three dimensions suggests that avian niches in these communities are both overdispersed in niche space and that bird species within the communities may have reached some sort of limiting similarity.

My own studies on the niche relationships of desert lizards along three dimensions also suggest overdispersion of niches (Pianka, 1973, 1974, 1975); however, niche separation among these lizard communities is not constant, but both average and maximal overlap vary inversely with the number of lizard species (Fig. 8.12). A decrease in overlap with increasing numbers of species is predicted by several theoretical formulations of limiting similarity and may be due to diffuse competition among coexisting species (see also below).

Guild structure

Another aspect of the pattern in which species are spread out in niche space concerns the extent to which functionally similar species exist in clusters of high overlap pairs. Members of such guilds (Root, 1967) would interact strongly with each other but weakly with other members of their community (see also section 9.5). In competitive communities, guilds would thus represent arenas of relatively intense interspecific competition.

Using the 'single-linkage' criterion of cluster analysis, a guild may be objectively defined as a group of species separated from all other such clusters by a distance greater than the greatest distance between the two most disparate members of the guild concerned. This definition allows complex hierarchical patterns of nesting of smaller guilds in the large ones (Pianka, 1978). Although techniques of measuring guild structure are embryonic, the concept has begun to attract increasing interest (Feinsinger, 1976; Inger and Colwell, 1977; Holmes *et al.*, 1979; Pianka *et al.*, 1979). Numerous intriguing questions can be raised about guilds, but very few answers can yet be supplied. For example, do guilds simply reflect gaps in resource space or can they be evolved when resources are continuously distributed? What are the effects of guilds on diffuse competition, community structure, and organization? What are components of guild structure (spacing, nesting, size, gaps, etc.) and how can they be measured? Some speculations about the effects of guild structure upon overall community stability are in section 9.5.

Experimental manipulations

Both Colwell and Fuentes (1975) and Connell (1975) have recently pointed out various limitations and shortcomings of natural experiments, especially the lack of a suitable control, and they make a strong case for experimental manipulation of densities in the field. Introduction and/or removal of species with concomitant monitoring of changes in population densities and/or niche shifts before, during, and/or after the experimental manipulation is potentially a fruitful avenue to studying competition in the field. Not all such experiments can be expected to produce results, however, because niche shifts and/or changes in population densities associated with the presence or absence of potential competitors need not necessarily occur in ecological time unless the populations concerned are periodically released from interspecific competition under natural conditions. A species that experiences strong interspecific competition pressures continually would not be expected to retain the capacity to use many of the resources and much of the niche space that is regularly exploited by its competitor; such a species might show only a slight niche shift in ecological time with removal of its competitor. Various such manipulative studies on competition have been undertaken with ants (Brian, 1952; Pontin, 1969), numerous marine invertebrates (Connell, 1961; Paine, 1966; Dayton, 1971; Menge, 1972; Vance, 1972; Stimson, 1970, 1973; Black, 1976; Haven, 1973), salamanders (Jaeger, 1970, 1971), lizards (Nevo et al., 1972; Dunham, 1980), birds (Davis, 1973), as well as with many rodents (DeLong, 1966; Koplin and Hoffman, 1968; Grant, 1972b; Joule and Jameson, 1972; Joule and Cameron, 1975; Rosenzweig, 1971, 1973; Schroder and Rosenzweig, 1975). Many of these experiments are reviewed by Connell (1975).

Some removal experiments have documented competitive exclusion in pairs of species too similar to coexist. For example, Connell (1961) demonstrated that the barnacle *Chthamalus* was able to persist in the intertidal rocky zone normally occupied only by larger *Balanus* barnacles when the larger barnacle species was removed. Noting that two lizards *Lacerta sicula* and *L. melisellensis* have mutually exclusive geographic distributions on small islands in the Adriatic, Nevo *et al.* (1972) report on the experimental introduction of small populations of each species to islands that supported only the other species; in two of three such experiments, the introduced species went extinct, while on a third island the introduced species appeared to be replacing the native form. These authors also performed a reciprocal transplant introduction experiment with the two species on two small but similar islands for future workers to monitor.

Field experiments on naturally coexisting species have also been informative. Vance (1972) studied competition for empty gastropod shells within and among three sympatric species of intertidal hermit crabs by measuring shell preferences and by manipulating the availability of empty shells. An index of 'shell adequacy' was generated by offering crabs a choice of empty uncontested shells; given such a selection, most crabs moved into larger shells than those in which they were found in nature. Under natural conditions, unoccupied shells tended to be in the smallest shell size categories. Very small hermit crabs tended to have shells close to their preferred size, while shell adequacy generally decreased with increasing crab size. Addition of empty shells to the natural environment resulted in increased crab densities, indicating that shells were indeed a limiting resource in short supply. Vance suggested that habitat specificity differences among the three species of hermit crabs could allow the observed coexistence in spite of considerable overlap in their utilization of a limiting resource (shells).

Dunham (1980) studied interspecific competition between two coexisting species of iguanid lizards in Texas by experimentally manipulating densities. During this four-year study, two years were below average in precipitation (and, presumably, in productivity) whereas the other two years were wetter than average. In the two dry years, removal of the larger lizard species (*Scelaporus merriami*) had numerous significant effects on the smaller species (*Urosaurus ornatus*), including increase in density, feeding success, growth rates, lipid levels and prehibernation body weights. In the two wet years, however, treatments did not differ from controls. Removal experiments demonstrated only one effect of the smaller species on the larger one: survival was significantly higher in one of the two dry years. Competition between these two lizard species is clearly not reciprocal and varies in intensity from year to year.

In another field experiment on competition between two broadly sympatric species of sea stars, *Pisaster ochraceus* and *Leptasterias hexactis*, Menge (1972) removed all individuals of the larger *Pisaster* from one small island-reef and added them to another similar reef; a third undisturbed nearby reef supporting both species was monitored as a control (the reciprocal experiment involving removal and addition of *Lep-*

tasterias was not done). Although these two species of sea stars differ greatly in size and reproductive tactics, their diets do overlap broadly (Menge
and Menge, 1974). Average weight of individual *Leptasterias* increased significantly with removal of *Pisaster* and decreased with its addition,
while the size of control *Leptasterias* did not change (Fig. 8.13a). Moreover, estimated standing crops (biomass/m²) of the two species varied

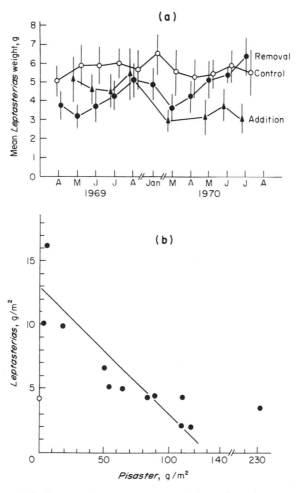

Fig. 8.13. (a) Changes in average wet weight of individual *Leptasterias*
with removal of *Pisaster* (●), addition of *Pisaster* (▲), and in controls
with no change in *Pisaster* densities (○). (Vertical lines represent 95 per
cent confidence intervals of means.) (b) Plot showing the inverse correlation between the biomass per m² of two species of sea stars, *Leptasterias
hexactis* and *Pisaster ochraceus* ($r = -0.64$, $p < 0.01$). From Menge (1972).

inversely over the areas sampled (Fig. 8.13b). Competition coefficients cannot be calculated from these data, but interspecific competition is clearly implicated. As indicated above, somewhat similar removal experiments have been undertaken with many other marine invertebrates and with small rodents (see Connell, 1974, 1975, for reviews of some of this work).

8.3   Neutral models of community structure

Until fairly recently, students of resource partitioning and community structure have been unable to accomplish much more than merely describe existing patterns of resource partitioning among various coexisting consumer species. Even such descriptive efforts seldom allow very useful comparisons with other studies of communities partially because there is 'no . . . standard protocol for community analysis' (Inger and Colwell, 1977). In some cases with low niche dimensionality, observed estimates of overlap have been compared with values of limiting similarity generated from various theoretical arguments such as those of MacArthur and Levins (1967) or May and MacArthur (1972). (For examples, see Orians and Horn, 1969 and/or May, 1974d.) But such comparisons may not be particularly revealing since values of limiting similarity depend strongly on the specific assumptions of models concerned (Abrams, 1975); this point was discussed more fully in section 5.3.

Sale (1974) responded to this dire need for null hypotheses by suggesting that communities might be compared to randomized versions of themselves. Overlap in observed communities of grasshoppers did not differ markedly from that in such randomized analogues, leading Sale (1974) to conclude that competition had not been a force in reducing overlap among or otherwise organizing these communities of insects. In a similar analysis using desert lizard communities, however, Lawlor (1980) found that average overlap was substantially lower in observed communities than in randomized replicates, suggesting that competition has shaped the organization of these lizard communities. This rather promising 'neutral model' approach has now been exploited in a number of studies of community structure (Caswell, 1976; Inger and Colwell, 1977; Pianka, Huey and Lawlor, 1979; Taylor, 1979; Joern and Lawlor, 1980; Connor and Simberloff, 1980).

8.4   Indirect effects of competition: competitive 'mutualists'

The traditional approach* to interspecific competitive interactions is strictly pairwise: only the *direct* effects of each species on any other target species are modelled. Indirect interactions, mediated through other members of the community, must also occur (Levine, 1976). Thus, two species with non-overlapping diets preyed upon by a common predator may nevertheless have a net negative effect on each other's population density. Similarly, two predatory consumer species with little or no dietary overlap can benefit each other indirectly if their prey species compete: an increase in either predator population depresses the density of its own resource population, hence ameliorating conditions for the other predator's major resource (for details, see Vandermeer, 1980). Under certain conditions, pairs of potential competitors can actually act to *increase* one another's densities if both

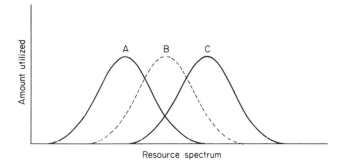

Fig. 8.14. Hypothetical illustration of the conditions leading to competitive mutualism between species A and C (here considered from the view of species C as the 'target' species—entirely analogous arguments can be made for species A). This example involves only a single resource dimension, but the phenomenon can also occur in cases of higher niche dimensionality. Resource requirements of species A and C overlap moderately, so that, in isolation, these two species are competitors. Both species A and C overlap more extensively with species B. When all three species coexist, the direct competitive effects of A on B are stronger than those between A and C; because these strong negative effects of A on B decrease the density of B, the negative impact of B on C is reduced by the presence of A. If this indirect effect mediated through species B is greater than the direct effect, A and C are in effect 'mutualists' in the context of the presence of species B. The arguments are easily extended to situations with more species.

* As for example in the 'alpha matrix' of competition coefficients for a community (Levins, 1968).

share other competitors (Lawlor, 1980). Such a competitive mutualism may arise when two weak to moderate competitors have a common strong competitor: because both species inhibit this third species markedly, each has a beneficial net effect on the other even though their direct pairwise interaction is detrimental (Fig. 8.14). Lawlor (1979) extends this approach to n-species communities, stressing that pairs of species with the potential for high overlap and strong interspecific competition might in fact interact only weakly in the context of an entire community. Further, he relates such indirect competitive mutualism to community assembly, noting that the addition of one of a pair of such species may make it easier for the other to invade a community. Members of different guilds may often be competitive mutualists (Pianka, 1980).

8.5 Some prospects and problems

Many possibilities still remain for important theoretical work on competitive interactions. Innovative new ways of modelling competition that depart from the traditional Lotka–Volterra equations and the concept of constant competition coefficients will be of great interest. Such approaches should provide fruitful insights into the actual *mechanisms* of competition. Even within the framework of the Lotka–Volterra equations, competition coefficients badly need to be treated as *variables* in both ecological and evolutionary time (see, for example, Leon, 1974, and Lawlor and Maynard Smith, 1976). As explained above, competitive effects between two species may frequently vary with the presence or absence of a third species; theory on interactive competition coefficients is virtually nonexistent. An attractive alternative to competition coefficients is to quantify species' interactions with partial derivatives that reflect the 'sensitivity' of each species' own density to changes in the density of the other ($\partial N_i/\partial N_j$ and $\partial N_j/\partial N_i$ terms measure the dynamics rather than the statics of competition). Alphas could also profitably be made density dependent; the actual *shapes* of resource utilization functions might ,be allowed to change (with an appropriate constraint such as holding the area under the curves constant) either in ecological time due to ethological release or in evolutionary time via directional selection conferring advantages on individual genotypes that deviate from the mean. Roughgarden (1974a, 1974b, 1974c, 1979) has made a start on such theoretical work. Niche overlap theory could be profitably expanded to incorporate effects of diffuse competition (Pianka, 1974, 1975); moreover, the theory of limiting similarity needs to be extended to many species and to multi-dimensional resource space.

Fundamental questions emerge from such theoretical considerations. Consider, for example, two competitive communities with similar numbers of species, one of which is composed of several distinct clusters of competitors with strong competitive interactions among themselves but very weak or nonexistent interactions between members of different clusters. Compare this community with another with the same number of species, but one in which all members interact moderately with all others (i.e., greater diffuse competition). Such a difference in guild structure between two communities might arise with a difference in niche dimensionality, for instance. What, if any, differences will there be between the two communities in community-level properties, such as stability? Will maximal tolerable overlap between pairs of species differ? One might find, for instance, stronger competitive interactions and greater niche overlap between *pairs* of competing species in the first community with fewer immediate neighbours in niche space, but *total* niche overlap summed over all interspecific competitors might well be greater in the second community with greater diffuse competition (see, for example, Pianka, 1974). Further work on indirect effects and competitive mutualism will obviously be of great interest.

Niche breadth theory could profitably be expanded to include age-specific phenotypic changes in resource use. Under such a selection regime, will population structure evolve to make most efficient use of available resources? What selective forces determine the optimal degree of within-phenotype versus between-phenotype components of niche breadth?

The neutral model approach has considerable promise. Algorithms for constructing various randomized analogues of real communities are by no means exhaustive. Families of such randomized communities with differing degrees of semblance to observed communities should help to elucidate the structure of the latter (see Lawlor, 1980, for the beginnings of such an approach). A major virtue of this methodology is that samples of analogue communities can be generated that are adequate for statistical comparisons with observed communities. The approach essentially allows 'experimentation' with real communities and is limited only by the imaginative powers of ecologists. A promising but as yet unexploited variant on these techniques is the artificial 'removal' and/or 'addition' of species to existing communities.

Numerous possibilities also exist for important empirical work, of course. For example, the within-phenotype and between-phenotype components of niche breadth have seldom been separated (but see Roughgarden, 1974b). As pointed out repeatedly earlier, field experiments clearly hold great promise, although their precise nature is not easily foreseen. Incomplete removal or addition experiments may well allow quantification of competition by monitoring rates of return to equilibrium densities, thereby measuring the dynamics of population growth (see Hallet and Pimm, 1979). Indeed, gathering useful new data on competition and niche relationships is probably far more challenging (and likely to be more significant) than adding to its existing theoretical foundation.

9
Patterns in Multi-Species Communities

ROBERT M. MAY

9.1 Introduction

For communities with many species, any description of the population dynamics of individual species in terms of the interactions between and within populations usually is very difficult. That is, the methods developed in chapter 2 for single populations and in chapter 5 for pairs of interacting populations are not easily extended to multi-species situations. As pointed out at the end of chapter 5, this is so both because of the proliferation of relevant parameters, and because the dynamical behaviour can become qualitatively more complicated.

One approach is to construct a large computer model, use observations or intuition to infer plausible parameter values, and then simulate the dynamics of the system. Helpful insights are more likely to emerge from such studies if they are subjected to exhaustive 'sensitivity analysis', in which the parameters are varied, one by one, with the aim of discovering how sensitively the dynamics depends upon the values of particular parameters. This is not an easy task. An interesting example of this work is by Harte and Levy (1975), who combine analytic techniques with numerical simulations. For an excellent review of current work in 'systems ecology', much of it motivated by the data amassed under the aegis of the International Biological Programme (IBP), see the collection of papers edited by Halfon (1979). I think that these studies will, in time, contribute greatly to our understanding of how ecosystems work; but I equally think that such efforts are in their formative stages, and I shall not discuss them further in this chapter.

An alternative approach is to abandon description of the behaviour of individual species, and focus instead on overall aspects of community structure. As we shall see below, there are many examples where the world appears chaotic and vagarious at the level of individual species, but nonetheless constant and predictable at the level of community organization. In this spirit, section 9.2 examines some generalities in patterns of energy flow in ecosystems, and section 9.3 considers the sorts of patterns of resource utilization and food web structure that underlie convergence in the structure and function of geographically separate but ecologically similar communities. Section 9.4 gives a somewhat more technical account of observed patterns in the relative abundance of species in different kinds of communities, and section 9.5 closes the chapter with some diffuse remarks on the relation between stability and complexity in ecosystems.

9.2   Patterns of energy flow

In the search for patterns, we may look at the total number of individuals in the different species or trophic levels, or at biomass, or at the way energy flows from one species to another or from one trophic level to another. Stated in this order, these quantities (numbers, biomass, energy flow) are progressively less amenable to casual observation; on the other hand, Table 9.1 suggests they are likely to be progressively more fundamental to the way the system is organized.

Table 9.1. Population density, biomass, and energy flow for five very different primary consumer populations (after Odum, 1968).

Population	Approximate density (m^{-2})	Biomass (g/m^2)	Energy flow $(kcal/m^2/day)$
Soil bacteria	10^{21}	0·001	1·0
Marine copepods (*Acartia*)	10^5	2·0	2·5
Intertidal snails (*Littorina*)	200	10·0	1·0
Salt marsh grasshoppers (*Orchelimum*)	10	1·0	0·4
Meadow mice (*Microtus*)	10^{-2}	0·6	0·7
Deer (*Odocoileus*)	10^{-5}	1·1	0·5

In the 1950s and early 1960s, studies of natural lakes (e.g., Cedar Bog Lake in Minnesota and Silver Springs in Florida) and of laboratory aquariums prompted the generalization that the efficiency at which energy is transferred from one trophic level to the next is around 10 per cent. That is, about 10 per cent of the net production (measured in calories) of plants ends up as net production of herbivores, and about 10 per cent of the nett production (in calories) of herbivores makes its way into net production in the first level of carnivores, and so on (see, e.g., Slobodkin, 1961).

Although this intriguing generalization tends to be true for lakes and aquariums (see, e.g., Kozlovsky, 1968, or the review in Krebs, 1972, ch. 22), subsequent research has shown it to be much less reliable for terrestrial and other kinds of aquatic communities.

To determine overall efficiencies of energy transfer, two questions must be answered. *First*, what fraction of the net production at one trophic level is actually assimilated by creatures at the next level? *Second*, how do these creatures apportion the assimilated energy between net production (growth and reproduction) and maintenance costs (respiration)? The second question is ammenable to fairly precise answers, but the first question is messier, as it can involve both particularities about the fraction of material that is assimilated rather than excreted by a given species, and generalities about the overall fraction of net production at one level that is actually used (consumed) by the next level. Some of these problems and ambiguities can be made more explicit by considering, say, mice and weasels. If we focus on the weasels, it is in principle straightforward to determine the efficiency with which 1 gm or 1 cal of eaten mouse is transformed into gms or cals of weasel. If we focus on the mouse population, it is hard to determine what fraction of their total biomass appears as net production in the next trophic level. Indeed, the answer ultimately depends on how we keep the books; the very notion of 'trophic level' does not stand up to close examination (where, for example, do the decomposers belong).

In pursuit of the second question, Humphreys (1979) has drawn together some 235 different studies of the 'production efficiencies' (net annual production as a ratio to production plus respiration, $P/(P + R)$) in natural populations of animals. Examining regression relations between P and R, Humphreys shows homeotherms (loosely, warm-blooded animals) can be separated into four significantly different groups: insectivores; birds; small mammal communities; and other mammals. Poikilotherms (cold-blooded animals) separate into three groups: fish and social insects; non-insect invertebrates; and non-social insects. The invertebrate groups further permit significant separation into trophic categories of herbivores, carnivores and detritus feeders.

These results are summarized in Table 9.2, which shows some interesting patterns. Both for non-insect invertebrates and insects other than social insects, the production efficiency is significantly lower for herbivores than for carnivores and detritus feeders. A plausible explanation is that biochemical conversion efficiencies are higher for animals eating other animals than for animals eating plants. Although there is scatter around the mean values of the production efficiencies for a given group, there is a significant overall distinction between homeotherms (with production efficiencies typically in the range 1–3 per cent) and poikilotherms (typically in the range 10–40 per cent).

Table 9.2. Mean production efficiency, $P/(P+R)$, for various groups of animals (from Humphreys, 1979, in which references are given).

Group	Mean production efficiency (per cent)	Sample size
Insectivores	0·9	6
Birds	1·3	9
Small mammal communities	1·5	8
Other mammals	3·1	56
Fish and social insects	10	22
Non-insect invertebrates	25	73
Non-social insects	41	61
Non-insect invertebrates		
herbivores	21	15
carnivores	28	11
detritivores	36	23
Non-social insects		
herbivores	39	49
detritivores	47	6
carnivores	56	5

Going some way toward answering the first question, Pimentel *et al.* (1975) have brought together several rough estimates of the percentage of net production of various terrestrial and aquatic plants that is consumed by their herbivores. This information is summarized in Table 9.3. It is notable that the three highest ratios are for aquatic plants and their herbivores, and for phytoplankton or algae and zooplankton; these aquatic systems will have overall plant–herbivore transfer efficiencies in the neighbourhood of 10 per cent. The terrestrial systems have consumption efficiencies which are lower, and sometimes considerably lower, than the marine ones. In part, this may be because in the sea much of the nett production at one level is indeed consumed by the next level, whereas on land a good deal of it goes directly to decomposers (rather than on up the trophic ladder). This point is made by Heal and MacLean (1975; see also Golley, 1968, and Petrusewicz, 1967) in their review of microbial and faunal production in terrestrial, freshwater and marine ecosystems: they discuss transfer efficiencies of different classes of organisms and at different levels in the trophic chain, and note that in terrestrial systems 90 to 99 per cent of secondary production is by decomposer organisms in the soil or leaf litter. A systematic attempt to take due account of the energy that flows through decomposers in terrestrial ecosystems is by Swift et al. (1980).

Table 9.3. The percentage of net productivity of various plant hosts that is consumed by feeding-animal species (from Pimentel et al., 1975, in which references are given).

Food-plant Host	Feeding-animal Species	Percentage of Productivity Consumed
Beech trees	Invertebrates	8·0
Oak trees	Invertebrates	10·6
Maple-beech trees	Invertebrates	6·6
Maple-beech trees	Invertebrates	5·9
Tulip poplar trees	Invertebrates	5·6
Grass + forbs	Invertebrates	4–20
Grass + forbs	Invertebrates	<0·5
Alfalfa	Invertebrates	2·5
Sericea lespedeza	Invertebrates	1·0
Grass	Invertebrates	9·6
Aquatic plants	Bivalves	11·0
Aquatic plants	Herbivorous animals	18·9
Algae	Zooplankton	25·0
Phytoplankton	Zooplankton	40·0
Marsh grass	Invertebrates	7·0
Marsh grass	Invertebrates	4·6
Meadow plants	Invertebrates	14·0
Sedge grass	Invertebrates	8·0

A very rough estimate of the 'food-chain efficiency', with which energy flows from one trophic level to the next, is obtained by multiplying together the appropriate entries in Tables 9.2 and 9.3. Such a multiplication assumes that all material consumed (Table 9.3) is indeed assimilated (Table 9.2 deals with the distribution of assimilated energy), with no fraction being excreted; Emlen (1973) cites figures of around or above 50 per cent for assimilation efficiencies, so that the approximation of taking it to be 100 per cent is not qualitatively in error.

Convolving Table 9.2 with Table 9.3, we see that food-chain efficiencies can vary over two or more orders of magnitude, from less than 0·1 per cent to more than 10 per cent. It is thus clear that there is no simple and universal rule setting energy transfer efficiencies between trophic levels at around 10 per cent, but instead there is an array of such figures, depending on the details of the environments and organisms involved. As this growing amount of information is put in order, there are two things to be done with it.

One is to understand it, as it were, from *below*. That is, to understand how the thermodynamics of metabolic processes, in conjunction with the bionomic strategy (in the sense of chapter 3) of the organisms involved, combine to put limits on the efficiency of energy transfer. For accounts of current work in this vein, see Morowitz (1968) or Gates and Schmerl (1975).

The other task is to make *upward* application of this knowledge, in an effort to understand the extent to which consideration of transfer efficiencies and the like may constrain the structure of real food webs. One aspect of this question is examined in section 9.5.3, below. The relation between energy flow and trophic structure can, however, be very complicated and non-obvious: mutualistic interactions such as plant and pollinator, or plant and seed-disperser, may appear inconsequential in terms of overall energy flow, yet play a central role in determining community diversity and/structure (for a penetrating review, see Paine, 1980).

An example of the way thermodynamic considerations can influence community structure is provided by comparison between warm-blooded and cold-blooded predators. Because their production efficiencies are significantly higher (Table 9.2), poikilothermic predators will have a higher ratio of predators to prey in the standing crop than will homeothermic predators, other things being equal (prey and predator sizes and lifetimes, hunting strategies, etc.). An interesting, if controversial, application of these general ideas is due to Bakker (1974, 1975a, b), who has used the systematic trends in predator–prey ratios deduced from fossil records as one tine on his multi-pronged argument to the effect that the later dinosaurs were warm-blooded; see Fig. 9.1. This is an example of the contemporary interplay between theoretical ecology and paleontology, a theme developed further in chapter 13.

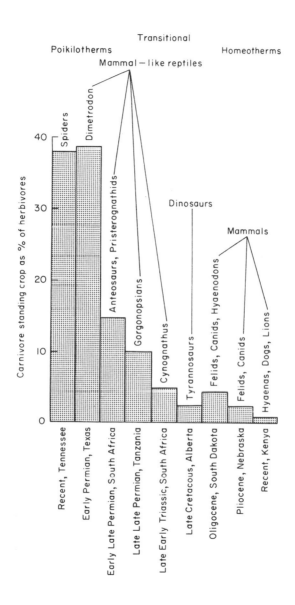

Fig. 9.1. The standing crop of predators expressed as percentage of the standing crop of potential prey, in various groups of living and fossil animals. The figure carries the implication that the later dinosaurs had predator/prey ratios more characteristic of warm-blooded than of cold-blooded animals. From Bakker (1974).

As a postscript to this section, it is revealing to consider current estimates of the world's yearly production of plant material. One of the earliest estimates of terrestrial primary productivity is due to Leibig around 1850: 100×10^9 metric ton per year. Ten years ago the widely accepted figure was $40–50 \times 10^9$ ton per year. The best contemporary estimate (Leith, 1975; Whittaker, 1975a) draws together many different strands of investigation, essentially all of IBP origin, to arrive at an estimate back near Leibig's, namely $110–120 \times 10^9$ ton per year. There are still uncertainties, mainly as to productivity in the tropics, and the Russians hold to a more generous figure of 170×10^9 (Rodin et al., 1975). Around 1960, the total marine primary productivity was thought to be about 140×10^9 ton per year; contemporary surveys (Leith, 1975; Rodin et al., 1975) agree on an estimate around $50–60 \times 10^9$. Thus the past decade has seen almost a factor of 10 revision in the terrestrial/ marine primary productivity ratio. Estimates of total secondary productivity in various types of ecosystems are less precise, and in general only order-of-magnitude accuracy can be claimed (Heal and MacLean, 1975). These uncertainties in such basic quantities emphasize just how short we are of the sort of detailed knowledge necessary for rational management of our environment.

9.3 Patterns in community structure

Patterns of ecosystem organization may also be discerned in terms of the actual species composing the community. To do this, we describe the constituent species according to the ecological functions they fulfil, rather than according to taxonomy or phylogenetic ancestry; we describe the villagers by the way they make their living, rather than by their family names or history. As mentioned at the beginning of section 9.2, such patterns have a somewhat more immediate appeal than the patterns in relatively abstract quantities such as energy flow.

A classic study of this kind is due to Simberloff and Wilson (1969), who eliminated the fauna from several very small mangrove islets in the Florida Keys, and then monitored the recolonization by terrestrial arthropods. In all cases the total number of species on an island returned to around its original value, although the species constituting the total were usually altogether different.

Heatwole and Levins (1972) have subjected these data to closer analysis, listing for each island the number of species in each of several

Table 9.4. Evidence for stability of trophic structure?

Island	Trophic classes								
	H	S	D	W	A	C	P	?	Total
E1	9 (7)	1 (0)	3 (2)	0 (0)	3 (0)	2 (1)	2 (1)	0 (0)	20 (11)
E2	11 (15)	2 (2)	2 (1)	2 (2)	7 (4)	9 (4)	3 (0)	0 (1)	36 (29)
E3	7 (10)	1 (2)	3 (2)	2 (0)	5 (6)	3 (4)	2 (2)	0 (0)	23 (26)
ST2	7 (6)	1 (1)	2 (1)	1 (0)	6 (5)	5 (4)	2 (1)	1 (0)	25 (18)
E7	9 (10)	1 (0)	2 (1)	1 (2)	5 (3)	4 (8)	1 (2)	0 (1)	23 (27)
E9	12 (7)	1 (0)	1 (1)	2 (2)	6 (5)	13 (10)	2 (3)	0 (1)	37 (29)
Total	55 (55)	7 (5)	13 (8)	8 (6)	32 (23)	36 (31)	12 (9)	1 (3)	164 (140)

The table is after Heatwole and Levins (1972). The islands are labelled in Simberloff and Wilson's (1969) original notation, and on each the fauna is classified
into the trophic groups: herbivore (H); scavenger (S); detritus feeder (D);
wood borer (W); ant (A); predator (C); parasite (P); class undetermined (?). For
each trophic class, the first figures are the number of species before defaunation,
and the figures in parentheses are the corresponding numbers after recolonization. The total number of different species encountered in the study was 231
(the simple sum 164 + 140 counts some species more than once).

(somewhat idiosyncratic) categories: herbivores; scavengers; detritus
feeders; wood borers; ants; predators; and parasites. The essentials of
their findings are contained in Table 9.4. This table suggests that, in
terms of trophic structure, the pattern is one of remarkable stability
and constancy. On the other hand, in terms of the detailed taxonomic
composition of the community of arthropod species on a particular
island, there is great variability. The total number of species encountered in the study was 231, whereas a glance at Table 9.4 shows
individual islands to have a subset of around 20–30 species; these 20 or
so species vary greatly from island to island, or before and after defaunation on the same island. Although the facts marshalled in Table
9.4 seem to tell a convincing story, a more rigorous approach is to test
them against the corresponding patterns generated under some appropriate 'null hypothesis', in which the species are drawn purely randomly
from the total pool. Such a re-examination has been carried out by
Simberloff (1976), who concludes that the apparent patterns shown in
Table 9.4 are not significantly different from those generated when the
species are put together randomly, disregarding trophic roles. Although
one can argue about the appropriateness of Simberloff's null hypothesis,
I think this analysis invalidates the conclusions of Heatwole and

Levins; indeed, I thought about removing Table 9.4 from this second edition, but decided to keep it as a Cautionary Tale. (Notice, however, that the species–area relations demonstrated by Simberloff and Wilson remain untouched by these revisionist exercises.)

Cody (1968) has analysed how ten grassland bird communities in North and South America are organized with respect to precentage of horizontal habitat selection, vertical habitat selection, and food selection: see Fig. 8.11 (page 187). Eight of the ten communities are clustered around one point in this 3-dimensional figure, which suggests an orderly and repeatable pattern for such bird communities, despite the particular species being very different (even to the extent that the avifauna in North and South America have different ancestry). Fager (1968) has shown that the community of invertebrates in a decaying oak log has a definite structure, although each log will have its own particular collection of species. MacArthur and MacArthur (1961) showed how the diversity of species within bird communities in North America is closely correlated with the amount of diversity of foliage (particularly foliage height), and Recher (1969) showed that this same relation applied directly to foliage height diversity versus bird species diversity in Australia. Southwood (1978) has outlined a tentative set of principles underlying the organisation of insect communities in Britain, illustrating his ideas with a variety of examples.

These are but a very few among a growing number of analogous studies of the structure of terrestrial, intertidal and aquatic communities.

The examples all point to an underlying community pattern, a trophic skeleton, which is stable and predictable. From the standpoint of the populations of individual species, one may gain an impression of ceaseless change and flux, dominated by environmental vagaries or the accidents of history. At the same time, from the standpoint of trophic structure, the picture may be one of steadiness and pattern.

The phenomenon of *convergence* is most familiar at the level of individual species, as displayed in museum showcases: the New World toucan and the Old World hornbill; hummingbirds and sunbirds; anteaters as represented in Neotropical, Oriental, Ethiopian and Australian realms by edentate antbears, pangolins, aardvarks and echidnas, respectively. But the considerations developed above find their expression in the convergence of the structure of entire communities of plants and animals in spatially separate but climatically similar regions.

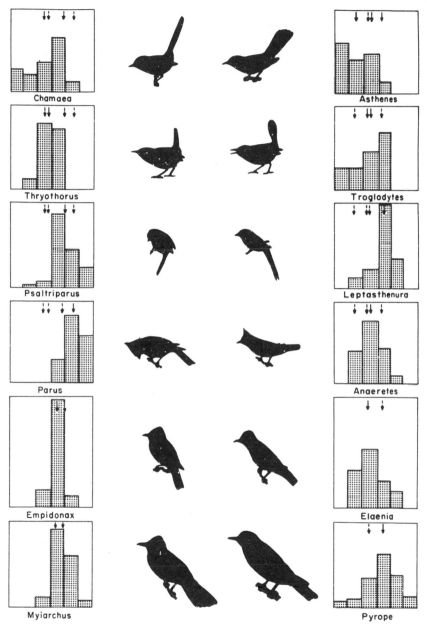

Fig. 9.2. Distributions of foraging heights, and silhouettes of four canopy insectivores (upper) and two sallying flycatchers (lower) in Californian chaparral (left) and Chilean matorral (right). The foraging height intervals are, from the left: ground; ground—6″; 6″–2′; 2′–4′; 4′–10′; 10′–20′. Arrows above each distribution indicate its mean, and those of the frequency distributions of neighbouring species. Silhouettes are all to the same scale. From Cody (1974).

One illustration is provided in Fig. 9.2, which shows the convergence in the physical appearance of members of guilds of insectivorous birds, one from Californian chaparral and the other from Chilean matorral. The physical similarities derive from underlying similarities in the way resources are divided up within the guild, as also indicated in Fig. 9.2. Cody (1975) has since shown that on a third continent, in South African macchia, equivalents of exactly the same six insectivores are found, including the four foliage species and the two sallying flycatchers. Such examples may be extended to encompass the fauna and flora of different regions not only at the present time, but even across the sweep of geological time. Thus Simpson (1949, 1969) has pointed to the convergence between the mammalian (and marsupial) communities in North and South America prior to the formation of the Panama land bridge in the mid-Pleistocene; up to this time the two faunal assemblies had followed separate evolutionary trajectories for 100 million years or so, each on its own island continent. In a more speculative fashion, Wilson (1975; see also Wilson *et al.*, 1973) has discussed the convergence in community structure between the dinosaur fauna of the Cretaceous, and typical mammalian faunas of today: see also chapter 13.

For a more detailed account of efforts at quantifying such community patterns, and using them to make predictive statements, see MacArthur (1972), Cody (1974, 1975) and the work discussed in chapter 10.

A different kind of community pattern, that has received surprisingly little attention, concerns the presence of many more species of small creatures than of big ones. An understanding of the way ecological systems are organized will need to account for these trends in the number of species in different size categories. Van Valen (1973) has compiled a vast amount of data to discuss the relation between number of species of mammals and birds and their physical size (specifically, weight), and Schoener and Janzen (1968; see also Janzen, 1977) have discussed the patterns observed in some communities of tropical insects. The problem of working with any one group (e.g., mammals, or beetles) is that ecological aspects of the species-size relation tend to be masked by the group blending into ecologically similar, but taxonomically different, groups at both low and high ends of its size range. Figure 9.3 therefore presents a very crude estimate of the overall number of species of terrestrial animals as a function of their physical size (specifically, length) (May, 1978b). There are many difficulties in this exercise: the

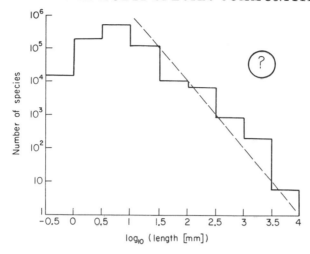

Fig. 9.3. The number of species of terrestrial animals, as a function of their characteristic body length. The question mark serves to emphasize that this is a tentative figure, based on very crude approximations. The dashed line illustrates the shape of the relation S ~ L⁻², eq.(9.1). (From May, 1978b.)

systematics of small arthropods and other invertebrates are in most cases in a rudimentary state ('We must take it for granted that a large part of the mite fauna of the world will remain unsampled, unnamed and unclassified for decades to come'; Mayr, 1969); and conventional taxonomic concepts have dubious validity once one goes below the one-millimeter size class. These caveats add up to the expectation that the number of species in the smaller size categories in Fig. 9.3 are seriously underestimated.

Essentially the only theoretical discussion of species-size relations is by Hutchinson and MacArthur (1959), which leads to the expectation that (at least for the larger size classes) the number of species, S, falls off with increasing linear dimension, L, roughly as

$$S \sim L^{-2}. \tag{9.1}$$

In essence, eq. (9.1) comes from the assumption that animals see their environment as a 2-dimensional mosaic (e.g., the total surface of a tree) to be partitioned up, on a scale set by their perceived grain size (which goes as L^{-2}). This speculation, eq. (9.1), is discussed further by May (1978b) and Southwood (1978), and is illustrated by the dashed line in Fig. 9.3.

9.4   The relative abundance of species

A more technical aspect of community patterns lies in the relative abundance of the various species.

For any particular group of S species, we may express the number of individuals in the ith species, N_i, as a proportion, p_i, of the total number of individuals, $N = \sum N_i$:

$$p_i = N_i/N. \tag{9.2}$$

This information may then be displayed on a graph of relative abundance versus rank, as exemplified by Figs. 9.6 and 9.8 below; such figures show the patterns in the relative magnitudes of constituent populations, from most to least abundant. Equivalently, the information may be incorporated in a distribution function, $S(N)$, as in Figs. 9.4 and 9.7 below: here $S(N)dN$ is the number of species comprising between N and $N + dN$ individuals.

Some of the main distributions which have been discussed in connection with natural systems are presented below. A more complete account, both of the distributions themselves and of the various equivalent ways they are conventionally displayed, is in Pielou (1975) or May (1975f). A separate source of statistical complication lies in estimating the magnitude of the various populations in the first place: the nuts and bolts of pertinent sampling techniques are well reviewed by Southwood (1976).

9.4.1   *Lognormal distributions*

Once the community consists of a relatively large assembly of species, the observed distribution of species relative abundance, $S(N)$, is almost always lognormal; i.e., there is a bell-shaped gaussian distribution in the logarithms of the species' abundances. This lognormal distribution has been documented for groups of organisms as disparate as diatoms, moths, birds or plants, provided always that the sample is large enough to contain a good number of species (see, e.g., Whittaker, 1972, 1975a). Figure 9.4 provides a typical example.

What is the explanation for these pervasive patterns? Given a largish group of species, it is likely that their relative abundances will be governed by the interplay of many more-or-less independent factors. It is in the nature of the equations of population dynamics that these

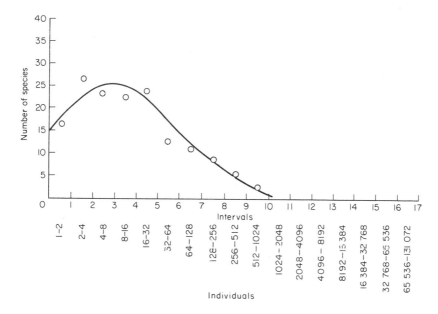

Fig. 9.4. Illustrating the lognormal distribution of relative abundance of diatom species in a sample taken from an undisturbed community in Ridley Creek, Chester County, Pennsylvania. The abundances are, as indicated, plotted as logarithms to the base 2: for further discussion, see the text. From Patrick (1973).

several factors should compound multiplicatively, and the statistical Central Limit Theorem applied to such a product of factors implies a lognormal distribution. That is, the lognormal distribution arises from products of random variables, and factors that influence large and heterogeneous assemblies of species indeed tend to do so in this fashion. This broad statistical argument similarly suggests lognormal patterns for the distribution of wealth in the USA, or for people or GNP among the nation states of the world. Such is in fact the case.

More specifically, ecologists usually write the lognormal distribution as

$$S(R) = S_o \exp(-a^2 R^2). \tag{9.3}$$

Here S_o is the number of species at the mode of the distribution, which species have populations around N_o; a is an inverse measure of the width of the distribution ($a = 1/(2^{\frac{1}{2}}\sigma)$, where σ is the standard deviation); and, following the convention established by Preston's (1948) early work, R expresses the abundance as a logarithm to base 2, so that successive 'intervals' or 'octaves' along the x-axis correspond to population doublings,

$$R = \log_2(N/N_o). \tag{9.4}$$

The two parameters S_o and a specify the distribution uniquely. In particular, the total number of species S is, to a good approximation, $S \simeq \sqrt{\pi} S_o/a$. One further useful parameter may be introduced. Define R_{max} as the expected R-value, or 'octave', for the most abundant species, and define R_N to be the location of the peak of the distribution in total numbers of individuals [i.e., the distribution of $N \times S(N)$]: the parameter γ is then defined to be

$$\gamma = R_N/R_{max}. \tag{9.5}$$

Any pair of the three parameters S_o, a and γ may now be used to characterize the distribution. The three are approximately related by $a^2 \gamma^2(\ln S_o) \simeq 0\cdot12$ (see May, 1975f).

The purpose of this brief notational frenzy is to discuss two empirical 'laws', which have provoked some speculation in the ecological literature. The first is the observation, made by Hutchinson (1953) and since supported by an increasing amount of data (see, e.g., Whittaker, 1972; Colinvaux, 1973, ch. 37), that the parameter a usually has a value around 0·2,

$$a \simeq 0\cdot2. \tag{9.6}$$

The second is Preston's (1962) 'canonical hypothesis', which says the usual value of γ is around unity,

$$\gamma \simeq 1. \tag{9.7}$$

This empirical rule holds true for a large body of data (Preston, 1962; see also Fig. 9.5). In addition to its intrinsic interest, the rule (9.7) is important in underpinning a theoretical explanation of observed species–area relationships (Preston, 1962; MacArthur and Wilson, 1967); this point is developed more fully in chapter 10. It is surprising that these two rules have not received more attention.

Before reading too much ecological significance into eq. (9.6), it is well to note that this rule also applies to the other lognormal distributions mentioned above (wealth, people and GNP of nations). To begin, assume that indeed $\gamma = 1$: the single parameter a now specifies the lognormal distribution, and there is a unique functional relationship between a and the total number of species S. The details of this relation (May, 1975f, Fig. 3) are such that as S varies from 20

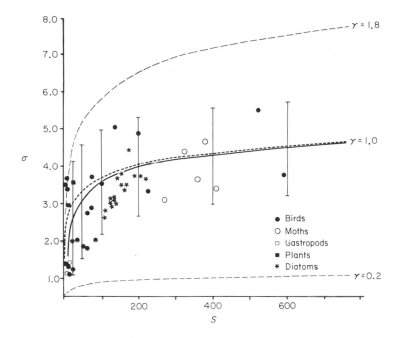

Fig. 9.5. The data points show the number of species, S, and the standard deviation of the logarithms of the relative abundances, σ, for various communities of birds, moths, gastropods, plants and diatoms (as labelled). The dashed lines show the relations between S and σ for communities in which relative abundance is distributed lognormally, with $\gamma = 1.8$, 1.0, 0.2; $\gamma = 1$ corresponds to Preston's 'canonical lognormal'. The solid line is the mean relation predicted by Sugihara's model of sequential niche breakage, and the error bars represent two standard deviations about this mean. (After Sugihara, 1980.)

to 10,000 species, a varies from 0·29 to 0·13. The ingenious reader can use the mathematical relations sprinkled two paragraphs above to establish the approximate proportionality $a \sim 1/\sqrt{\ln S}$, which makes explicit the very weak dependence of a on S. More generally, for lognormal distributions it can be seen that there is a relation between the total number of species and individuals, S and N, and the values of the parameters a and γ, and that the enormous range of communities with S ranging from 20 to 10,000 and N ranging from 10S to 10⁷ S is characterized by values of a in the range 0·1 to 0·4 and of γ in the range 0·5 to 1·8 (May, 1975f, Fig. 5).

These facts have led May (1975f) to conjecture that the enigmatic rules (9.6) and (9.7) are merely mathematical properties of the lognormal distribution for large S (with the lognormal itself simply reflecting statistical generalities).

Although this is probably the correct explanation of the rule (9.6), Sugihara (1980) has shown that the 'canonical hypothesis', eq. (9.7), is obeyed too accurately to be explained by May's arm waving. Using all the available data on natural communities, Sugihara plots the number of species, S, against the standard deviation of the logarithmic abundance, σ, to get Fig. 9.5. For a lognormal distribution, any specified value of the parameter γ leads to a unique relation between S and σ; these relations are shown for $\gamma = 1\cdot8$, $\gamma = 1\cdot0$ and $\gamma = 0\cdot2$. It appears that natural communities conform more closely to the canonical relationship, eq. (9.7), than can be explained by mathematical generalities alone.

Sugihara has also suggested a biological mechansim that will produce the observed patterns. He imagines the multidimensional 'niche space' of the community as being like a hypervolume, which is *sequentially* broken up by the component species, such that each of the S fragments denotes the relative abundance of a species. This is both biologically and mathematically very different from MacArthur's model in which a 'stick' is broken *simultaneously* into S pieces; the sequential breakage pattern (with any fragment being equally likely to be chosen for the next breakage, regardless of size) seems more in conformance with evolutionary processes, and the patterns of relative abundance thus generated are unlike those of the 'broken stick' model. The solid line in Fig. 9.5 shows the mean relation between S and σ predicted by Sugihara's model, and the error bars show the range of two standard deviations about the mean. Of course, the fact that this model fits observed distribution patterns does not prove the model to be correct; other biological assumptions could presumably also fit the distribution (although none such have yet been proposed).

9.4.2 *Other distributions*

In simple communities, with attention focussed on a relatively small and homogeneous group of species, distributions of species relative abundance other than lognormal are often seen.

One example is the patterns that various people have described for early successional plant communities (see, e.g., Whittaker, 1975a; McNaughton and Wolf, 1973). In the early stages of succession, plant communities will tend to consist of a handful of weedy, pioneer,

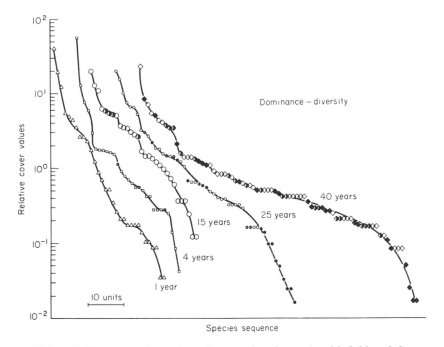

Fig. 9.6. Patterns of species relative abundance in old fields of five different stages of abandonment in southern Illinois. The patterns are expressed as the percentage that a given species contributes to the total area covered by all species in a community, plotted against the species' rank, ordered from most to least abundant. Symbols are open for herbs, half-open for shrubs, and closed for trees. From Bazzaz (1975).

r-selected species that happened to get there first. The species arriving first will tend to grow rapidly, pre-empting a fraction (*k* say) of the available space or other governing resource before the arrival of the next species, which in turn will tend to pre-empt a similar fraction of the remaining space before the arrival of the third, and so on. The consequent pattern will show up, very roughly, as a geometric series when the relative abundances are arranged hierarchically, from most to least abundant. A typical illustration is provided by the earlier successional stages in Fig. 9.6; in this figure, relative abundance is plotted logarithmically against rank, so that the geometric series 'niche pre-emption' pattern shows up as a straight line.

As succession proceeds, things become more complicated, and a multitude of ecological dimensions are likely to be relevant to the ultimate composition of the community. As discussed above, this leads to a lognormal distribution of species relative abundances: on an abundance-versus-rank plot this produces the sort of S-shaped curve seen for later successional stages in Fig. 9.6, with a preponderance of 'middle class' species. The later, lognormally distributed community tends to be an egalitarian socialist society compared with the feudal hierarchy characteristic of early succession.

Something akin to a reversal of these successional patterns takes place when mature communities become polluted.

This has been shown particularly for diatoms in streams and lakes subject either to 'enrichment' by waste heat, sewage, or other organic materials, or to toxic pollution by heavy metals or other poisons (Patrick, 1973, 1975). The initial equilibrium diatom community essentially always shows the classic lognormal distribution. When polluted, the community typically shows a pattern in which a few species become exceptionally common, with their relative abundances tending to exhibit a geometric series on a rank-abundance plot; it is plausible that the cause is an early successional type of r-selection for success in handling the all-important new factor. Closer analysis may, or may not, reveal a lognormal distribution of relatively rare species

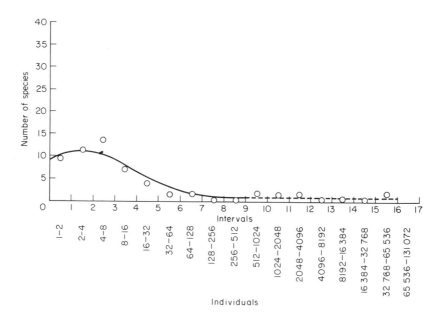

Fig. 9.7. The distribution of relative abundance of diatom species in a community subjected to organic pollution (Back River, Maryland). The features are as discussed in the text. From Patrick (1973).

hanging on down in the tail of the distribution. Figure 9.7 gives a typical example, which is to be contrasted with the undisturbed community of Fig. 9.4.

Another beautiful example is provided by the experimental 'Parkgrass' plots at the Rothamsted Experimental Station. These plots were set aside in 1856, and each was subjected to some specified treatment, such as the withholding or overapplication of certain fertilizers. The resulting changes in the relative abundances of the grass species present have been monitored over the past century. One set of results is shown in Fig. 9.8. Tilman (1980) has made a thorough and perceptive analysis of the Parkgrass data, showing the patterns exhibited in Fig. 9.8 to be typical. Figure 9.8 illustrates the arguments developed above, with the overfertilized plot showing changing patterns of relative abundance of species, strikingly like succession in reverse.

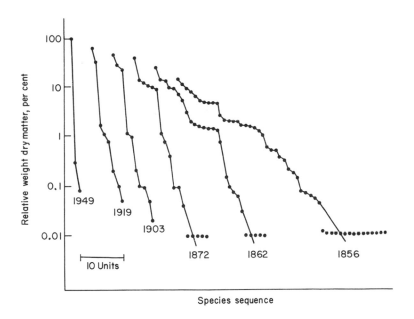

Fig. 9.8. Changes in the patterns of relative abundance of species in an experimental plot of permanent pasture at Parkgrass, Rothamsted, following continuous application of nitrogen fertilizer since 1856. Species with abundance < 0·01 per cent were recorded as 0·01 per cent. Notice that time runs from right to left; the patterns look like the successional patterns of Fig. 9.6, run backwards in time. (From Kempton, 1979, following Brenchley, 1958.)

9.4.3   *Richness and diversity indices*

One single number that goes a long way toward characterizing a biological community is simply the total number of species present, S. This is sometimes called 'richness'.

Often one will wish to go beyond this, yet stop short of describing the full distribution $S(N)$, by constructing some second number to characterize the 'diversity' or 'evenness' of the community; some single number to distinguish between the community with 30 equally abundant species and that with one common species and 29 rare ones.

The statistically-minded will see all such numbers simply as moments of the $S(N)$ distribution, and will be inclined to choose the *variance* of the distribution as a sensible measure of its 'diversity' (as was done, for example, in constructing Fig. 9.5). Several of the indices which have been proposed in the ecological literature are of this kind, involving $\sum p_i^2$ in one form or another [as $\sum p_i^2$, or as $1 - \sum p_i^2$, or as $1/\sum p_i^2$, with p_i defined by eq. (9.2)]. More fashionable is the Shannon diversity index,

$$H = -\sum p_i \ln p_i, \tag{9.8}$$

which is linked by an ectoplasmic thread to information theory. H is a very insensitive measure of the underlying distribution of relative abundance (see, e.g., May, 1975f, Figs. 6, 7). If some such diversity index is to be used, I much prefer one of the 'variance' measures (involving $\sum p_i^2$ in some shape or form), or even the naive 'dominance' index provided by the ratio of the most abundant population to the total population (i.e., the value of p_i for the most common species).

The connections between patterns of relative abundance and degree of pollution, which we noted in section 9.4.2, have sparked interest in the possibility of using some index of diversity to measure how much an ecosystem has been perturbed by human activity. One such study has used the diversity index H of eq. (9.8) as a phenomenological measure of gross toxicity in San Francisco Bay (Armstrong *et al.*, 1971). Clearly, an engineering-style index of this kind would simplify tasks of environmental management.

Taking diatom communities in polluted streams as one among many possible examples, we may note that any such single number will be dominated by the distribution of the handful of common species. But the time scale for recovery of the pristine ecosystem (and, indeed, whether it can ever recover) may depend on the lognormal tail of species which are uncommon in the polluted community, and whose presence will not show up in any overall diversity index. This is a cogent argument for describing the community by its full distribution of species relative abundance, and not trying to condense information into a single diversity index, which may mislead (see also Williamson, 1973).

9.5   Stability and complexity

One aspect of community ecology that has attracted much attention over the years is the relation between the complexity of a food web and its stability. Here increased 'complexity' is usually loosely taken to mean more species, more interactions, and hence more parameters characterizing the interactions. Increased 'stability' may be identified with relatively lower levels of population fluctuation, or with ability to recover from perturbations, or simply with persistence of the system.

Elton (1958), among others, has drawn together a collection of empirical and theoretical observations in support of the conclusion that complexity implies stability. This notion has tended to become part of the folk wisdom of ecology, and has sometimes been elevated to the status of a mathematical theorem (e.g., Hutchinson, 1959). But the empirical evidence is at best equivocal: there are many examples of simple natural systems that are stable (in all of the above senses), and of complex ones that are not. Among the growing number of contemporary reviews, I particularly like that by Watt (1968, pp. 39–50).

Mathematical models are useful as one among many tools for working on this question. In what follows, we will explore the relation between stability and complexity in various kinds of model ecosystems, and, where possible, see how the theoretical conclusions match observed properties of real ecosystems.

9.5.1   *Complexity and stability in general models of ecosystems*

In assessing the stability character of any ecosystem model, various factors enter. *First*, equilibrium configurations will exist only within certain restricted ranges of the interaction and environmental parameters (such as birth rates, assimilation efficiencies, predator attack rates, etc.; the r_i, K_i and α_{ij} in eq. (8.1), for example). Outside this restricted region of parameter space, the equations will describe a collapsing ecosystem in which some, or all, of the populations are fated for extinction. As was illustrated in chapters 2, 5, 6 and 7, mapping out this stable region of parameter space is a standard mathematical exercise. *Second*, even when the parameters are fixed at values which do admit of an equilibrium solution, this equilibrium may be unstable to large perturbations in the population numbers: this involves the ideas of 'resilience', and of global versus local stability, as discussed in section 5.2.3.

These two questions are intertwined, both mathematically and biologically. For parameter values near the centre of the domain of stable values, the dynamical landscape in population space may in general be thought of as a comparatively wide and deep valley; all but the most extreme disturbances to the populations will eventually return to the equilibrium configuration. As the parameter values are chosen nearer the edge of the domain of stable values, the dynamical landscape in population space becomes more like a shallow valley nestled in the top of a volcano; modest disturbances to the populations will see the system spill out from the volcano top. Finally, for parameter values outside the stable domain, there is no valley at all; the volcano tip has given way to a rounded hilltop. Moreover, in the real world, we do not deal with fixed parameter values, but rather the environmental and interaction parameters are themselves fluctuating, in turn driving the population perturbations.

In short, a system which is stable only within a comparatively small domain of parameter space may be called *dynamically fragile*. Such a system will persist only for tightly circumscribed values of the environmental parameters, and will tend to collapse under significant perturbations either to environmental parameters or to population values. Conversely, a system which is stable within a comparatively large domain of parameter space may be called *dynamically robust*.

When these kinds of studies are made, a wide variety of mathematical models suggest that as a system becomes more complex, in the sense of more species and a more rich structure of interdependence, it becomes more dynamically fragile.

This is an important theoretical result, and it comes from models at several different levels of abstraction.

At the most concrete level, we may take specific models such as the multispecies generalization of the simple Lotka–Volterra prey–predator model [eqs. (5.1) and (5.2)], and contrast the typical dynamical behaviour of systems with few and with many species: the 'complex' systems are never more, and are usually less, stable than the 'simple' ones (May, 1975a). Other studies have used generalizations of the Lotka–Volterra equations of chapter 8 [eq. (8.1)]:

$$dN_i/dt = r_i N_i \left[1 - \sum_{j=1}^{S} \alpha_{ij} N_j \right]. \qquad (9.9)$$

Assigning the coefficients α_{ij} random values that are equally likely to be positive or negative (except for the diagonal elements, α_{ii}, which are kept negative), Roberts (1974) has shown the probability to get a 'feasible' equilibrium, such that all S species are present with positive population values, decreases as S increases. These results have been extended, and some analytic understanding added, in further studies by Goh and Jennings (1977), Gilpin (1975c) and Tregonning and Roberts (1979). A particularly interesting numerical study of these Lotka–Volterra eco-systems is by Gilpin and Case (1976), who show the typical 'randomly assembled' S-species system will collapse to one or other of many alternative simpler systems, usually comprising only two to four species. This is a brief summary of a large and growing literature, all bringing the message that the more species that are stirred into the pot in these abstract general models, the harder it is to end up with a stable ecosystem.

Another approach to the stability–complexity question is by Caswell (1976). He frames a 'neutral hypothesis' by studying the likely distribution of relative abundance of individuals among S species, assuming *no* biological interactions among them. Comparing the 'neutral' distributions of species' relative abundance with real ones for birds, fish, insects and plant species in tropical and temperate zones, Caswell finds the real communities to be less diverse (either in the sense of fewer species or of greater dominance by a few common species) than would be the case in the absence of interspecific interactions. The discrepancy is greatest in the tropics, where biotic effects are most pronounced. Caswell concludes 'the diversity of natural communities may be maintained in spite of, rather than because of, such [biological] interactions [between species]'.

A further class of very abstract models deal only with general expressions for the way a population's fluctuations about its equilibrium value [$x_i(t) = N_i(t) - N_i^*$ for the ith population] are governed by its interaction with other populations, as described by interaction co-efficients b_{ij}:

$$dx_i/dt = \sum_j b_{ij} x_j. \qquad (9.10)$$

These coefficients b_{ij} measure the influence of the jth species upon the rate of change of the ith, in the neighbourhood of equilibrium. Numerical (Gardner and Ashby, 1970; McMurtrie, 1975) and analytic (May, 1972b) studies of randomly assembled food webs of this kind have involved 3 parameters: the number of species, S; the average connectance of the web (the number of food links in the web as a fraction of the total number of topologically possible links), C; and the average magnitude (disregarding sign) of the interaction between linked species, b. If all self-regulatory terms are taken to be $b_{ii} = -1$, for large S these systems will tend to be stable if

$$b \, (SC)^{\frac{1}{2}} < 1, \tag{9.11}$$

and unstable otherwise. Thus increasing complexity, in the sense of an increasing number of species S, or of increasing connectance C, or of increasing average interaction strength b, works against dynamical stability.

At a yet more abstract level, similar conclusions emerge from studies which describe only the topology or 'loop structure' of the food web, independent of the actual magnitude of the interactions (Levins, 1975; May, 1975a).

Thus, as a mathematical generality, increasing complexity makes for dynamical fragility rather than robustness.

This is *not* to say that, in nature, complex ecosystems need appear less stable than simple ones. A complex system in an environment characterized by a low level of random fluctuation and a simple system in an environment characterized by a high level of random fluctuation can well be equally likely to persist, each having the dynamical stability properties appropriate to its environment.

Moreover, if we regard evolution as an existential game, where the prize to the winner is to stay in the game (Slobodkin, 1964), we may conjecture that ecosystems will evolve to be as rich and complex as is compatible with the persistence of most populations. In a predictable environment, the system need only cope with relatively small perturbations, and can therefore achieve this fragile complexity, yet persist. Conversely, in an unpredictable environment, there is need for the stable region of parameter space to be extensive, with the implication that the system must be relatively simple.

In brief, a predictable ('stable') environment may permit a relatively complex and delicately balanced ecosystem to exist; an unpredictable ('unstable') environment is more likely to demand a structurally simple, robust ecosystem.

In addition, it must be emphasized that real communities are not assembled randomly (May, 1975a, pp. 3–4 and 75–77; Lawlor, 1978; Tregonning and Roberts, 1979). They are the winnowed products of the long workings of evolutionary processes. Consequently we should be alert for those special structural features of real ecosystems that help reconcile species richness and apparent complexity with dynamic stability. Some possible such mechanisms are discussed below.

9.5.2 *Testing ideas about stability and complexity*

The Eltonian view that complexity confers stability carries the corollary that, in the biologically crowded and species rich tropics, populations should be more stable than in temperate and boreal regions. On the other hand, the ideas set out above suggest population fluctuations are likely to be at much the same level in all regions. Earlier attempts to determine the facts were somewhat anecdotal, partly because of the paucity of tropical data: Owen and Chanter (1972) found that a few species of *Charaxes* butterflies in Sierra Leone had fairly constant abundance, but emphasized their belief that most populations of tropical insects fluctuated a lot; Smith (1972) documented dramatic population explosions of a moth, *Urania fulgens*, in tropical America; and Bigger's (1976) review of tropical crop pests suggested they fluctuate as much as temperate ones. Leigh's (1975) survey of the available information on the amplitudes of fluctuation of herbivorous vertebrates showed no consistent tropical–temperate differences. Wolda (1978) has recently drawn together his own tropical data with that of other people, to give a systematic compilation and analysis of 138 studies providing information about the average level of year-to-year fluctuation exhibited by groups of tropical and temperate insects. Summarizing this synoptic study, Wolda says 'The data suggests rather strongly that tropical insects have, on the average, the same annual variability as have insects from the temperate zone. There is no indication of a greater stability of tropical insects.'

A different class of empirical study has sought to test the ideas embodied in eq. (9.11), by examining the extent to which the product SC tends to remain constant as species richness S varies among a set of otherwise broadly similar communities. McNaughton (1978; see also the critical discussion in Lawton and Ralliston, 1979, and McNaughton,

1979b) collected data on plant species in 17 grassland stands in Serengeti National Park; for these communities, in which the interactions were purely competitive, the product SC was remarkably constant at $4.7 \pm$ 0.7 (at the 95 % confidence level, with S ranging from 4 to 19). Rejmanek and Stary (1979) reconstructed food webs for 31 plant–aphid–parasitoid communities (in which S ranged from 3 to 60) in Central Europe, and also found connectance varied inversely with species richness (with SC ranging from 2 to 6, and the relationship $SC \simeq 3$ covering all the data well). Yodzis (1980) has analysed the 24 food webs catalogued by Cohen (1978); unlike those in the relatively homogeneous systems studied by McNaughton and by Rejmanek and Stary, these webs exhibit many different kinds of interactions. Yodzis shows that again SC remains roughly constant ($SC \sim 4$) as S ranges from 8 to 64, although there is a tendency here for SC to decrease slightly with increasing S (which could be explained by a systematic decrease in the average interaction strength, b, as S increases).

The authors of these three studies all point out that the underlying mechanism keeping SC roughly constant is the tendency for the larger communities to be organized into relatively small subunits or 'guilds' of species, with most interactions taking place within guilds. This ties in with the theoretical suggestion (first made by May, 1972b, and subsequently refined by McMurtrie, 1975, Siljak, 1974, 1975 and Goh, 1979a) that, for specified S and C, dynamic stability may be enhanced by assembling the food web as a set of loosely coupled subunits.

Other authors have recently explored the extent to which real eco-systems avail themselves of this trick of reconciling species richness with dynamical robustness by being made up of loosely coupled sub-systems. Gilbert has argued that his *Heliconius–Passiflora* systems in particular (Gilbert, 1975), and that tropical plant–insect food webs in general (Gilbert, 1977), are so constituted. Lawton and Pimm (1978) and Beddington and Lawton (1978) have argued the case more broadly, observing, for example, that most insect herbivores are monophagous or oligophagous, giving rise to relatively discrete food chains even in species-rich plant communities.

In addition to having a limited number of trophic levels (see below), and possibly being organized into loosely coupled blocks (as just dis-cussed), the interactions among species in real food webs are constrained in many other ways that make the system very different from a ran-domly constructed one. Such constraints, which can affect the relation between complexity (increasing S and C) and dynamic stability, include 'donor control' (related to conversion of biomass within the system; DeAngelis, 1975; Mazanov, 1978), the nature of predators' functional responses (Nunney, 1980), the hierarchical structure of food webs (such that food loops in which A eats B, B eats C, and C eats A are uncommon), and other things. A recent review of these many important and unresolved issues is by May (1979).

9.5.3   How many trophic levels?

One conspicuous property of the structure of natural ecosystems is that food chains are typically short, rarely consisting of more than four or five trophic levels (Hutchinson, 1959).

A quantitative elaboration of this point has been given by Cohen (1978) and Pimm (1980), who analysed data compiled by Cohen for 19 food webs. These webs, which include terrestrial, freshwater and marine examples, contain a total of 102 top predators (animals themselves free from predation). Cohen and Pimm have independently traced out all the food chains connecting top predators to basal species (plants, detritus, or arthropods that fall into freshwater systems). The number of trophic levels is fairly consistently around three, and only for one of the 102 top predators can a food chain involving more than six species (five links) be found.

Such steady patterns in the number of trophic levels are in pronounced contrast to the great variability in the amount of energy flowing through different ecological systems. Primary productivity varies over three to four orders of magnitude in both terrestrial and aquatic ecosystems, and the productivity of fish and of terrestrial animal populations varies over five or more orders of magnitude. As we discussed in section 9.2, there are further variabilities in the efficiency of energy transfer from one level to the next, with such efficiencies typically being much lower for warm-blooded than for other animals.

The conventional explanation for the number of trophic levels is that they are determined by energy flow; if only, say, 10 per cent of the energy entering one level is effectively transferable to the level above it, the number of levels is clearly limited. As Pimm and Lawton (1977) have recently observed, however, this explanation is not easily reconciled with the number of trophic levels being essentially independent of enormous variations in the amount of energy flow and in the transfer efficiencies: 'Food chains are not noticeably shorter in barren Arctic and Antarctic terrestrial ecosystems compared with a productive tropical savannah or the fish guilds of a tropical coral reef.'

Pimm and Lawton alternatively suggest the explanation may lie in the dynamics of the various populations in the community. They use numerical studies of the stability properties of variously structured Lotka–Volterra models to argue that long food chains may typically result in population fluctuations so severe as to make it hard for top predators to persist. This suggestion has been further developed by several people (Saunders, 1978; Vincent and Anderson, 1979; Pimm and Lawton, 1978); in particular, DeAngelis *et al.* (1978) have shown that long food chains may be kept relatively stable provided the transit time for a molecule or a unit of energy through the web is fast.

Pimm and Lawton's work is provocative, and certainly correct in calling attention to the inadequacy of the conventional wisdom in this area. Their notion that the number of trophic levels may be limited by considerations of dynamic stability is, however, debatable, deriving as it does from the properties of rather special mathematical models.

The discussion of the length of food chains typifies much of the material in this chapter: the empirical patterns are widespread and abundantly documented, but instead of an agreed explanation there is only a list of possibilities to be explored.

9.5.4 *Some general implications*

The general arguments given in section 9.5.1, which see the paradigms of trophic complexity such as the tropical rainforest or Lake Baikal as being the evolutionary products of stable environments, march with the more down-to-earth discussions of r- and K-selection in chapter 3. r-selection is associated with a relatively unpredictable environment and a simple ecosystem, K-selection with a relatively predictable environment and a complex, biologically crowded community: a system comprising r-selected species, with their relatively large clutch size or seed set, is well adapted to bounce back after disturbance, whereas a K-selected assembly may be unable to recover after an unusually severe disturbance. In a similar vein, we saw in section 5.4 both that mutualism tends to be a dynamically fragile relationship, and that it is a much more common phenomenon in stable tropical environments. Such biological particularities provide the detailed mechanisms which underlie the dynamical generalities enunciated above. These themes are pursued further in some of the papers in the volumes edited by van Dobben and Lowe-McConnell (1975) and by Farnworth and Golley (1974).

On this view, there is no reason necessarily to expect simple natural monocultures to be unstable. And, indeed, there are many instances of robustly enduring natural monocultures. The marsh grass *Spartina* is one conspicuous example. Another is bracken which in recent years, partly as a result of hilly areas grazed by cows being given over to sheep, has shown itself to be a robust, even aggressively invasive, natural monoculture over increasing areas in Britain (Lawton, 1974). The instability of so many man-made agricultural monocultures is likely to stem not from their simplicity, as such, but rather from their lack of any significant history of coevolution with pests and pathogens.

An important general conclusion is that the large and unprecedented perturbations imposed by man are likely to be more traumatic for complex natural systems than for simple ones. This inverts the naive, if well-intentioned, view that 'complexity begets stability', and its accompanying moral that we should preserve, or even create, complex systems as buffers against man's importunities. I would argue that the complex natural ecosystems currently under siege in the tropics and subtropics are less able to withstand our battering than are the relatively simple temperate and boreal systems.

10

Island Biogeography and the Design of Natural Reserves

JARED M. DIAMOND and ROBERT M. MAY

10.1 Introduction

The flora and fauna of islands have played a central role in the development of ecological thought, from the early formulations of evolution and biogeography by Darwin and Wallace, through Mayr's demonstration of the role of geographic isolation in speciation, to the analytical theory of island biogeography pioneered by MacArthur and Wilson. Some reasons why islands have been well suited to provoking or testing theoretical ideas are that they have definite boundaries, come in many different sizes and heights and remotenesses, often have relatively simple communities of plants and animals, and serve as ready made evolutionary laboratories offering replicate 'natural experiments' in community assembly.

Islands may be real islands in the ocean, or they may be virtual islands such as hilltops (where for many species the surrounding lowland presents a distributional barrier), lakes, or wooded tracts surrounded by open land. In particular, the natural reserves and wildlife refuges that are set aside from large areas bent to man's purposes may be thought of as islands in a sea of altered habitat. In view of the manifest destiny of much of the world's tropical rain forest, we may ask such questions as: How many species of Amazonian plants and animals will survive if only 1 per cent of the Amazonian rain forest can be preserved? At what rate will species be extinguished? Which species will be likely to survive in the reserve, and which will most likely be lost? These are pressing questions, to which the theory of island biogeography holds at least some of the answers.

This chapter first treats 'static' aspects of the equilibrium biota, discussing empirical and theoretical relationships between the island area, A, and the number of species present, S. Second, 'dynamic' aspects of the equilibrium are examined; equilibrium is seen as a balance between immigration and extinction, and the rates at which the system approaches the equilibrium configuration from below, from the neighbourhood, and from above are discussed. Next we discuss *which* species tend to be present on a given island at a given stage in its history, and conclude with some speculations as to the emergent principles for the design of natural reserves.

10.2   Species–area relations

10.2.1   *Empirical relations*

There have been many studies which compare the number of species, S, on islands of different area, A, but with similar habitat and in the same archipelago or island group. For both plants and animals, and for a variety of taxonomic groups from birds to beetles and ants, such

Table 10.1. The species–area exponent z.

Organism	Location	z	Source
beetles	West Indies	0·34	Darlington
reptiles and amphibians	West Indies	0·30	Darlington
birds	West Indies	0·24	Hamilton, Barth, Rubinoff
birds	East Indies	0·28	Hamilton, Barth, Rubinoff
birds	East-Central Pacific	0·30	Hamilton, Barth, Rubinoff
ants	Melanesia	0·30	MacArthur and Wilson
land vertebrates	Lake Michigan Islands	0·24	Preston
birds	New Guinea Islands	0·22	Diamond
birds	New Britain Islands	0·18	Diamond
birds	Solomon Islands	0·09	Diamond and Mayr
birds	New Hebrides	0·05	Diamond and Marshall
land plants	Galapagos	0·32	Preston
land plants	Galapagos	0·33	Hamilton, Barth, Rubinoff
land plants	Galapagos	0·31	Johnson and Raven
land plants	World-wide	0·22	Preston
land plants	British Isles	0·21	Johnson and Raven
land plants	Yorkshire nature reserves	0·21	Usher
land plants	California Islands	0·37	Johnson, Mason, Raven

Values of z in eq. (10.1), as deduced from observations on various groups of plants and animals in various archipelagoes. For original references, see May (1975f).

studies commonly lead to a relation of the form

$$S = cA^z \qquad (10.1)$$

The dimensionless parameter z (the slope of the regression line on a $\log S$ versus $\log A$ plot) typically has a value in the range 0·18–0·35; c is a proportionality constant, which depends *inter alia* on the dimensions in which A is measured and on the taxonomic group studied. The data are sometimes better fitted by a relation of the form $S = a + b \log A$: for a fuller discussion see, e.g., Diamond and Mayr (1976).

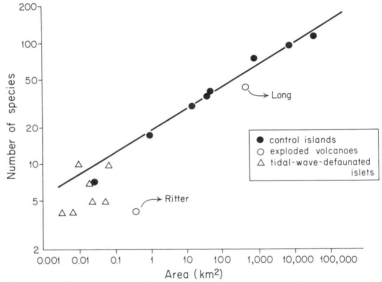

Fig. 10.1. An example of the relation between species number and island area in an archipelago: number of resident, nonmarine, lowland bird species S on islands in the Bismarck Archipelago, plotted as a function of island area on a double logarithmic scale. The solid circles represent relatively undisturbed islands, and the straight line $S = 18 \cdot 9\, A^{0 \cdot 15}$ was fitted by least-mean-squares through the points for the seven largest islands. The open circles refer to the exploded volcanoes, Long and Ritter, where species number is still below equilibrium, especially on Ritter, because of incomplete regeneration of vegetation. The open triangles refer to coral islets inundated by the Ritter tidal wave in 1888. (From Diamond, 1974.)

Table 10.1 gives a list of the values of z in the S–A relation (10.1) for groups of plants and animals in various parts of the world. Figure 10.1 illustrates the relation, for number of bird species on islands of the Bismarck Archipelago near New Guinea.

A rough rule, which summarizes the *S–A* relation (10.1) with values of *z* in the range typically observed, is that a tenfold decrease in area corresponds to a halving of the equilibrium number of species present.

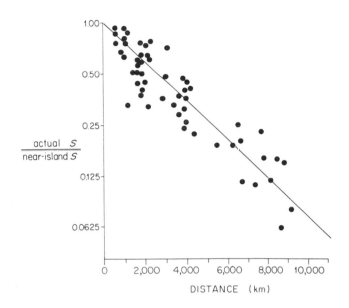

Fig. 10.2. An example of the relation between species number and island distance from the colonization source, for birds on tropical islands of the Southwest Pacific. The ordinate (logarithmic scale) is the number of resident, nonmarine, lowland bird species on islands more than 500 km from the larger source island of New Guinea, divided by the number of species on an island of equivalent area close to New Guinea. The abscissa is island distance from New Guinea. The approximately linear relation means that species number decreases exponentially with distance, by a factor of 2 per 2600 km. (From Diamond, 1972.)

Thus, to answer one of the paradigmatic questions in the introduction, saving 1 per cent of the Amazonian rain forest might correspond, very roughly, to saving 25 per cent of the original species. Such relations are admittedly crude and neglectful of detail, but they provide an informed first guess at the relation between the area of a reserve and the number of species which are eventually likely to be preserved in it.

A further empirical rule is that, if one compares islands of similar area, S decreases with increasing distance D from the colonization source. Figure 10.2 illustrates this trend for birds on tropical Pacific islands colonized from New Guinea.

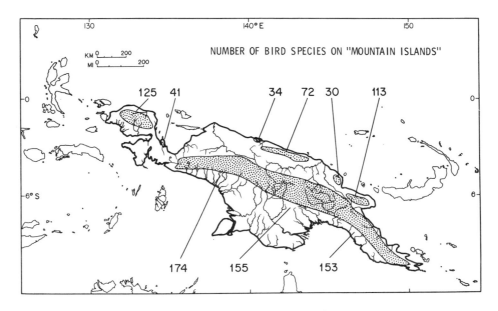

Fig. 10.3. An example of the relation between species number and area of a 'habitat island'. New Guinea consists of a large central mountain range plus about six smaller mountain ranges along the north coast (dotted areas), separated from each other by a 'sea' of intervening lowlands. There are many New Guinea bird species that are confined to higher elevations in the mountains, and for which New Guinea itself therefore behaves as an island archipelago. Numbers on the map give the number of such montane bird species on each small range and at three different locations on the central range. Note that the larger ranges have more montane species. Most of the variation in S not correlated with variation in A is correlated with variation in altitude.

Dependence of S on A is also observed for 'habitat islands' within continents or other islands. Figures 10.3 and 10.4 are two from among the many illustrative examples that could be chosen; see also MacArthur and Wilson (1967), Vuilleumier (1970), Brown (1971), and Diamond (1975a).

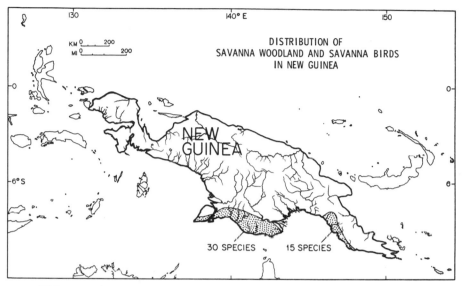

Fig. 10.4. An example of the relation between the area *and* degree of isolation of a habitat island, and the number of species it contains. Most of New Guinea is covered by rain forest, but two separate areas on the south coast (shaded in the figure) support savanna woodland. The characteristic bird species of these savannas are mostly derived from Australia (the northern tips of which are just visible at the lower border of the figure). The Trans-Fly savanna (left) not only has a larger area than the Port Moresby savanna (right), but is also closer to the colonization source of Australia. As a result, the Trans-Fly supports twice as many bird species characteristic of savanna woodland (about 30 compared with 15) as does the Port Moresby savanna. (From Diamond, 1975a.)

10.2.2 *Theoretical Explanations*

The phenomenological eq. (10.1) is based on observational data, and is useful as such. This section attempts to provide the equation with a theoretical underpinning.

In the preceding chapter, we saw that for any large and fairly heterogeneous assembly of species the distribution of relative abundance is likely to be of lognormal form. Going further, Preston (1962) observed that this lognormal is commonly of 'canonical' form, corresponding to the parameter γ (see section 9.4) having the special value $\gamma = 1$. In this event, there is a unique relation between the number of species, S, and the total number of individuals in the biota, N say. This was noted by Preston (1962) and by MacArthur and Wilson (1967), who both went on to add a second assumption, namely that there is an approximately linear relationship between the total number of individuals, N, and the island area, A. The underlying implicit biological assumption, that total density of individuals is independent of area (and of S), is not strictly valid, but the deviations do not greatly affect the predicted species–area relation. When put together with the assumption of a canonical lognormal species relative abundance (which relates S to N), the assumption relating N to A leads to a unique relationship between S and A. This mathematical relationship between S and A is complicated (May, 1975f, p. 112), but for large S ($S > 20$ or so) it is increasingly well approximated by eq. (10.1) with $z = 0.25$ (May, 1975f; this analytic result is to be compared with the numerical curve-fitting previously employed by Preston, 1962, to get $z = 0.262$ and by MacArthur and Wilson, 1967 to get $z = 0.263$). At low values of S, the exact S–A relation obtained under the above assumptions exhibits a downturn, of just the kind exhibited by the data (solid circles and open triangles) in Fig. 10.1.

The details of these derivations depend on the rather mystical 'canonical' assumption that $\gamma = 1$. More generally, as we explained in section 9.4, we expect some lognormal distribution of species relative abundance, with the parameter γ in the neighbourhood of unity (say 0.6 to 1.7) for a wide range of values of S and N: see May (1975f, Fig. 4). When coupled with the assumption that N is proportional to A, this leads to a 1-parameter family of S–A curves, depending on the explicit value of the parameter γ. As for the special 'canonical' case, these relations are well approximated by eq. (10.1) once S is relatively large, with z now a function of γ. Specifically, $z = 1/(4\gamma)$ for $\gamma > 1$ and $z = (1 + \gamma)^{-2}$ for $\gamma < 1$, so that the plausible range of variation of γ leads to eq. (10.1) with values of z in the range 0.39 to 0.15 (May, 1975f).

This provides a detail-independent explanation of the empirical eq. (10.1), based on statistical generalities along with the assumption that biomass is roughly proportional to area. However, as discussed in section 9.4, the statistical arguments which lead to the lognormal distribution and a z value around 0.25 are no more than plausible generalities, and there are many circumstances (e.g., early succession, disturbed habitat, etc.) where those arguments do not apply. Any attempt to explain the systematic differences in z-values among the groups of species listed in Table 10.1, or other fine details of S–A relations, will demand that more attention be paid to biological details. One such more fundamental approach is to derive the S–A relation from an understanding of extinction and immigration rates (see below): this work has been initiated by Schoener (1976b) and Gilpin and Diamond (1976). It emerges from such a treatment that z decreases with increasing immigration rates, as exemplified by the progressive decrease in z for birds of New Guinea islands, New Britain islands, Solomon islands and the New Hebrides (Table 10.1), and increases with increasing extinction rates.

10.3   Rates of approach to equilibrium

10.3.1   *Extinction and immigration*

Preston (1962) and MacArthur and Wilson (1963, 1967) have pointed out that the number of species on an island is set by a dynamic balance between immigration and extinction. For any particular island, the nett extinction rate will increase as the total number of species present increases; conversely, the nett rate at which new species are added — the immigration rate—will decrease as S increases. This situation is illustrated schematically in Fig. 10.5 (see also Figs. 10.6 and 10.7). The equilibrium number of species, S^*, is that at which extinction and immigration rates are equal.

This perception gives insight into the relation between an island's size and degree of isolation, and the equilibrium number of species on it. Species immigrate onto an island as a result of dispersal of colonists from continents or other islands: the more remote the island, the lower the immigration rate (i.e., the shallower the dashed curve in Figs. 10.5, 10.6 and 10.7). Species established on an island run the risk of extinction due to fluctuation in population numbers: the smaller the island, the smaller is the population and the higher the extinction rate (i.e., the more steeply rising the solid curve in Figs. 10.5 and 10.7). Area also affects immigration and extinction in several other ways: through its relation to the magnitude of spatial and temporal variability in resources; by being correlated with the variety of available habitats, as stressed by Lack (1973); and by being correlated with the number of 'hot spots', or sites of locally high utilizable resource production for a particular species (Diamond, 1975b). All in all, the larger and less isolated the island, the higher is the species number at which it should equilibrate.

To illustrate how these ideas can be elaborated in more quantitative fashion, we consider the unrealistically simple case where all species have the same, constant immigration rate, μ, and extinction rate, λ.

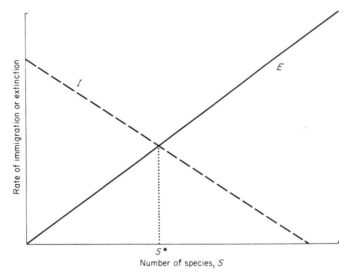

Fig. 10.5. Schematic illustration of an island's extinction curve E (solid line), and immigration curve I (dashed line), as functions of the number of species on the island, S. This figure is for the unlikely case when each species has the same extinction rate and immigration rate, so that the nett rates (expressed as number of species per unit time) are linear functions of S: see eqs. (10.2) and (10.3). The island equilibrium number of species is S^*, where the curves intersect.

Needless to say, this circumstance never prevails in the real world, but it provides a simple model which is useful for exposition; it may also be a sensible approximation to more general nonlinear models (e.g., Fig. 10.7 below) in the neighbourhood of equilibrium. If all species have the same extinction rate, λ, the nett extinction rate, E, expressed as number of species extinguished in unit time, is

$$E = \lambda S. \tag{10.2}$$

Similarly, the nett immigration rate, I, again expressed as number of species newly immigrating in unit time, is

$$I = \mu(S_T - S). \tag{10.3}$$

Here S_T is the total number of potential immigrants in the mainland pool, so that $(S_T - S)$ is the number of candidates for immigration to an island on which S species are already present. Equations (10.2) and (10.3) give the curves illustrated in Fig. 10.5. At equilibrium $E = I$, which gives the equilibrium species number, S^*, as

$$S^* = [\mu/(\lambda+\mu)]\, S_T. \tag{10.4}$$

If the island is not in equilibrium, the change in the number of species in unit time is given by the difference between immigration and extinction:

$$dS(t)/dt = I - E. \tag{10.5}$$

That is, from eqs. (10.2) and (10.3),

$$dS(t)/dt = \mu S_T - (\lambda+\mu)S(t). \tag{10.6}$$

Equation (10.6) may be integrated. If the initial number of species on the island is $S(0)$ at $t = 0$, then

$$S(t) = S^* + [S(0) - S^*]\, e^{-(\lambda+\mu)t}. \tag{10.7}$$

This expression describes the rate at which the system approaches the equilibrium value S^*. Setting aside the excessive simplicity of the underlying assumptions, this expression could be fitted to observational data to obtain estimates of λ, μ and S^*.

However, it is grossly unrealistic to assume that all species have the same λ and μ values, hence that extinction and immigration rates vary linearly with S [eqs. (10.2) and (10.3)]. In reality, λ and μ values of the species coexisting in a community range over many orders of magnitude and have an approximately lognormal distribution (Gilpin and Diamond, 1980). There are two major reasons why nett extinction rates should be expected to increase faster than linearly with S, and likewise why nett immigration rates should fall faster than linearly as S increases: species differ greatly in their values of λ or of μ; and competition causes, for each species, μ to decrease and λ to increase with S. An extreme view, which is implicit in some of the writings of Lack (1973, 1976), is that there are a certain number S^* of species permanently resident on the island, and that any other immigrants will fail to breed; there are a particular S^* species that are ecologically appropriate to the island. This situation is illustrated in Fig. 10.6, which shows the corresponding ultimately steep extinction curve: the extinction rates for the S^* species which, as it were, 'belong' on the island are effectively zero; for any other species, the extinction rate is effectively infinite.

The real situation almost invariably lies between the extremes depicted in Figs. 10.5 and 10.6. In general, as more and more species are packed in, the island approaches ecological saturation, and the

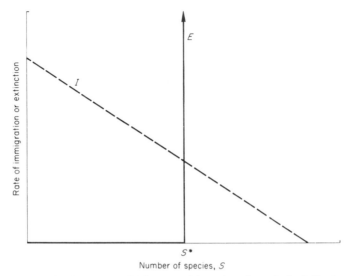

Fig. 10.6. Extinction (solid line) and immigration (dashed line) curves for an island, as functions of S, as in Fig. 10.5. The difference is that here there are S^* species which 'belong' on the island, having effectively zero extinction rate, and *no* other species are capable of breeding on the island. This results in an infinitely steep extinction curve E, as illustrated [i.e., eq. (10.8) in the limit $n \to \infty$].

overall extinction curve, E, steepens rapidly. This may be represented by writing

$$E(S) = \varepsilon(S/S_0)^n, \tag{10.8}$$

where n is now a parameter which describes the steepness of the curve. The simple case of eq. (10.2), constant extinction rate for species, corresponds to the limit $n = 1$; the opposite extreme of Fig. 10.6 corresponds to the limit $n \to \infty$ (whence $E \to 0$ for $S < S_0$, $E \to \infty$ for $S > S_0$). Observational data for birds on real islands suggests values of n around 2 to 4 (Gilpin and Diamond, 1976; Schoener, 1976b; see also Terborgh's, 1974, work below). Analogous expressions may be used to parameterize the nonlinear behaviour of most curves for nett immigration, $I(S)$, which for birds prove to be even steeper than the extinction curves ($n \sim 6$: Gilpin and Diamond, 1976). Figure 10.7 illustrates such extinction and immigration curves for lowland bird species on a typical island in the Solomon Archipelago, in the tropical Southwest Pacific (Gilpin and Diamond, 1976).

The comparison between the realistic Fig. 10.7 and the idealized extremes of Figs. 10.5 and 10.6 sheds light on what has sometimes been misperceived as a conflict between the views of MacArthur and Wilson and those of Lack. Species equilibrium is indeed a dynamic thing (as suggested by MacArthur and Wilson). However, as stressed from the start by MacArthur and Wilson (e.g., Fig. 4 of their 1963 paper), the actual extinction and immigration rates which describe species turnover are, in reality, given by curves that are closer in character to Fig. 10.6 (in the spirit of Lack) than to Fig. 10.5.

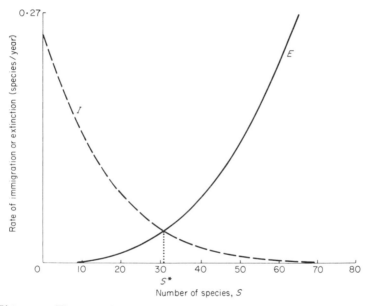

Fig. 10.7. The actual extinction (solid line) and immigration (dashed line) curves for the avifauna on Three Sisters, one of the smaller Solomon islands: the rates are plotted as relative number of species to go extinct, or to immigrate, per year as a function of the number of species on the island. For this island, the extinction curve is of the form of eq. (10.8) with $n = 2.75$, and the equilibrium number of species is $S^* = 32$. For further details, see Gilpin and Diamond (1976). These actual curves are to be compared with the idealized extremes depicted in Figs. 10.5 and 10.6.

We now confront these ideas with data on the way the number of species approaches the island equilibrium value from below ($S(0) < S^*$), in the neighbourhood of equilibrium $(S(0) \sim S^*)$, and from above $(S(0) > S^*)$.

10.3.2 *Species number increasing toward equilibrium*

One type of study involves observing the increase in species number on islands where the flora and/or fauna have been removed either by natural catastrophe or by experimental manipulation. In these situations, the immigration term I is of predominant importance in eq. (10.5), at least in the initial stages.

The most famous such study was provided by a 'natural experiment', the recolonization of the volcanic island of Krakatoa after its biota had been destroyed by an eruption in 1883 (Docters van Leeuwen, 1936; Dammerman, 1948; see MacArthur and Wilson, 1967, pp. 43–51). Here the number of bird species returned in a relatively short time to the value appropriate to the island's area and isolation, whereas the number of plant species is still rising; the rate of approach to equilibrium obviously depends on the plant or animal group under consideration. Similar 'natural experiments' are provided by the birds of Ritter and Long Islands near New Guinea, whose faunas were destroyed by volcanic eruptions in 1888 and about two centuries ago, respectively (see Fig. 10.1); by the birds of seven coral islets in the same area, where the tidal wave following the Ritter eruption destroyed the fauna in 1888 (Diamond, 1974); and by the initial colonization of a newly created volcanic island such as Surtsey, off Iceland.

As discussed in section 9.3, Simberloff and Wilson (1969) created an analogous 'artificial experiment' by fumigating several tiny mangrove islets off the coast of Florida and observing the recolonization by arthropods. Similarly, A. Schoener (1974) suspended plastic sponges and measured the rate at which they were colonized by marine invertebrates.

10.3.3 *Species number around equilibrium*

Another type of test is provided by turnover studies at equilibrium. According to the above interpretation, although the *number* of species on an island may remain near an equilibrium value, the *identities* of the species need not remain constant, because new species are continually immigrating and other species are going extinct. Estimates of immigration and extinction rates at equilibrium for birds have been obtained by comparing surveys of an island in separate years; among such studies are those for the birds of the Channel Islands off California (Diamond, 1969; Hunt and Hunt, 1974; Jones and Diamond, 1976), Karkar Island off New Guinea (Diamond, 1971), Mona Island off Puerto Rico (Terborgh and Faaborg, 1973), islands off the coast of Britain and Germany (data of Jones and Diamond, summarized by Diamond and May, 1977), and European mainland census plots (Järvinen, 1979). Two practical problems in turnover studies are: (a) to estimate man's effect on natural turnover rates (see Jones and Diamond, 1976); and (b) to correct for the precipitous decline in apparent turnover rate with increasing interval between censuses, as a result of extinctions and reimmigrations in the interval remaining unnoticed and cancelling each other's effect (Diamond and May, 1977).

All these turnover studies found that a certain number of species present in the earlier survey had disappeared by the time of the later survey, but that a similar number of other species immigrated in the intervening years, so that the total number of species remained approximately constant unless there was a major habitat disturbance. As expected from considering the risk of extinction in relation to population size, most of the populations that disappeared had initially consisted of few individuals. The turnover rates per year (extinction and immigration rates) observed in these bird studies were in the range of 0·2 to 20 per cent of the island's bird species for islands of area ranging from 400 to 0·4 km².

10.3.4 Species number decreasing towards equilibrium

The situation of greatest relevance to floral and faunal conservation arises when some fraction of a habitat is set aside as a reserve, and the rest destroyed. Such a reserve will at first be 'supersaturated', containing more species than are appropriate to its area at equilibrium. The situation is the exact converse of an island which has had its biota destroyed (section 10.3.2): equilibrium of species number will be approached from above, and the extinction term E will be of predominant importance in eq. (10.5), at least in the early stages.

A natural experiment of this kind is provided by so-called landbridge islands. During the most recent ice age, which lasted for an extended period of thousands of years and ended about 10,000 years ago, enough water was locked up in glaciers to make the ocean levels about 100 m. lower than at present. Consequently islands that are now separated from continents or larger islands by water less than 100 m. deep were once attached, and shared the continental biota. Examples of such land-bridge islands are Britain off Europe, Aru and other islands off New Guinea, Trinidad off South America, Fernando Po off Africa, and Borneo and Japan off Asia. Subsequent to these islands being created by rising sea levels about 10,000 years ago, their continental complement of species has slowly decreased towards the equilibrium value appropriate to an island of their modern area.

Terborgh (1974) has made a quantitative such study for birds on five neotropical land-bridge islands. The number of bird species currently on each island is known, and the number present on each island before it was cut off 10,000 years ago may be estimated from the neighbouring mainland species numbers. Terborgh notes that in this relaxation process extinction is predominant [so that I is neglected in eq. (10.5)], and he chooses to describe the nett extinction rate by eq. (10.8) with $n = 2$; thus he fits the data to

$$dS/dt = kS^2. \tag{10.9}$$

The extinction parameter k thus deduced is shown, as a function of island area, in Fig. 10.8. Similar results, and in particular the tendency for extinction rates to decrease as island area increases, have been obtained by Diamond (1972, 1973) for birds on land-bridge islands off New Guinea and elsewhere in the southwest Pacific, and by Wilcox (1980) for mammals on islands of the southeast Asian continental shelf. Brown (1971) has made an analogous study of the distribution of small mammals in forests which are now isolated on mountaintop 'islands' rising out of the 'sea' of desert in western North America, but which were connected by a continuous forest belt or 'habitat land-bridge' during times of cooler Pleistocene climates (see Table 10.2).

Terborgh (1974) has made a dramatic application of the calculations summarized in Fig. 10.8, by showing that they correctly predict the extinction rates observed within the present century on Barro Colorado Island. This island was created from a former hilltop by the flooding of adjacent valleys to create Lake Gatun when the Panama Canal was constructed, and since 1923 it has been carefully protected as a wildlife preserve. A meticulous account of the bird species present on the island over the past 50 years has recently been published by Willis (1974). From the k-versus-area patterns for post-Pleistocene relaxation rates on neotropical land-bridge islands, Fig. 10.8, Terborgh makes an

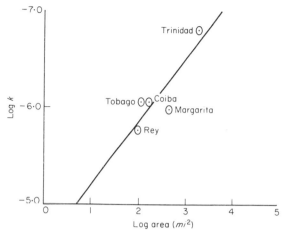

Fig. 10.8. The extinction paramotor k of eq. (10.9) as a function of island area, for birds on land-bridge islands in the West Indies. (From Terborgh, 1974.)

estimate of the k value for Barro Colorado Island. Equation (10.9) can then be used to get an estimate of the decline in the island's number of bird species since 1923. This theoretical estimate of 16–17 forest bird species lost is gratifyingly in accord with the actual number of 15.

10.4 Which species survive?

Up to this point the problem of survival has been discussed in statistical terms: what fraction of its initial biota will a reserve eventually save, and how rapidly will the remainder go extinct? We now go on to consider the survival probabilities of individual species.

These considerations bear directly on conservation strategies. If each species had a roughly equal probability of survival, then large numbers of small reserves could be a satisfactory policy: each vest-pocket reserve would lose most of its species before reaching equilibrium, but with enough reserves any given species would be likely to be among the survivors in at least one reserve. The flaw in this strategy is that different species usually have very different area requirements for survival, arising from very different rates of immigration and of extinction.

Consider first the question of immigration. Even if there are many small reserves, a species that is incapable of dispersing from one reserve to another across the intervening sea of unsuitable habitat is doomed to eventual extinction: its lights will wink out, one by one, with no chance of reignition. Conversely, a species capable of dispersing from one reserve to another may persist by virtue of a dynamic balance between local extinction and reimmigration, provided recolonization rates are high enough or extinction rates low enough. Dispersal ability obviously differs enormously among plant and animal species. Flying animals tend to disperse better than non-flying ones; plants with wind-borne seeds tend to disperse better than plants with heavy nuts. The more sedentary the species, the more irrevocable is any local extinction, and the more difficult will it be to devise a successful conservation strategy. These conservation problems will be most acute for the sort of slowly dispersing species found in normally stable habitats, such as tropical rain forest. Even the power of flight cannot be assumed to guarantee high dispersal ability. For example, 134 of the 325 lowland bird species of New Guinea are absent from all oceanic islands more than a few km from New Guinea, and are confined to New Guinea plus associated land-bridge islands. Similarly, many neotropical bird families with dozens of species have not one representative on a single New World island lacking a recent land-bridge to the mainland; and not a single member of many large Asian bird families has been able to cross Wallace's line separating the Sunda Shelf land-bridge islands from the oceanic islands of Indonesia. Such bird species are generally characteristic of stable forest habitats, and have insuperable psychological barriers to crossing water gaps. In short, low recolonization rates may mean either that a species cannot, or that it will not, cross unsuitable habitats.

Given this variation in ability to recolonize, we turn to consider how species vary in extinction rates of local populations. The New Guinea land-bridge islands alluded to in section 10.3.4 offer a convenient test situation (Diamond, 1972, 1975a). As mentioned in the preceding paragraph, there are 134 New Guinea lowland bird species that do not cross water gaps, and consequently post-Pleistocene extinction of these species on land-bridge islands cannot have been reversed by recolonization. Because extinction rates on relatively small islands are high (cf. Fig. 10.8), virtually all these species are now absent from all land-bridge islands smaller than 50 km². On the other hand, these 134 species vary greatly in their distribution on the seven larger (450–8000 km²) land-bridge islands. At one extreme, some species (e.g., the frilled monarch flycatcher, *Monarcha telescophthalmus*) have survived on all seven islands; at the other extreme, 32 species have disappeared

from all seven islands. Most of these 32 extinction-prone species fit into
one or more of three categories: birds whose initial populations must
have numbered few individuals because of very large territory requirements (e.g., the New Guinea Harpy Eagle, *Harpyopsis novaeguineae*);
birds with small initial populations because of specialized habitat
requirements (e.g., the swamp rail, *Megacrex inepta*); and birds which
are dependent on seasonal or patchy food sources, and which normally
go through drastic population fluctuations (e.g., fruit-eaters and
flower-feeders). These observations tend to be confirmed by other bird
studies, and by the work of Brown (1971), referred to above, on differential extinction rates among mammal species isolated on mountain
tops (Table 10.2).

Another natural experiment in differential extinction is provided by
New Hanover, a 1200 km² island which in the late Pleistocene was connected by a land-bridge to the larger island of New Ireland in the Bismarck Archipelago. Although today New Hanover has only lost about
22 per cent of New Ireland's species, among these lost species are 19 of
the 26 New Ireland species confined to the larger Bismarck islands,
including every endemic Bismarck species in this category. That is,
those species most in need of protection were differentially lost: as a
faunal reserve, New Hanover would rate as a disaster. Yet its area of
1200 km² is not small by the standards of many of the tropical rain
forest parks that one can realistically hope for.

Just as rising post-Pleistocene sea levels caused area fragmentation
and hence population extinctions by dissecting formerly continuous
land masses into islands, so (on a shorter time scale) have man's activities caused population extinctions by fragmenting formerly continuous
forests into isolated wood-lots. Among bird populations disappearing in
fragmented Brazilian forests, the best predictor of extinction is low
abundance (data of Willis, analysed by Terborgh and Winter, 1980). A
similar conclusion applies to bird extinctions in an isolated Ecuadorean
forest (Leck, 1979). Low abundance is also strongly correlated with
risk of extinction in turnover studies at equilibrium (Jones and Diamond, 1976).

The survival prospects for a particular species may be quantified by
determining its 'incidence function', $J(S)$ (Diamond, 1975b). We have
noted, at least for birds and mammals, that some species occur only on
the largest and most species-rich islands, other species occur also on
medium-sized islands, and others also occur on small islands. These
patterns may be displayed graphically by grouping the islands into

Table 10.2. Mammals on mountain islands of the Great Basin.

Number of islands per species	Species	Habitat and diet	Weight (g)	Ruby	Toiyabe	Toquima	White-Inyo	Snake	Schell Creek	Deep Creek	White Pine	Oquirrh	Roberts Creek	Diamond	Stansbury	Grant	Spruce	Spring	Pilot	Panamint
14	Neotoma cinerea	GH	317	×	×	×	×	×	×	×	×	—	—	×	×	—	×	×	×	×
14	Eutamias umbrinus	GH	57	×	×	×	×	×	×	×	×	×	×	×	—	×	×	×	—	—
13	Spermophilus lateralis	GH	147	×	×	×	×	×	×	×	×	—	—	×	—	×	×	×	×	—
12	Microtus longicaudus	GH	47	×	×	×	×	×	×	×	×	×	×	—	×	×	—	—	—	—
9	Marmota flaviventer	GH	3000	×	×	×	×	×	×	×	×	—	—	—	×	—	—	—	—	—
8	Thamomys talpoides	GH	102	×	×	×	×	—	×	×	×	—	—	×	—	—	—	—	—	—
6	Sorex vagrans	C	6·7	×	×	—	—	×	×	—	—	×	×	—	—	—	—	—	—	—
6	Sorex palustris	C	14	×	×	×	×	×	—	—	—	×	—	—	—	—	—	—	—	—
4	Zapus princeps	SH	33	×	×	—	—	—	—	—	—	×	×	—	—	—	—	—	—	—
4	Ochotona princeps	SH	121	×	×	×	×	—	—	—	—	—	—	—	—	—	—	—	—	—
3	Mustela erminea	C	58	—	×	—	×	×	—	—	—	—	—	—	—	—	—	—	—	—
3	Spermophilus beldingi	SH	382	×	×	×	—	—	—	—	—	—	—	—	—	—	—	—	—	—
1	Lepus townsendi	SH	2500	×	—	—	—	—	—	—	—	—	—	—	—	—	—	—	—	—
	Number of species per island			12	12	9	9	8	7	6	6	5	4	4	3	3	3	3	2	1

classes containing similar numbers of species (e.g., 1–4, 5–9, 10–20, 21–35, 36–50, etc.), calculating the incidence J or fraction of the islands in a given class on which a particular species occurs, and plotting J against the total species number S on the island. Some such incidence functions for birds in the Bismarck Archipelago are shown in Fig. 10.9; a much more full discussion is in Diamond (1975b). Since S in the Bismarcks is closely correlated with area, these graphs in effect represent the probability that a species will occur on an island of a particular size.

For most species, J goes to zero for S values below some value characteristic of the particular species, meaning there is no chance of long-term survival on islands below a certain size. These incidence functions can be interpreted in terms of the 'bionomic strategy' of the species, along the lines of chapter 3. Figure 10.9a is typical of species in the so-called 'super-tramp' category (Diamond, 1974, 1975b); as the figure shows, such birds tend to be found on small islands, but *not* on large ones (because of competitive exclusion from species-rich faunas). Figure 10.9b illustrates the incidence function for an intermediate category of species, while Fig. 10.9c is typical of species found only in

Footnote to Table on facing page.

Out of the Great Basin desert of the western United States rise 17 mountain ranges to elevations above 10,000 ft. Boreal habitats on the summits of these 'mountain islands' are now isolated from each other by the surrounding sea of desert, but were connected to each other and to the source boreal habitats of the much more extensive Rocky Mountains and Sierra Nevada during cooler Pleistocene periods. On these mountaintops live 13 species or superspecies of small flightless mammals, which cannot cross the intervening desert today but which reached the mountain islands over Pleistocene bridges of boreal habitat. Since the Pleistocene these mammal populations have been subject to risk of extinction without opportunity for recolonization. The table illustrates the regular distributional pattern produced by this differential extinction. The presence (×) or absence (−) of each named species is indicated for each named mountain island, along with the total number of islands inhabited by each species (left-most column) and total number of species inhabiting each island (bottom row; this number correlates well with island area). C = carnivore; GH = generalized herbivore (present in most habitats); SII = specialized herbivore (present in only certain habitats); body weight is given in grams. Herbivores can maintain higher population densities than carnivores, small animals higher densities than large animals, and species of generalized habitat preference higher densities than habitat specialists. Thus, the main patterns of differential extinction emerging from the table are that: rare species survive on fewer islands than abundant species; rare species become confined to the larger islands; and different small islands tend to end up with the same group of abundant species. From Brown (1971).

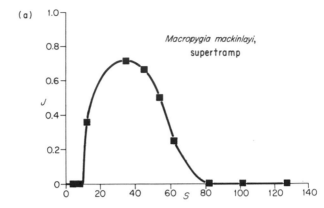

Fig. 10.9a. The incidence function $J(S)$ for the 'supertramp' pigeon *Macropygia mackinlayi*, in the Bismarck Archipelago. The incidence J represents the fraction of those islands, with S values in a given small range, on which the species occurs, as a function of the number of species S on the island. Each point is typically based on 3–13 islands. For a more detailed discussion, see Diamond (1975b).

the presence of many others (i.e., on large islands). In general terms, one may recognize Figs. 10.9a–c as depicting the spectrum from an extreme r-strategy through to an extreme K-strategy (see chapters 2 and 3); the J-versus-S curves, however, go further to give quantitative substance to these notions.

Incidence represents either the fraction of a set of similar islands which a certain species inhabits at a given time, or else the fraction of time that a certain species inhabits a particular island. These fractions

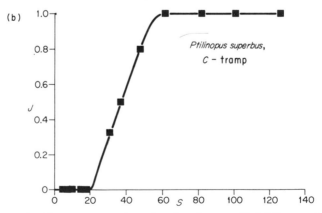

Fig. 10.9b. The incidence function for the so-called 'C-tramp' pigeon *Ptilinopus superbus*, in the Bismarcks. (From Diamond, 1975b.)

approach equilibrium probabilities determined by a balance between immigration and extinction rates for the particular species, just as total species number approaches an equilibrium value determined by the immigration and extinction rates of all species (section 10.3.1). Hence incidence J equals $\mu/(\mu + \lambda)$, where μ and λ are now the immigration and extinction probabilities of the particular species. This simple model accounts for the bent-S-shaped form of the incidence functions of Figs. 10.9b and 10.9c, as well as for species differences in the abscissa position of the rising portion of the curve (Diamond and Marshall, 1977; Gilpin and Diamond, 1981).

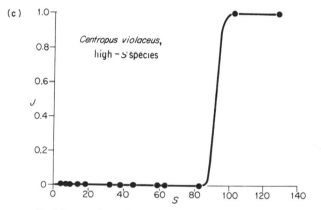

Fig. 10.9c. Incidence function for the 'high-S' cuckoo *Centropus violaceus*, in the Bismarcks. (From Diamond, 1975b.)

'Assembly rules' furnish additional information as to which species are likely to be present on a given island. These rules (Diamond, 1975b) codify the observation that, of all possible combinations that could be formed from within a group of related species, only certain combinations exist in nature. The rules are evocative of those of the early days of atomic spectroscopy, with its empirical catalogues of allowed and forbidden spectral lines. Thus there are some pairs or groups of species that never coexist; others that form an unstable combination by themselves, but may form part of a stable larger combination; still others constitute stable combinations that resist invaders that would transform them into a forbidden combination; and so on. For the Bismarck avifauna, a detailed account of the assembly rules for such guilds as the cuckoo-dove, the gleaning flycatcher, the myzomelid-sunbird, and the fruit-pigeon guild has been given by Diamond (1975b, pp. 393–411). Faaborg (1976) has made similar detailed studies of the incidence and assembly of the birds on West Indian islands. In some cases these rules clearly derive from competition for resources, and from harvesting of resources by permitted combinations so as to minimize the unutilized resources available to support potential invaders; other cases are less transparent.

These detailed observations as to the extinction, immigration and overall incidence $J(S)$ of particular species of course provide the underlying explanation for the patterns of curvature in extinction and immigration functions, as illustrated in Fig. 10.7 (Gilpin and Diamond, 1981). Ultimately, the incidence functions and assembly rules are presumably determined by considerations of limits to similarity and niche overlap (see chapters 5, 8) and by the constraints imposed by food web structure (see section 9.3, and in particular the study by Heatwole and Levins, Table 9.3).

The sort of quantitative information that is embodied in incidence functions and assembly rules holds the promise of providing the basis for a predictive science of environmental management.

10.5   Design principles for natural reserves

The work described in this chapter clearly is relevant to questions about the properties of natural reserves: how is the eventual number of species in the reserve related to its area; how do extinction rates vary with area; how do area-dependent survival probabilities vary among species? The answers to these questions prompt the enunciation of certain 'general design principles' for floral and faunal preserves (see, e.g., Willis, 1974; Wilson and Willis, 1975; Diamond, 1975a and 1976a; Terborgh, 1974; May, 1975g; and the chapters in Soulé and Wilcox, 1980, for a book-length treatment). As for most generalities, such design principles are useful, but must be applied with caution in any one specific instance. They are no substitute for a painstaking study of extinction and immigration rates, of incidence functions and assembly rules, for each particular conservation project.

In general, a large reserve is better than a small reserve for two reasons: the large reserve can hold more species at equilibrium, and it will have lower extinction rates. A less straightforward question is whether to prefer one large reserve or else several smaller reserves whose areas add up to the same total as the single large reserve. If one considers only total number of species without regard to how species differ in their need for protection, either arrangement may have a slight advantage depending on the particular situation (Simberloff and Abele, 1976a). However, recall that different species have different area requirements and that species requiring large areas are often the ones most threatened by man's activities and in need of protection (section 10.4). On these grounds, fragmenting a large reserve into several smaller reserves is a bad rather than a good policy (see Diamond, 1976b, Terborgh, 1976, Whitcomb, Lynch, Opler, and Robbins, 1976, and Simberloff and Abele, 1976b, for various views on this question). Many species, especially those of tropical forests, are stopped by narrow dispersal barriers; for these species, major roads or power lines may effectively fragment a park and hence cause increased risk of extinction.

If one must settle for several smaller parks, a way to raise the equilibrium number of species in any one such park is to raise the immigration rate into it. This can be done by careful juxtaposition of the scattered parks, and by providing corridors or stepping-stones of natural habitat between them.

Any given reserve should be as nearly circular in shape as other circumstances allow. Such maximization of the area-to-perimeter ratio minimizes dispersal distances within the reserve, and avoids 'peninsular effects' whereby dispersal rates to outlying parts of the reserve from more central parts may be so low as to perpetuate local extinctions, thus diminishing the reserve's effective area.

These general design principles are subject to many qualifications and equivocations. *First*, several smaller reserves may have the compensating advantage that in an inhomogeneous region each reserve may favour the survival of a different group of species. Even in a homogeneous region, separate reserves may save more species of a set of vicariant similar species, one of which would ultimately exclude the others from a single reserve. *Second*, the above principles ignore epidemiological aspects of park management: many scattered parks are less susceptible to the ravages of an epidemic disease or analogous disaster. *Third*, there is the obvious point that some 'edge' species, that thrive at the interface between habitats, will prefer several smaller parks, or parks with high perimeter-to-area ratios; conversely, edge-intolerant species will be differentially worse off with several smaller areas, and will be unable to survive once the reserves become too small. (However, many edge species will fare well in non-park habitats transformed or dissected by man, while edge-intolerant species will furnish a disproportionate number of the species most dependent on parks for survival.) *Fourth*, one must consider the dynamical features such as the spatial and temporal patterns of stable oscillation that so frequently characterize populations both in mathematical models and in the real world. The evidence for a roughly 50-year cycle in elephant population eruptions in the area that is today Tsavo National Park in Kenya suggests that such dynamical aspects of natural populations can create management problems when even a very large area is enclosed as reserve.

All these questions need to be illuminated by further research, particularly field work on organisms other than birds. There is clear and present need for the development of techniques to estimate the size and other properties a reserve should possess if it is to fulfil its designated conservation purpose.

11
Succession

HENRY S. HORN

11.1 Introduction

Plant succession has often been idealized as a process whereby a community that has suffered an episodic devastation slowly regenerates a semblance of its former self in the absence of disturbance. These regenerative changes in the community were interpreted by Clements (1916) as emergent properties analogous to the recovery of an organism from injury, and the analysis of properties of the whole community has been continued (Margalef, 1968; Odum, 1969). A recent trend views succession in the context of the adaptations of individual species, independent of any transcendent properties of the whole community (Connell, 1972; Drury and Nisbet, 1971, 1973; Horn, 1974; Connell and Slatyer, 1977). Tradition attributes the origin of this individualistic attitude to Gleason (1926), but Thoreau (1860) talked about it much earlier as though it were already commonplace.

Over the past 20 years, several workers have converged on representing succession as a Markovian replacement process. A table is set up to show the probability that a given plant will be replaced in a specified time by another of its kind or by another species. With the aid of this table, the current composition of a community can be 'projected' into the future. Markovian models can be applied not only at the level of plant-by-plant replacements (Anderson, 1966; Horn, 1975a, b), but also to transitions among trees of different sizes in a growing stand (Usher, 1966; Leak, 1970; Moser, 1972), local groups of trees (Botkin *et al.*, 1972), woodlots of varying composition (Stephens and Waggoner, 1970), groups of species (Williams *et al.*, 1969), regional forests (Shugart *et al.*, 1973), and even vegetation whose composition is only defined qualitatively (MacArthur, 1961).

An important assumption of most of these models is that the table of replacement probabilities does not change as time passes. Under this assumption, the composition of the modelled community approaches a particular steady state. This state is dependent on the configuration of the table of replacement probabilities, and thus on the biological interactions within and among species, but it is independent of the community's initial composition, and hence independent of the historical vagaries of the devastation that started the succession. Such steady states are notoriously difficult to establish in nature. When very old and 'stable' forests are examined closely, there is always evidence for a history of episodic disturbance (Connell, 1978). Many trees may be of a particular age, dating from a past devastation by wind, water, landslide, or fire (Maissurow, 1941; Jones, 1945; Loucks, 1970; Heinselman, 1973). The forest may preserve evidence of a sequence of fires (Buell *et al.*, 1954; Houston, 1973) or of hurricanes (Stephens, 1955; Henry and Swan, 1974), or a wide variety of natural disturbances (White, 1979). Even in the absence of devastation, trees may occur in evenly aged groups dating from episodes that favoured seedling establishment, either climatically (Cooper, 1960) or due to the absence of predators (Peterkin and Tubbs, 1965). Where succession is fast enough to be observed directly, as among herbs or the encrusting beasties of rocky sea shores, the succession may be dominated by chronic predatory devastation (Harper, 1969; Paine, 1974; Connell, 1978; Fox, 1979), and alternative quasi-stable communities may develop, whose structure can only be explained by reference to specific historical events (Sutherland, 1974).

The ideal of convergent and undisturbed reestablishment of a stable community is not general in nature, if indeed it exists. Hence there have been periodic controversies over whether succession should imply anything more than the historical description of individual examples (unusually lucid reviews are: Gleason, 1927; Whittaker, 1957; Drury and Nisbet, 1971, 1973; and Miles, 1979). McIntosh (1980) provides an excellent historical review of ideas about succession, spiced with journalistic rhetoric.

In this chapter I analyse the basic linear model of succession to show how chronic disturbances, interspecific competition and local regeneration affect the rate of convergence of succession. This analysis discloses some of the obstacles to previous theories of succession. The effect of several biologically interesting non-linearities is discussed intuitively, but nonetheless rigorously. In particular, if the presence of vigorous young plants is strongly dependent on nearness to mature plants, or on a highly seasonal rain of seeds, then succession may not be convergent at all. If a community is devastated on a sufficiently regular schedule, its species will be those whose lifespans match the length of the less eventful interludes. These species may then evolve increased sensitivity to devastation late in life, and thereby perpetuate the environmental cycle to which they are adapted.

Although many of the ideas in this chapter arose from a mathematical analysis, I have found that they can be supported by direct appeal to biological assumptions without the intervening mathematical machinery. My examples betray my fondness for trees and forests, and they are illustrative rather than critical or exhaustive. However, the ideas are equally applicable to any communities of sessile plants or sessile animals.

11.2 Forest succession as a tree-by-tree replacement process

By making an accurate map of every tree in a forest and then mapping the same area 50 years later, one can estimate the probability that a tree of each species will live for 50 years, and for each tree that dies, the probability that it will be replaced by another of its own species or by another species. This is a generalization of the familiar process whereby a population's demographic characteristics are summarized in a 'life table' of survivorship and fecundity at each age (see chapters 2, 3). Namkoong and Roberds (1974) have analysed such a life table to support recommendations for management of Californian coastal redwood forests. Table 11.1 shows probabilities of replacement, calculated in a roundabout way, for Gray Birch, Blackgum, Red Maple, and Beech, four species that can successively dominate abandoned farmland near Princeton, New Jersey (Horn, 1975a).

From Table 11.1 and the current composition of the forest, it is easy to calculate the expected composition of the forest 50 years hence. For example:

$$RM_{50 \text{ yrs hence}} = 0.5 GB_{now} + 0.25 BG_{now} + 0.55 RM_{now} + 0.03 BE_{now}.$$
$$(11.1)$$

Analogous equations can be written for each species, combining the current composition of the forest with the appropriate column of Table 11.1. Then the composition of the forest can be predicted 100 years hence by combining the table according to the same recipe with the calculated composition 50 years hence; and so on for 150 years, 200 years and more. The recipe consists of the rules for multiplication of a row vector (the current composition of the forest) by a matrix (the table of tree-by-tree replacement probabilities). Once the composition is projected far enough into the future, further multiplications by the successional matrix will leave the species' abundances unchanged.

Table 11.1. Fifty year tree-by-tree transition matrix for Gray Birch, Blackgum, Red Maple and Beech.

	50 years hence			
Now	Gray Birch	Black-gum	Red Maple	Beech
Gray Birch	5 + 0	36	50	9
Blackgum	1	37 + 20	25	17
Red Maple	0	14	37 + 18	31
Beech	0	1	3	61 + 35

Diagonal is percentage of trees still standing plus percentage that have been replaced in 50 years by another of their own kind. Off-diagonal terms are percentage of trees replaced by another species in 50 years. The percentage of trees still standing was estimated by assuming that trees die at a constant rate such that 5 per cent are left standing after 50 years for Gray Birch, 150 years for Blackgum and Red Maple, and 300 years for Beech. For those trees that die in 50 years, replacement of one species by another is assumed to be proportional to the number of saplings of the latter under a large sample of canopy trees of the former. The forests in which these data were gathered have many more species (Horn 1975b); so this matrix is only an illustrative caricature.

Table 11.2. Predicted composition of a succession.

Age of Forest (years)	0	50	100	150	200	...	∞	Very Old Forest
Gray Birch	100	5	1	0	0		0	0
Blackgum	0	36	29	23	18		5	3
Red Maple	0	50	39	30	24		9	4
Beech	0	9	31	47	58		86	93

The succession starts with a field full of Gray Birch and then obeys the transitions specified by the matrix of Table 11.1. The stationary composition after an infinite amount of time is gotten by solving the set of equations in which the composition equals the composition multiplied by the matrix. The composition of the 'very old forest' is the proportion of trees of the four species in the canopy of an actual forest that may not have suffered an extensive disturbance for 350 years or more. Several species have been left out of the analysis for simplicity (See Horn 1975b for more details).

Table 11.2 shows an example of these calculations. For a succession starting with Gray Birch several predictions are made. Red Maple should dominate quickly and Beech should slowly increase to predominate later. Gray Birch should be lost rapidly. The forest should become stationary and predominantly Beech, but Blackgum and Red Maple should persist at low abundance, so that the stationary forest is a mosaic of successional patches. All of these predictions are consistent with what apparently happens in a real forest (Horn, 1975a). Similar tests have confirmed the predictive value of similar models for short-term changes in Wisconsin hardwood forests (Moser, 1972), stands of varying composition in New England (Stephens and Waggoner, 1970; Botkin et al., 1972), and regional vegetation patterns in Wisconsin (Shugart et al., 1973).

11.3   Linear models and their consequences

The fundamental property of a model like the preceding one is that it converges on a stationary composition after enough time. The proportional representation of species in this stationary composition is independent of the initial composition of the forest; it depends only on the matrix of replacement probabilities. Peden et al. (1973) have shown this for an elaborate computer simulation of forest growth, but it is true for a much wider class of models (section 11.4.2; Horn, 1975b). Hence there is no uniquely biological significance to the convergence itself. The biologically interesting questions involve the structure of the matrix of replacement probabilities. Three idealized structures are shown in Fig. 11.1, along with a schematic representation of the quasi-real case of Table 11.1. The idealized structures, chronic and patchy disturbance, obligatory succession and competitive hierarchy, differ markedly in their patterns and rates of convergence to the steady state. An outline that is strikingly similar to mine has been independently proposed by Connell and Slatyer (1977) in the context of an extensive review of the machinery of succession. They also discuss preemption by the first species to invade, which prevents further succession from taking place.

11.3.1   *Chronic, patchy disturbance*

Where disturbances that result in the deaths of individual organisms and a few neighbours are frequent, localized, and uniformly distributed in time and in space, interactions between species take the form of a race for unchallenged dominance in recent openings, rather than direct competitive interference. An individual of any species may succumb to a given disturbance and its place may be taken by an individual of any other species. Therefore, if an episodic devastation changes the relative abundances of any two species, they are capable of adjusting their abundances to accord with the stationary distribution in as little time as a single generation. Where a community is dominated by chronic

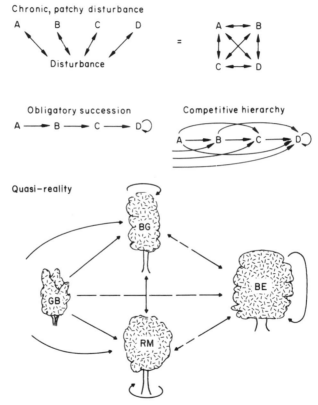

Fig. 11.1. A, B, C and D represent hypothetical species. Where disturbance is chronic and patchy, any species is likely to invade an opening that results from the death of any other species; all conceivable replacements are possible. An obligatory succession results if later species require preparation of their environment by earlier species. In an ideal competitive hierarchy each later species can outcompete earlier species, but can also invade in their absence. The quasi-realistic case illustrates the replacements of Table 11.1, with obvious abbreviations for Gray Birch, Blackgum, Red Maple and Beech.

and patchy physical disturbance, succession should be rapid, convergent, and relatively independent of the historical peculiarities of the devastation that started it.

I suspect that examples of this kind of succession will be found among flood plain forests of the temperate zone, but the most thoroughly studied example that I know is among invertebrates of the rocky intertidal, where some communities are shifting successional mosaics of patches that were battered at various times by gigantic driftwood (Dayton, 1971). Furthermore, Dayton (1975) has found that succession among intertidal algae reaches a stationary state more quickly in areas exposed to waves than in sheltered areas or deep tidal pools.

Patchy disturbance followed by invasion of any of several species makes an important contribution to the celebrated local diversity of tropical and sub-tropical forests (Knight, 1975; Connell, 1978). Some of this spatial diversity is due to patchy soil (Austin *et al.*, 1972), and some perhaps to deterministic micro-successions that have started at different times, but much can only be attributed to probabilistic accidents of just which species colonizes a recent opening at the appropriate time (Webb *et al.*, 1972). Other mechanisms are involved in the regeneration of these forests. Knight has shown that many of the species that are characteristic of older forests on Barro Colorado Island in Panama are also invaders of recently vacated fields (a 'competitive hierarchy' of section 11.3.3 or the 'initial floristic composition' of Egler, 1954). Webb and friends have found several species whose regeneration is only effective in the immediate vicinity of mature trees of the same species (section 11.4.1), and Hubbell (1979) has documented many examples of clumped spatial distribution among trees of a tropical deciduous forest. The relative importance of these mechanisms can only be discovered by future studies of the explicit dynamics of tropical forests.

11.3.2 *Obligatory succession*

The obligatory succession is characterized by the necessity of early successional species to pave the way for later species, a prime ingredient in Clements' (1916) theory of the cause of succession. This has been called 'relay floristics' by Egler (1954) and 'facilitation' by Connell and Slatyer (1977). Later species must await the dominance and demise of early species, and the convergence on a stationary state that includes the later species is exceedingly slow, even though it is

certain. The classic example of this type of succession is the convergence of communities on sandy or boggy soil. Accumulation of organic
matter from appropriately adapted plants increases the water reten-
tion of the sandy soil or increases the depth and drainage of peat,
in either case modifying the soil in favour of plants with intermediate
tolerance of aridity and dampness. Despite the logic of this textbook
example, a careful look at the communities that are cited to support
it shows that differences between communities are due largely to secular
changes in the level of the water table (Olson, 1958; Heinselman, 1963;
H. Raup, 1975). A major role of successional plants remains to be demonstrated critically, beyond a change in composition of only two species.
A very clear and well-documented example is the mineral soil that is
left by retreating glaciers, which gains fertility from nitrifying bacteria
in root nodules of early successional alders (Crocker and Major, 1955).
Other early successional plants have this effect and in addition insulate
the soil, allowing permafrost to form, which in turn prevents drainage
and allows the development of tundra tussocks of sphagnum and
sedge (Viereck, 1966).

11.3.3 Competitive hierarchy

In the case of the competitive hierarchy, late successional plants
are increasingly capable of dominating eventually in crowded competition with early successional species, but the late successional
species are also capable of invading the earliest stages of succession.
This succession may or may not reflect the historical accidents of a
particular devastation, depending on how early in the succession the
later species invade. The succession can begin to converge on dominance
by the later species as soon as they do invade. Indeed, careful studies
of many north temperate forest successions have shown that late
successional species are present almost as soon as early successional
species invade a vacated field (Drury and Nisbet, 1973; Oliver, 1975;
Heinselman, 1973). The early successional species grow quickly to
dominate early, but they cannot survive crowding. Of course crowding
also suppresses the later species, but they are at least able to endure
to dominate later. Thus this succession proceeds not by the invasion
of later species, but by a thinning out of the 'initial floristic composition' (Egler, 1954).

I have examined the mechanism behind a competitive hierarchy
relating growth rate of trees to light intensity (Horn, 1971). Most leaves

are able to photosynthesize at 90 per cent of their maximal rate with as little as 25 per cent of full sunlight. Therefore a tree with leaves scattered throughout its volume can have a very large total leaf area, with interior leaves that pay for themselves as long as the tree is in the open. When such a multilayered tree is in shade, however, the interior leaves may respire more than they photosynthesize. In shade, the optimal tree has a monolayer of leaves in a shell about its periphery, and all leaves intercept light at its highest intensity. A monolayered tree can grow in the sunlight, but not as quickly as a multilayered tree. If a monolayered and a multilayered tree simultaneously colonize an abandoned field or an opening in the forest, the multilayered tree will grow faster to dominate until it is crowded by its neighbours, when the monolayered tree gains a competitive advantage. If a monolayered tree reaches the opening first, it grows slowly, but its shade prevents invasion by multilayered trees. If a multilayered tree reaches the opening first, it grows quickly, but its understory is open to invasion by a monolayered tree. Although the monolayered tree is suppressed in the understory, it eventually dominates. As a result, the multilayered tree, like Diamond's 'supertramp' birds (chapter 10), must send its seeds flying to attain unchallenged dominance in newly vacant areas.

11.3.4 *Quasi-reality*

The tree-by-tree replacements inferred for four species in a New Jersey forest (Quasi-reality of Fig. 11.1) have the overall pattern of a competitive hierarchy. Patchy disturbance plays a role as well, as when a Red Maple or a Blackgum replaces a Beech, or a Blackgum replaces a Red Maple. There is also a hint of obligatory succession, since Beech has not been commonly observed invading open fields in this locale. An additional feature is the tendency of individuals of three species to be replaced by new individuals of their own species. Such self-replacement can be added to the successional species in the idealized obligatory succession and in the competitive hierarchy, where it further slows convergence since early species give way less readily to later species. Even among late successional species, copious self-replacement slows convergence on the stationary state. The recovery of the stationary state from even small disturbances in the relative abundances of late successional species will be sluggish if those species tend to perpetuate themselves locally, rather than sorting out statistically with each other.

There is the possibility of local cycles of replacement among Blackgum, Red Maple and Beech. More dramatic explicit cycles of replacement involving Beech have been documented by Forcier (1975), Fox (1977) and Woods (1979).

11.3.5   *When is succession rapidly convergent?*

There are three paradoxes in the analysis of Fig. 11.1 that help to explain some of the difficulties of previous theoretical approaches to succession. Natural selection favours copious self-replacement whenever it is possible. Therefore evolution tends to slow successional convergence and to destroy the environmental context of selection for adaptations to a particular successional status for a particular species. Drury and Nisbet (1973) go so far as to argue that there are no adaptations to succession itself, but that succession occurs because species that are adapted to geographically local niceties of soil and climate attain short-lived dominance in a temporal gradient of physical conditions that is sharpened by interspecific competition. This argument has been developed further by Pickett (1976).

Not only is convergence slowed by self-replacement, it may even be destroyed if the amount of self-replacement depends heavily on a species' local abundance. This effect will be defended in section 11.4.1. It has the consequence that succession is certainly slow and dependent on historical accidents, perhaps not even convergent, where the supposed biological cause of succession is most apparent, namely interspecific facilitation in the obligatory succession or interspecific competition in the competitive hierarchy.

On the other hand, succession is rapid and predictably convergent in communities that are dominated by frequent, patchy, and essentially random disturbances. This underscores the fact that successional convergence, where it occurs, is a statistical phenomenon, rather than a uniquely biological one. Only in the competitive hierarchy is succession rapid and convergent enough to be observed and blatant enough to be biologically interpreted. The very reason that this is so is that a prime ingredient of Clements' (1916) successional recipe is missing; early successional species are not needed to pave the way for later species.

Although these results have been presented in the context of tree-by-tree replacements, they are applicable at any other level from size classes within a species to regional blocks of vegetation. Indeed they are applicable to any successional patterns that are generated by linear replacements of one state by another with characteristic probability, whether this probability is measurable in practice or not. Furthermore, the contrasts of Fig. 11.1 can be largely inferred from the topology of arrows connecting the species and the number of heads on each arrow. The potential convergence of a succession can be predicted from primitive and qualitative observations of which species are capable of locally replacing which other species.

11.4 Biologically interesting non-linearities

11.4.1 *Recruitment dependent on local density*

The rain of seeds or of other propagules at any stage of succession depends on what species are within the normal distance of dispersal. If in addition the numerical success of propagules is increased in proportion to the local abundance of conspecific plants, succession need not converge at all, even in the absence of disturbance. I have presented an analytical model of this situation (Horn, 1975b), and Acevedo (1978) has explored it with computer simulations, but its major consequence can be argued intuitively. Any species whose abundance has increased as a result of a devastation becomes more efficient at replacing both itself and other species, especially another species whose abundance has decreased below what it was prior to the devastation. Therefore any departures from a stationary distribution tend to be not only self-perpetuating, but even self-augmenting. The composition of the community depends on the historical vagaries of which species were made temporarily more abundant during the last episode of major disturbance.

Since most temperate trees set far more seed than ever reach even the sapling stage (Harper and White, 1974), recruitment of a particular species depends on its presence, but may not be strongly influenced by its local abundance. Nevertheless if a sufficiently large area is clearcut, recruitment to the opening will be less representative of the previous composition of the forest, than will be recruitment to the smaller openings of a patch-cut forest. (Foresters call this 'group selection', but the term has been usurped by other ecologists in an entirely different context.) The extensive removal of all merchantable trees of one or more species, so called high-grading, may also alter the pattern of recruitment. Indeed foresters recognize that patch-cutting should produce a faster and more dependable regeneration of the previous forest (Twight and Minckler, 1972), but the process may be immediately uneconomical, or it may invite further devastation by deer or by wind.

If recruitment of young plants is generally proportional to the local abundance of conspecific adults, the consequences for successional theory are profound. Succession would not be necessarily convergent. Alternative stable communities, dependent on accidents of history, could persist side by side in an otherwise uniform environment. The dynamic study of recruitment in a diverse forest is a very promising area for future research. The studies reviewed by Harper and White (1974) and the ideas reviewed by Whittaker and Levin (1977) are still insufficient to allow a guess of whether forest succession should or should not generally be convergent.

11.4.2 *Direct and persistent effects of devastation*

Devastation itself can be represented as a plant-by-plant replacement matrix in the same way that patchy disturbance is shown in Fig. 11.1. Then succession over a long time interval can be represented by multiplying the initial composition of the forest by a matrix like *SSASSSBSSSSASCSS*, the product of many successional matrices (*S*) interspersed with various devastational matrices (*A*, *B*, and *C*). Such a succession might be dominated by the pattern of devastation or it might be dominated by a community whose composition is nearly stationary. The balance between these extremes is set by three properties of the matrices and their pattern of occurrence: the frequency of devastation relative to the speed of convergence in the interludes, the number of distinctly different communities resulting from devastation (*A*, *B*, *C*), and whether the effects of a devastation are persistent (represented, e.g., by *SSABBBBSS*). The first two of these have already been analysed intuitively in section 11.3; and the next paragraph is an optional technical note that distills the essence, or at least the fragrance, of that analysis.

The initial composition of a forest can be represented as a linear combination of the eigenvectors of S (also called latent or characteristic vectors of S). Then succession, which multiplies this composition (c_o) by successively higher powers of S, is represented as the same linear combination with eigenvectors (v_i) multiplied by their eigenvalues (λ_i, the latent or characteristic roots of S), each raised to a power (t)

that measures the passage of time since the last devastation. Thus:

$$c_t = c_o S^t = (\textstyle\sum_i a_i\, v_i) S^t = \textstyle\sum_i a_i\, v_i\, \lambda_i{}^t \qquad (11.2)$$

where c_t is the composition after t of the periods of time over which S is measured, the a_i are appropriate constants, and the i range over the number of rows or columns in S. As time passes, the composition of the forest will be increasingly dominated by the eigenvector corresponding to the largest eigenvalue, which is unity for an exhaustive probability matrix like S, because $\lambda^t = 1$ forever if $\lambda = 1$. The contribution of other eigenvectors to community composition depends on the magnitudes of their eigenvalues, decaying rapidly for those with small eigenvalues, and decaying slowly for those with large eigenvalues, because λ^t tends toward zero as t gets very large for $|\lambda| < 1$. The size of the second largest eigenvalue is therefore an inverse measure of the speed with which succession converges on the stationary distribution (Usher, 1979). Since the eigenvectors of most matrices like S are orthogonal, the number of eigenvalues that are nearly as large as the largest is a measure of the number of 'degrees of freedom' with which Nature may perturb a community so that subdominant eigenvectors put up a lengthy fight against the dominant one. Hence the number of markedly different and possibly persistent devastations (the A, B, C above) is measured by the dispersion of eigenvalues of S. I have not been able to develop a uniformly interpretable and simple measure of this dispersion. However, as R. Levins has pointed out to me, a whole class of measures might be based on the sum of squares of the eigenvalues, which is the trace of S^2 and thus the sum of products of the elements of S with their transposes. The sum receives contributions only from the self-replacements and two-headed arrows of Fig. 11.1, which, happily, have the biological interpretations already discussed in section 11.3. The only seemingly counter-intuitive notion of this analysis is the demonstration that the number of significantly different kinds of devastation is a property of the successional matrix, rather than a property of either the cause or the effect of devastations.

   This analysis shows that important and interesting dynamic properties of succession are inherent in a diagram with arrows between species that replace one another. Dale (1977) and Carleton and Maycock (1978) give examples of respectively simple and complex techniques for generating such a diagram.

Of course if the direct effects of devastation are themselves persistent or self-reinforcing, then understanding the structure of a community requires a knowledge of its ancient history. A clear and yet subtle example is Blydenstein's (1967) interpretation of some patchy savannah-like vegetation on the Llanos of Columbia. Apparently fires have swept the area in the distant past, leaving small patches unscathed. The soil drained readily in severely burned areas, which now support only grassy vegetation which dries and burns frequently, intensifying the drainage of the soil, and so on. The less severely burned areas retained their moisture, and they have developed dense patches of forest, which retain enough moisture to remain relatively resistant to the fires that frequently sweep the plains around them.

11.4.3 *Adaptation to regular devastation*

If devastation occurs on a sufficiently regular schedule, this schedule gives a competitive advantage to early successional species whose lifespans match the intervals between catastrophes. Examples span a wide spectrum of intervals. Eastern White Pine is adapted to fill the interval of about 100–300 years between fires (Loucks, 1970; Heinselman, 1973) or severe hurricanes (Henry and Swan, 1974). Chaparral in California (Hanes, 1971) and Pitch Pine in New Jersey (McCormick and Buell, 1968) thrive with a more frequent burning on a cycle of about 10–50 years. Annual weeds are adapted to tillage (Harper and White, 1974), and flood plain herbs to spring flooding. Giant Sequoia has the characteristics of an early successional species, despite the great age of remaining stands (Biswell *et al.*, 1966; Rundel, 1971). Several of the famous groves are on the moraines and outwash of mountain glaciers of the past (Rundel, 1972; Matthes, 1965). I enjoy the fantasy of a venerable Sequoia biding its time in anticipation of the next local ice age. At the opposite end of the spectrum, the colon bacterium is marvellously adapted to devastations of 24 hour regularity.

The occurrence of physical disturbance need not be regular to cause a highly regular pattern of devastation. This is easiest to demonstrate for fires, but comparable models could be developed for other disturbances. Figure 11.2 shows a hypothetical model in which the intrinsic inflammability of vegetation increases as plant material accumulates after a fire. When the inflammability reaches a threshold, which is high in wet periods and wet climates and low during drought,

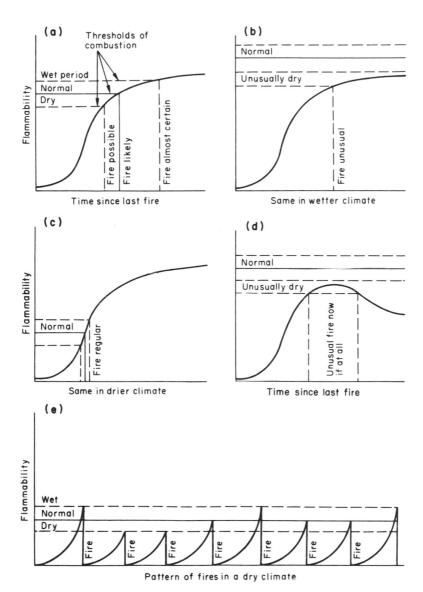

Fig. 11.2. Graphical model for predicting the temporal pattern of natural fires; see text for a discussion.

common accidents of ignition result in fires. In a very wet climate (Fig. 11.2b), only an unusual drought results in a fire, and the temporal distribution of fires is irregular. In a very dry climate (Fig. 11.2c), the timing of a fire may be little changed by wide variations in the timing of weather, and a very regular pattern of fires may result (Fig. 11.2e). The details and predictions of Fig. 11.2 correspond closely to interpretations of cyclic occurrence of fires by Loucks (1970), Rowe and Scotter (1973), and Heinselman (1973), though Heinselman also found that geographically distant fires were synchronized by extensive droughts. Johnson (1979) has explicitly analysed the intervals between fires in boreal woodland, finding that their frequency depends on the structure of the vegetation and their regularity on the roughness of the terrain.

The temporal pattern of fires depends as much upon the intrinsic shape of the curve of inflammability as it does on the temporal pattern of droughts that favour fires. In particular, the regularity of fires is determined by the rate of increase in inflammability near the average threshold for conflagrations in a given environment. If the fires occur on a sufficiently regular schedule, early successional species whose lifespans are matched to the interval between fires may increase their local persistence if they became yet more inflammable late in life. This in turn would steepen the curve of inflammability, and the period between fires would become still more uniform. Mutch (1970) has measured the combustibility of vegetation of several species whose persistence in nature depends upon fires, and found that these species are more combustible than are related species that inhabit communities that are rarely burned. Biswell (1974) notes that Chamise chaparral in California remains vigorous and stable only if it is burned every 15–20 years, and he describes a mechanism whereby 15 year old Chamise increases its own inflammability. The terminal twigs die and then proliferate by lateral sprouting, and the sprouts repeat this behaviour until the crown of the bush is a well laid bed of tinder.

If accumulated vegetation has a sufficient ability to hold water, inflammability may eventually decrease with an increase in the standing vegetation. Such a situation is shown in Fig. 11.2d. In this case there is a limited period during which the community risks fire. After this period has passed, it is exceedingly unlikely that fire will sweep the community. Historical accidents will then determine which portions of the community appear to be maintained by fire and which appear to escape entirely. Such a mechanism is implied in Blydenstein's (1967) description of the patches of forest in the Llanos of Colombia, and in the early development of some pine forests (Weaver, 1974). Connell (1975) has interpreted alternative and apparently stable associations of intertidal beasties in the context of a model like Fig. 11.2d, with one species of barnacle either being eaten by snails and replaced by another, or else escaping predation by growth, and dominating the stationary community. Here is another class of models in which regular and biologically interpretable patterns of succession may be driven by historical and more or less random events.

Although this model has been developed in the context of fire as the catastrophic disturbance, the same model could be used to discuss patterns of devastation by wind, flood, landslide, predation, disease, or even economic management. The degree of regularity of devastation in each case depends on the intrinsic increase in sensitivity of the vegetation as it grows back following devastation, measured at the threshold of sensitivity at which devastation becomes likely. Such measurements remain to be made (Wright and Heinselman, 1973), even in the case of fire, where appropriate general models would have immediate practical applications in the management of parks and preserves whose major vegetation is naturally dependent on fire (Wright, 1974). Although the techniques for prescribed burning are highly enough developed to allow generalizations about applications to a wide range of environments (Kayll, 1974), there are no general principles that help to predict how a given community will respond to a particular pattern of burning.

11.5   Conclusion and prospect

The only sweeping generalization that can safely be made about succession is that it shows a bewildering variety of patterns. Succession may be rapidly convergent on a stationary vegetational composition, or it may be slow and apparently dependent on accidents of history. Succession may lead to alternative stationary states, depending on the initial composition. The vegetation may continually change in response to more or less random changes in the physical environment, or it may change cyclically by responding to and perhaps reinforcing an environmental cycle.

Analytical models dispel some of the bewilderment by showing that the general pattern of a succession is largely determined by biologically interpretable properties of the individual species that take part in the succession. Where successional convergence is found, it is a statistical phenomenon, rather than a biological one. Whether convergence is rapid or slow depends upon how effectively each species locally regenerates itself and replaces other species. Succession is rapid and convergent where each species can more or less randomly replace any other species directly. Succession is slow and dependent on history if early successional species must modify the environment before later species can invade. Succession will only be directly observable, convergent, and biologically interpretable in those rare cases where the most important species form a competitive hierarchy. Deaths among the competitive dominants may still leave room for the persistence of early successional species, and the 'undisturbed' forest may actually be a mosaic of successional patches.

Whether succession is convergent or not depends critically on how strongly the amount of recruitment of young plants is determined by the local density of mature plants of the same species. This relationship should be explored for a complex community. Until it is known, one cannot predict whether the community will remain stable under human exploitation or will shift radically to a new stationary state in response to a small perturbation.

Some communities are adapted to a more or less regular pattern of natural devastation. More work is needed at both theoretical and empirical levels before generalizations can be made about human management of such communities. In particular the dynamics of a community's developing sensitivity to natural devastation deserve as much attention as the temporal pattern of the physical causes of devastation.

The goals of forest management have recently been extended, from the traditional farming of an economically valuable harvest of wood, to constructing and maintaining forests with a wide variety of specified amenities, to the repair of de facto devastation by exploitation, and even as Raup (1964) argues persuasively, to managing for flexibility of unanticipated future use. A sound conceptual approach to these problems is a prerequisite to developing experiments in management. It is equally important to discover the conditions under which the preservation or reconstruction of a natural area requires the absence of management.

So far the linear models of forest succession have made gratifying predictions of what was already known. From the general consequences of linear models, I can even predict the range of surprises that remain to be discovered in them. Sections 11.3 and 11.4.2 show that the broad properties of succession depend on the topology of arrows that show the competitive interactions within and between species. Therefore the effect of any conceivable management depends on whether one helps or hinders these competitive interactions. The only simple treatments that will have dramatic effects are just those practices of seeding, weeding and patch-cutting that enlightened foresters have recommended for centuries. If there are real surprises, they must be hiding in biologically interesting non-linearities like the effect of local density on recruitment or the dynamics of sensitivity to disturbance.

12

Sociobiology

HENRY S. HORN

12.1 Introduction

Sociobiology is the study of the biological basis of patterns of social behaviour in animals. Much of sociobiology is reviewed in a companion volume *Behavioural Ecology* (Krebs and Davies, 1978), and the subject is treated in depth in E. O. Wilson's (1975) *Sociobiology: The New Synthesis*. Representative selections of recent papers have been drawn together in volumes edited by Clutton-Brock and Harvey (1978) and by Hunt (1980). A state-of-the-art appraisal has emerged from a Dahlem Workshop in Berlin (Markl, 1980). A good review of subjects related to sexual behaviour has been written by Daly and M. Wilson (1978). In the preface to a republication of Darwin's (1871) *The Descent of Man, and Selection in Relation to Sex*, Bonner and May (1971) review the many ideas of sociobiology that he presaged.

This chapter is an advertisement of an idiosyncratic selection of topics that use techniques and insights drawn from theoretical population biology, with deliberate emphasis on questions and difficulties rather than on proven results.

One of the most intriguing problems of sociobiology is the evolution of altruistic behaviour. Darwinian selection favours strength, fecundity and selfishness, so that selflessness in behalf of the welfare of others seems paradoxical. Yet apparent altruism is a common feature of many social systems. Several mechanisms have been proposed to account for the evolution of altruism. Each presents its own problems and paradoxes, in addition to the problems of disentangling the roles of the various mechanisms in any particular instance.

The classic case of altruism, recognized as a 'minor difficulty' by Darwin (1859), is that of worker ants and bees, who forgo their own reproduction entirely in order to help their mother to store pollen and honey and to raise sisters and brothers. Hamilton (1964) seemed to resolve this difficulty by emphasizing that, because of the peculiar mechanism of sex determination in bees, sister bees are more closely related genetically to each other than they are to their daughters, so that helping to raise sisters pays greater genetic dividends than raising an equal number of daughters. Hamilton's argument seems to account for much of the organization of the society of bees, but there are problems in applying it to the evolutionary origin of sociality, and plausible alternative suggestions have been proposed.

A second major paradox in sociobiology is the pervasiveness of sex and sexual behaviour, given that sex dilutes the genetic contribution of an individual to its offspring by a factor of one half in each generation. Despite recent and penetrating theoretical analyses, chiefly by Williams (1975) and Maynard Smith (1978a), the adaptive significance of sex is not definitively known.

The adaptive features of life histories are also the subject of sociobiology. Much of this has been discussed by May (Chapter 2) and Southwood (Chapter 3) in this book and by Horn (1978) in the companion volume. Many of the details of reproductive social behaviour can be interpreted as adaptations to patterns of inevitable juvenile and adult mortality and as coadaptations with patterns of dispersal.

A wide range of other topics in sociobiology has no direct connection with population biology, but is beginning to profit from quantitative approaches similar to those of population biology, particularly game theory and optimal decision-making. These topics again are covered in detail in the companion volume (Krebs and Davies, 1978).

Finally, there has been a recent furore over the application of the ideas of sociobiology to human behaviour. Papers advocating various extreme positions have been gathered together and reprinted by Caplan (1978). The debate is polarized about the issue of whether human behaviour is largely learned, with very little genetic component, or strongly affected by genetic predispositions and even genetic determinants. The polarization and acrimony tend to draw attention away from the important and interesting interactions between genetic and cultural evolution. A strong propensity for learning, teaching, and culture is the part of the inheritance of all humans that sets us apart from the rest of the animal world, and therefore its selective value must have been critical in the evolution of humans (Bonner, 1980). Conversely, the genetic evolution of humans has been entirely in a social and cultural context; so that models of genetic change in humans must include cultural parameters (Cavalli-Sforza and Feldman, 1981; Lumsden and Wilson, 1980).

12.2   Evolution of altruistic behaviour

Four kinds of mechanism have been suggested as a basis for the evolution of altruistic behaviour. The first is an extension of the idea of 'mother love' to cover not only the sacrifices that parents make on behalf of their own offspring, who are in fact their Darwinian fitness, but also the lesser sacrifices made on behalf of more distant relatives. Modern understanding of this mechanism, which has been called kin selection, is largely due to Hamilton (1964), though suggestions of the mechanism can be found in Fisher (1930) and Haldane (1932) if one looks hard enough. The second mechanism, called group selection, is run by the suggestion that groups composed of altruists should live with greater economy than groups of selfish individuals, and should therefore be more resistant to extinction or should competitively replace selfish groups (Wright, 1945). Wynne-Edwards (1962) applied this mechanism to explain the constancy of many animal populations as a result of socially imposed self-regulation. The third mechanism, reciprocal altruism, formalized by Trivers (1971), can only operate in social groups of relatively fixed composition because it depends on social mechanisms to enforce return payments for altruistic behaviour. In cases of both kin selection and reciprocal altruism, an apparently altruistic act is interpreted as selfish when viewed in a broader context (Dawkins, 1976).

It is worth noting explicitly that each of these evolutionary mechanisms requires that altruistic behaviour be dispensed in a discriminatory manner. Kin selection requires the benefits of selfless behaviour to be conferred preferentially to relatives. Group selection requires isolated groups and therefore altruistic behaviour is directed *de facto* at home group members. Reciprocal altruism is directed toward those individuals who are likely to reciprocate. Under all three mechanisms, social relations with respectively non-kin, members of other groups, and non-reciprocators should be selfish and competitive.

Kin selection and group selection are often invoked to interpret altruism and thus to counter Darwinian selection among individuals. This does not mean that either mechanism precludes individual selection, nor that either is a logical alternative to individual selection. Indeed, both mechanisms can reinforce individual selection for traits that do not involve a decrease in individual fitness (Dawkins, 1979).

The fourth mechanism, suggested by Alexander (1974), is parental enforcement of altruistic behaviour on the part of offspring. If altruism among offspring results in greater effective reproductive output by parents, natural selection will favour parents who enforce such altruism. This mechanism may play a particularly important role in the social behaviour of highly social insects (section 12.3).

12.2.1 *Kin selection*

The term 'kin selection' is not found in Hamilton's initial papers on the subject (Hamilton, 1963, 1964). Instead he defined the 'inclusive fitness' of an individual to include not only Darwinian fitness, namely the number of its own offspring who survive to pass its genes to the next generation, but also collateral relatives who could pass on genes that were identical by descent with those of the focal individual. Relatives are of course suitably discounted as they become more distantly related. Hamilton then argued that natural selection would favour any genetically influenced behaviour that increases inclusive fitness, even if that behaviour decreases an individual's likelihood of survival or decreases the number of its own offspring. As a corollary, altruistic behaviour would be favoured whenever the direct loss in reproductive value (as defined by Fisher, 1930; see Horn, 1978) to the behaver is less than the gain in reproductive value to the recipients of the behaviour, discounted as the recipients are more distantly related to the behaver (West Eberhard, 1975). The more distant a relative and the more costly the behaviour, the less likely it is that this criterion can be met. Thus self-sacrificial acts are expected only on behalf of several close relatives, acts of low risk are expected in behalf of close and distant relatives, and selfish behaviour is expected toward non-relatives.

Hamilton's (1963) quantitative criterion for kin selection of altruism is easy to define and to interpret. Suppose that the performance of an altruistic act decreases the fitness of the altruist by a factor $(1-c)$ relative to a non-altruist, but increases the fitness of the beneficiary by $(1+b)$. The altruistic behaviour will be favoured by natural selection provided that the cost in fitness (c) to the performer is less than the fitness benefit (b) to the recipient, discounted by the relatedness (r) of the recipient to the altruist. An appropriate measure of relatedness is the probability that a gene in the altruist is identical by descent with its homologue in the recipient. The cost of the altruistic act to the altruist is then c, and the benefit, in currency of the altruist's genetic fitness, is rb. Selection favours the altruistic act if the benefit outweighs the cost; that is, if $rb > c$. Cavalli-Sforza and Feldman (1978) have examined this simple criterion in the context of genetic models for evolution at a single locus, and found that it is technically correct for only a narrow range of conditions; they discuss the conditions under which it may be seriously wrong. However, May and Robertson (1980) discuss a wide range of conditions under which the criterion, though technically incorrect, is a good approximation.

Although the criterion for kin selection of altruistic behaviour seems simple, it is exceedingly difficult to measure in practice. The costs and benefits of behavioural acts should be measured in terms of marginal decreases and increases in reproductive value, and reproductive value is itself such a challenge to measure that its change in response to the presence or absence of a given behaviour seems unobtainable. Furthermore, a detailed quantitative test of the criterion of kin selection requires that all animals involved in social behaviour have known pedigrees; and humans have trouble enough working out their own pedigrees! Most testable predictions of kin selection are therefore qualitative, though, for reasons discussed in section 12.2.4, definitive tests of kin selection must be quantitative. Altruistic behaviour should be preferentially dispensed to close relatives, or to close neighbours if the population's structure is viscous enough for neighbours to be likely to be close relatives. Very distant relatives, non-relatives, and strangers should be ignored or exploited in a selfish fashion.

The best evidence for the operation of kin selection comes from comparative studies of social structure in insects with different mating systems (Wilson, 1971; Hamilton, 1972; also section 12.3). More indirect evidence comes from the fact that alarm calls in response to predators are given preferentially when among relatives by ground squirrels (Dunford, 1977; Sherman, 1977) and prairie dogs (Hoogland, 1980). Patterns of infanticide in polygynous and socially stable families of mammals show discriminatory dispensation of selfish behaviour as predicted by kin selection. For example, in lions (Bertram, 1975) and langur monkeys (Hrdy, 1977), when the dominant male is replaced by a new male from outside of the group, the new male may attempt to kill youngsters that he did not sire.

Arguments about kin selection are deceptively simple and are often loosely phrased or misinterpreted. Sample errors are exposed and discussed by Dawkins (1979).

12.2.2   *Group selection*

Group selection is a purported mechanism for the evolution of altruism by the following argument (after Wynne-Edwards, 1962). Although selfish behaviour is favoured by selection among individuals within a group, those groups with many selfish individuals exterminate themselves, usually by overreaching their resources. The former domains of selfish groups are then recolonized by altruistic individuals from more provident and hence long-lived groups. An obvious requirement of this mechanism is that the population be divided into isolated groups between which there is little migration or dispersal, and within which group members suffer a common fate.

The restriction on migration between groups is so stringent that group selection seems an implausible mechanism for the evolution of altruism. To illustrate this, make the assumption that selfishness and altruism are simple, heritable alternative behaviours. Furthermore assume a population that is initially composed of groups with only altruistic individuals, so that altruism need only be maintained by group selection, not originated. Each time that a selfish mutant arises within a group it is favoured by individual selection over the altruists in that group, but it and its selfish spawn must drive that group to extermination before a single selfish individual is exported to another, altruistic, group. Otherwise the selfish type will persist, and if more than one escapes to another group, the selfish type will increase. A sufficiently low rate of migration may prevent the escape of selfish individuals to altruistic groups, but the rate of dispersal from altruistic groups must at least be sufficiently high to colonize every vacancy left by an exterminated selfish group. Thus for group selection to maintain altruistic behaviour in the face of individual selection for selfishness, rates of intergroup movement must be sufficiently low for selfish individuals and sufficiently high for altruists.

More elaborate and more realistic models of group selection have been reviewed by Wilson (1973). They, too, involve elaborate restrictions on population structure and movements for selection among groups to counter the results of selection within groups; it is arguable whether these restrictions are likely to be commonly met in practice. However, these models also show that group selection can be an important mechanism over a wide range of conditions when it is working in concert with individual selection.

A different approach to group selection for altruism is taken by Gilpin (1975), after May (1973, p. 102). Gilpin examines a wide range of population dynamic models of predator–prey interaction, finding that they can switch from a regime of quasistable oscillations to rapid extinction of the predator, and that an abrupt switch is caused by a small increase in the predator's efficiency of killing prey under certain plausible conditions. Gilpin infers that a group of predators with only a few selfish gluttons may extinguish itself so quickly that it may indeed be unable to export a single glutton to a more altruistic and provident group. Thus the criterion for maintenance of altruism by group selection is fulfilled. Of course Gilpin's ideas only apply to altruistic behaviour that has appropriate dynamical consequences for populations of predator and prey, and it remains to be seen how widespread the necessary special conditions are in nature. The mechanism is most plausible for predator–prey systems of pathogens and their hosts, though of course pathogens are also subject to the traditional form of group selection because all members of a group of pathogens inhabiting a diseased host may share a common fate. In either case, the evolution of such systems deserves study.

Alexander (1974) and others have suggested that human social structures provide the setting for a special kind of group selection. Humans are capable of forming groups of individuals who are not necessarily all closely related, groups which interact as groups, competing, extinguishing, and replacing each other. The importance of such group selection in early human evolution must remain entirely speculative.

12.2.3 *Reciprocity*

Trivers (1971) formalized a theory of reciprocal altruism and used it to interpret apparently altruistic behaviour in a wide variety of social groups of relatively fixed composition. His suggestion is that altruistic behaviour is differentially expressed in behalf of those individuals who are likely to reciprocate in the future, and that many other social acts have coevolved as mechanisms for detecting and discriminating against the cheaters who accept favours but do not dispense them. The theory has yet to receive a convincing critical test, though it has produced many ingenious and intriguing *ad hoc* interpretations of social behaviour. It requires small social groups of relatively stable composition so that opportunities exist for reciprocation and for detection of cheaters.

12.2.4 *Altruism and population structure*

All of the above mechanisms for the evolution of altruism either require or are favoured by a population that is fragmented and that has limited dispersal. In such a population neighbours are likely to be more closely related than distant individuals. Thus a pattern of altruism toward neighbours but not strangers can evolve by kin selection without any need for explicit identification of kin and calculation of their relatedness. Such a population may begin to meet the stringent criteria of migration and dispersal for group selection of altruism. Such a population may produce the stable association of given individuals that are prerequisites for accounted reciprocity. Population parameters that favour one of the mechanisms for the evolution of altruism favour two others as well, making it difficult to disentangle them and to evaluate them by any broad comparison of population structure with social behaviour.

Patterns of dispersal are critical to resolving this dilemma. D. S. Wilson (1980) points out that many populations are either fragmented or spatially structured by territoriality or homing behaviour, with the result that most individuals interact with only a small number of neighbours over the greater part of their lives. Mates may or may not be chosen from a wider array of individuals, but very often the individuals of a given neighbourhood suffer a common fate that depends upon the genetic composition of the neighbourhood. This situation is formally equivalent to group selection, with the groups being the local neighbourhoods of interacting individuals. As Wilson further argues, these neighbourhood groups may be much smaller than genetically closed populations that have traditionally been considered to be the units of group selection. Hence they have higher rates of both extinction and recolonization, both of which are rapid, and neighbourhood groups might often meet the stringent criteria for group selection of altruism. The distribution of genetic types among such neighbourhoods is important. If each contains an exactly representative sample of the population as a whole, then group selection and individual selection are indistinguishable. If each group contains only one genetic type, and if this uniformity is a result of common ancestry, then group selection is indistinguishable from kin selection. Even a random distribution of genetic types among neighbourhood groups will produce some groups that are statistically representative of the whole population, some that are internally uniform, and many that are intermediate. Thus there will always be some mixture of group selection, individual selection, and kin selection in any population with a moderately stable spatial structure, and the balance of this mixture will depend on arcane details of the pattern of dispersal between neighbourhoods.

12.3   Haplodiploidy and the eusociality of ants, bees and wasps

One of the early and impressive successes of theory in sociobiology is Hamilton's (1964) explanation of the genetic advantage gained by worker bees who forgo their own reproduction to help raise sisters, in a colony described as eusocial (= 'truly social') to distinguish it from more casual social groups. Ants, bees, and wasps are peculiar, though not unique, among insects in that the sex of an individual is determined

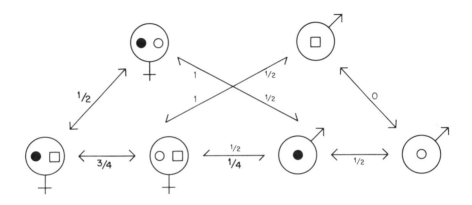

Fig. 12.1. Relations among family members in a single, outbred mating between haplodiploid insects.

Parents are shown above, the four possible types of offspring below. Homologous genes from the diploid pair of the mother are represented respectively by light and shaded circles; the homolog from the father is shown by a light square. Offspring from fertilized eggs are females; from unfertilized eggs, male. Double-headed arrows show symmetrical relationships, half-headed arrows show asymmetrical relationships. Each fraction is the probability that the relative pointed to shares a gene identical by descent with the individual whence the arrow originates. These probabilities are calculated as follows for the examples marked in boldface; the genes referred to are homologous alleles identical by descent at a given locus. Mother to daughter: half of mother's genes are carried by half her daughters; the other half of mother's genes are carried by the other half of her daughters; $(\frac{1}{2} \times \frac{1}{2}) + (\frac{1}{2} \times \frac{1}{2}) = \frac{1}{2}$. Daughter to mother: half of daughter's genes are carried by her mother; the other half are not; $(\frac{1}{2} \times 1) + (\frac{1}{2} \times 0) = \frac{1}{2}$. Sister to full sister: the maternal half of a female's genes are carried by half her sisters; the paternal half are carried by all her sisters; $(\frac{1}{2} \times \frac{1}{2}) + (\frac{1}{2} \times 1) = \frac{3}{4}$. Sister to brother: the maternal half of a female's genes are carried by half her brothers; the paternal half are not carried by any brothers; $(\frac{1}{2} \times \frac{1}{2}) + (\frac{1}{2} \times 0) = \frac{1}{4}$.

by whether it develops from an egg that has been fertilized, in which case it is a female, or from an egg that has not, in which case it is a male. This genetic system of haplodiploidy produces a number of asymmetries in relatedness among family members, which are illustrated in the diagram (Fig. 12.1). Hamilton argues that the female workers are more closely related to their sisters than to their own offspring, and that therefore staying at home to help raise reproductive sisters is genetically more productive than setting up a new home and raising an equal number of reproductive daughters. Trivers and Hare (1976) have shown that many other aspects of colony organization in ants, bees, and wasps are interpretable as optimal responses to these asymmetrical relationships. For example, in outbred monogamous species, reproductive sisters are worth three times as much as reproductive brothers from a genetic point of view, and indeed workers expend about thrice as much effort on behalf of sister queens as on behalf of brothers. Trivers and Hare's data and interpretations have been severely criticized by Alexander and Sherman (1977), but the latter's more careful comparative analysis shows an even more accurate pattern of differential investment related to differential degrees of relatedness. Alexander and Sherman show that conflicting patterns of optimal investment for mothers versus daughters are more often resolved in favour of mothers, and they infer that parental manipulation (Alexander, 1974) of the behaviour of offspring plays an important role in the pattern of social behaviour in ants, bees, and wasps.

Simplified accounts of the role of halpodiploidy in the origin of eusocial behaviour often make the misleading statement that it pays a worker bee more to help raise siblings than to raise her own offspring. The statement is true for outbred monogamous families only if the sex ratio among reproductive siblings is biased in favour of females; were the sex ratio 50:50, then the average relation of a sister to her sibling would be $(1/2 \times 3/4) + (1/2 \times 1/4) = 1/2$, the same as her relation to potential offspring (West Eberhard, 1975; Charnov, 1978; Wade, 1979). Such a bias could come about by adjustment of the bias toward fertile eggs by the mother (a form of parental manipulation, Alexander, 1974), or differential care of sisters by daughters, or a combination of both. However the bias originates, optimal behaviour by both mother and daughter should reinforce it, making it genetically worthwhile for a daughter to forgo reproduction and to help in her mother's home. Haplodiploidy must play a strong role in the evolutionary origin of eusocial organization in ants, bees and wasps, but that role cannot be played except in the context of some already complicated social behaviour. Thus the role of haplodiploidy in the origin of eusociality is catalysis rather than predisposition.

Ultimately the origin of eusociality must have to do with likelihood of colony establishment by different numbers of starting individuals. The otherwise lowly paper wasp is ideal for studying colony formation. New nests may be founded by one female or by several, though instances of more than five foundresses are rare. Multiple foundresses are usually sisters or at least former nestmates (West Eberhard, 1969; Metcalf and Whitt, 1977a; Klahn, 1979), with the tendency to be nestmates being almost universal among foundresses who settle near their own natal nest (Klahn, 1979). One of several foundresses usually dominates the others and does most of the reproduction (West, 1967; Metcalf and Whitt, 1977b); the subordinates are often less vigorous than the dominant and may not produce as many offspring as does the average single foundress. However, the total production of offspring of a subordinate plus the discounted offspring of her sister usually equals or exceeds that of the average single foundress. Moreover, when populations are high and nest sites are rare, nests founded and guarded by two sisters are more resistant to usurpation by competitors than are nests of single foundresses (Gamboa, 1978). Colonies of multiple foundresses may be re-established after predation; hence where predation is frequent, colonies of multiple foundresses produce more young than those of single foundresses, both per colony and per female; where predation is unlikely, colonies of multiple foundresses still produce more offspring than singly founded colonies, but not more per foundress (Gibo, 1978). Worker wasps invest equally in brothers and sisters, suggesting that their mother's interests are favoured over their own (Noonan, 1978). Paper wasps provide clear evidence that altruistic behaviour is favoured by individual selection, kin selection among parents or among offspring, parental manipulation, and/or haplodiploidy. The balance among these mechanisms depends on the particular species of wasp, the locality, patterns of dispersal, population density, and regime of predation. Obviously a complex of selective forces, but an interpretable complex, is responsible for the maintenance of social behaviour in paper wasps. Their social behaviour is supposedly rudimentary, but at least one species stores honey for the winter (Strassman, 1979).

Darwin (1859) cited the evolution of sterility among workers of ants, bees, and wasps as a 'minor difficulty', but he was much more worried about the evolution of morphologically specialized castes among insects.

This problem has recently been analysed by Oster and Wilson (1978), who borrow control theory from the discipline of engineering to show the increased reliability of systems with multiple pathways from resources to products, the increased energetic efficiency due to multiple morphologies adapted to a diversity of tasks, and the optimal timing of a colony's switch from production of more workers to production of reproductive individuals.

Although understanding of the evolution of eusociality in ants, bees, and wasps is far from complete, it has profited greatly from studies in the analytical context of theoretical population biology. These studies have emphasized the importance of Hamilton's (1964) inspired suggestions about haplodiploid genetics, but they have also shown the critical roles played by the energetic efficiency of colonies and the ecology of their establishment, roles classically recognized by Darwin.

12.4   Sex

The adaptive significance of sex is fully discussed by Maynard Smith (1978b), in the companion volume; he gives an excellent summary of the arguments in his book (Maynard Smith, 1978a) and in that by Williams (1975). For more detail, see these and Ghiselin (1974).

12.4.1   *The paradox of sex*

In higher organisms, the process of sexual reproduction involves the production of gametes by meiosis, which halves the chromosome number. The gametes then fuse with those from another individual to form the zygotes which develop into the next generation of diploid organisms. Thus sex involves dilution of an individual's genetic contribution to the next generation by one half, when compared to production of a comparable number of cloned diploid individuals. Williams (1975) calls this the 'cost of meiosis', but Maynard Smith (1978b) suggests that it is really the 'cost of producing males' because it does not arise for simultaneous hermaphrodites or isogamous species, where both parents contribute equally to the nourishment of offspring. It is perhaps more accurate to say that there is indeed a two-fold cost of meiosis, but that it is immediately repaid by the doubled output of having both parents raise offspring in hermaphrodites and isogamous species. Similarly, the cost is immediately repaid in all species in which the male contributes substantially to the nourishment of the offspring, either directly by supplying food to them or to his mate, or indirectly by defending an exclusive territory on which his mate or offspring live.

Maynard Smith makes the argument for the cost of producing males as follows. If males contribute little to the nourishment of offspring, then a given female should produce the same number of offspring whether she does this parthenogenetically, by cloning or by self-fertilization, or sexually. All of the parthenogenetic female's offspring are productive females, like herself, but only half of the sexual female's offspring are productive females, whereas the other half are relatively unproductive males. Thus the parthenogenetic female has double the reproductive output of the sexual female. This argument is correct but facile. The cited twofold advantage of the parthenogenetic female cannot be realized in crowded populations (K-selected populations, Southwood, Chapter 3), where there is not room for the doubled number of females. Furthermore, although males may themselves be unproductive, they are able to parasitize productive females, who nourish offspring half of whose genetic complement comes from their father. Daughters of a sexual female nourish offspring half of whose genetic complement is their own. Therefore male offspring are as effective as female offspring as a means of passing genes to future generations, and the twofold cost of sex is not due to the low productivity of males compared to females, but rather to the 'cost of meiosis' for both sexes.

However the twofold cost of sex is expressed, it makes the prevalence of sex somewhat paradoxical. Long-term advantages to sex that accrue to a population or a species do not fully resolve this paradox, even if they produce an advantage that averages over several generations to twofold per generation. These long-term advantages occur as a result of a change in environment; yet between environmental changes, the cost of meiosis is paid in every generation. Thus the selective machinery behind the long-term advantages of sex shares the restrictions of the maintenance of altruistic behaviour by group selection.

Reviewers of comparative sexual systems in animals have yet to remove from the paradoxical category all those instances in which the male contributes substantially to the nourishment of offspring through indirect means, or those in which the female makes a minimal contribution that is nearly matched by the male's. However, there would certainly be enough instances left for the prevalence of sexual behaviour to remain paradoxical.

12.4.2 *Advantages of sex*

The long-term evolutionary advantages of sex are reviewed by Maynard Smith (1978a, 1978b). As yet there is no direct evidence for their importance, though computer simulations have shown their plausibility for theoretical populations with realistic dynamics. The most striking advantages are conferred when the environment changes once in every generation. Of course, such changes are really short-term changes (Chapter 3) from a genetic point of view. The advantage arises because the environment that offspring face is always different from that which their parents faced, and requires the different genetic constitution that sexual recombination creates. Spatial variation in the environment can also produce an environment that changes in every generation if the modal distance of dispersal of offspring exactly matches the scale of a regular pattern of spatial variation. Whether environmental variation is spatial or temporal, an exact match between the scale of that variation and generation time seems unlikely *a priori*. Empirical tests of these ideas have yet to be made. A likely testing ground would be a critical comparison between univoltine insects, which take a full year to develop, and closely related bivoltines, which have two broods in a season and thus face environmental conditions that alternate every other generation.

The most convincing short-term advantage of sex occurs when it is advantageous for a parent to have a variety of genotypes among its offspring. This advantage is discussed in detail by Williams (1975) and is argued most succinctly by Maynard Smith (1978b). If offspring are broadcast sparsely and widely over a variety of habitats, to each of which a different genotype is best adapted, and if populations are so sparse that competition is rare, then the optimal family consists of individuals of a single genotype, namely that which is best adapted to the commonest habitat. If other genotypes are added to the family they will land more often on the commonest habitat, to which they are less well adapted, than they will on the less common habitats appropriate to their genotype. Moreover, faced with little competitive challenge, the particular genotype that is best adapted to the commonest habitat, although able to flourish in only a limited number of habitats, might be able to survive in many. Asexual propagation is favoured, just as it is when the whole environment is completely uniform. However, if the dispersal of offspring is local and to a wide variety of habitats and if competition is moderate or severe, then genetic variety in a family, and thus sexual reproduction, may be favoured for either of two reasons, both involving competition among siblings. If offspring are capable of moving about so that each may choose the habitat to which its genotype is best adapted, then siblings may reduce competition among themselves by occupying a variety of habitats, rather than being crowded into a single one. Sexual reproduction and hence a genetic variety among offspring allows reduced sibling competition by dispersal among a variety of habitats. If offspring are not able to move about and choose appropriate habitats, and if competition is locally severe, then a parent that sexually produces a genetic variety of offspring and spreads them among a variety of habitats is likely to have at least one best adapted genotype in patches of several types of habitat. In this case the best adapted genotype competitively eliminates its genetically different siblings, but it has an even chance in competition with the genetically similar offspring of other parents.

It seems paradoxical, or at least confusing, that sex can be favoured either as a mechanism for reducing competition among siblings or because local competition among siblings is particularly intense. The difference between these cases depends on subtle differences in patterns of dispersal and competition. The latter case requires high fecundity to produce the large number of losing tickets in the genetic raffle among siblings (Williams' 1975 apt analogy). Both cases require intense competition; otherwise a given genotype would be able to survive in a variety of habitats, and genetic variety within a family would not be necessary. Thus the short-term advantages of sex that derive from competition among siblings apply most forcefully to K-selected species (Chapter 3) that produce many young, few of whom survive to reproductive adulthood. Fortunately, such species include many vertebrates and insects for which sexual behaviour would otherwise appear to be maladaptive. Again these interpretations of the adaptive significance of sex have yet to be tested critically in Nature.

The best critical evidence for the adaptive significance of sexual recombination comes from comparative studies of hermaphrodites (Ghiselin, 1974; Charnov, Maynard Smith and Bull, 1976; Charnov, 1979), creatures that can propagate both sexually and parthenogenetically (Williams, 1975), and plants (Charlesworth and Charlesworth, 1978). In broad outline, the theories presented above are confirmed. Hermaphrodites repay the cost of meiosis by reciprocal fertilization, which may occur immediately, by shifts in sex role from time to time, or by a change in functional gender over the course of life. The cost of producing males is minimized by bizarre mating systems that allow sex ratios that are heavily biased in favour of females. And facultative changes from asexual to sexual propagation occur at times in the season or in the life history when environmental conditions are changing or when offspring are dispersed copiously, but for the most part locally.

Most discussions of sex are biased toward 'higher' (i.e., bigger) organisms; most pathogens, parasites, slime molds, and other free-living microbes have both sexual and asexual forms within their life histories. This is a fact that theory can easily be stretched to cover. However, the adaptive advantages of sex in those species that always have it are far from being understood. In particular, there is no critical evidence from field work to show that the twofold cost of meiosis is repaid by the theoretical mechanisms that have been proposed.

12.4.3 *Inbreeding*

It is tempting to argue that inbreeding ameliorates the cost of meiosis, because an inbreeding individual is matching its own genes with genes that are identical by descent. However, by inbreeding, an individual may be depriving both itself and a related mate of an opportunity to parasitize another unrelated individual to help raise related offspring. Stated in another way, it is numerically immaterial whether two related individuals pass on their genes by mating with each other or by both mating with unrelated individuals, as long as they both mate and produce the same number of viable offspring in either mating system. It is the relative viabilities of inbred versus outbred offspring that determine the costs and benefits of inbreeding; the cost of meiosis is usually not affected by inbreeding.

It is often assumed that inbred offspring are necessarily of lower viability than outbred, because of the pairing of deleterious recessive genes of common origin. Definitive evidence from the field for this assumption is scant but persuasive. Lowered fecundity, lowered resistance to disease, lowered viability, and/or lowered viability of offspring have been observed for humans (Cavalli-Sforza and Bodmer, 1971), titmice (Greenwood, Harvey and Perrins, 1978), baboons (Packer, 1979, based on early deaths of all four offspring of only one male), and ungulates bred in zoos, some of which were wild-caught females (Ralls, Brugger and Ballou, 1979). There are many examples of such inbreeding depression among livestock and other captive breeds, but there are a few spectacularly successful inbred lines as well (Ralls *et al.*, 1979). Bengtsson (1978) and Shields (1979, 1981) argue that deleterious recessives will not be allowed to accumulate in populations that have a long history of some degree of inbreeding, and Shields balances the traditional argument by pointing out that excessive outbreeding may cause the disruption of locally coadaptive gene complexes. Bengtsson and Shields suggest that an equilibrium will be established in which an optimal balance of inbreeding and outbreeding will reinforce the current genetic structure of a population, and will be maintained by appropriate patterns of dispersal, particularly of juveniles. Bateson (1980) reviews another behavioural mechanism for attaining this balance, namely an individual's choice of a novel, but not unusual, mate.

Brother–sister matings are a special case. They may not ameliorate the cost of meiosis for the argument given above. In particular, it has been argued that a male can mate with his sister and then can go on to mate with other less related females; but it is not inbreeding alone that has given this male a reproductive advantage, but the combination of his inbreeding and his beating other males to their sisters. Brother–sister matings can, however, favour sex by lowering the cost of producing males in a round-about way. If brothers routinely mate with sisters, then a parent may produce many daughters and only a small number of sons, because one son can fertilize many daughters (Hamilton, 1967, 1972). As Alexander and Sherman (1977) observe, female haplodiploid insects can control the sex ratio of their offspring by the proportionate fertilization of eggs, and when brother–sister matings are the rule, female biased sex ratios are adaptive. The female biased sex ratio in turn allows daughters to gain a genetic advantage from helping to raise siblings, rather than having their own offspring. This is yet another plausible evolutionary route to eusociality in ants, bees, and wasps (section 12.3).

Although the cost of meiosis is not generally affected by differing degrees of inbreeding, the advantages of sex are. The genetic variety attainable among offspring is limited by the heterozygosity of their parents and the genetic difference between their parents, both of which are decreased where inbreeding is more prevalent.

12.5 Life histories

The adaptive function of life histories has been described by Southwood (Chapter 3) as well as by Horn (1978) in the companion volume. Life histories have critically important implications for sociobiology for two reasons. First the ultimate biological basis of any behaviour must lie in its effect on net reproductive output, which can only be calculated in a meaningful way from the parameters of life histories (namely age-specific rates of survival and fecundity). Second, the life history parameters set the context in which animals interact socially; social interactions can only be of the most ephemeral sort unless rates of survival are high. Thus sociobiology has far more to do with K-selected (Southwood, Chapter 3) than r-selected animals.

12.5.1 *Patterns of mortality and adaptive life history*

Life history can be viewed as an adaptive response to patterns of unavoidable mortality (Southwood, Chapter 3). High rates of mortality favour high fecundity, but the pattern of mortality within the life history is as important as its overall rate. At any stage in life, energy can be devoted to growth and survival or to reproduction. If death is very likely before the next reproductive season, then reproduction is favoured. Juvenile mortality has just the opposite effect. In a sense the prereproductive mortality of offspring should be viewed less as mortality than as a discount on the fecundity of their parents, because only offspring that survive to produce their own offspring will pass on their parents' genes. Therefore a high rate of unavoidable juvenile mortality is effectively a low rate of fecundity, favouring the direction of energy into growth and survival. The balance between reproduction versus growth and survival in an optimal life history is thus determined by a ratio between adult and juvenile survival (Charnov and Schaffer, 1973; Horn, 1978; Waller and Green, 1981).

12.5.2 *Dispersal*

Dispersal is obviously favoured when the local environment is deteriorating while better conditions persist elsewhere, but Hamilton and May (1977) have argued that it is adaptive for parents to enforce dispersal of some of their offspring even from a stable and favourable environment. In a crowded population, non-dispersive offspring compete among themselves for a limited number of openings, as well as competing with dispersive offspring of other parents. By dispersing its offspring, a given parent has a chance of 'seeding' several openings, and this benefit may exceed the cost in mortality due to dispersal itself and the numerical advantage that non-dispersers might have in competing for the few local openings near their parents. Comins, Hamilton and May (1980) develop these points in detail. The study of dispersal from crowded populations is of great theoretical importance because the pattern of a creature with many young who suffer a high mortality in dispersal and establishment, but with an otherwise stable population structure, cuts across the canonical classification of r-selected species versus K-selected species (Horn, 1978). Furthermore, such a mixture of large families, dispersal, and competition for few openings provides the ideal setting for testing the proposed advantages of sex that are due to competition among siblings (section 12.4.2).

The importance of patterns of dispersal has already been mentioned several times in this chapter. Dispersal is the critical parameter in the stringent criterion for the evolution of altruism by group selection (section 12.2.2). The pattern of dispersal determines whether neighbours are close or distant relatives in comparison to strangers, and hence determines the possibility of kin selection for a pattern of cooperation with neighbours but competition with strangers (section 12.2.1). Limited dispersal of long-lived adults produces an ideal setting for the evolution of reciprocal altruism (section 12.2.3). Patterns of dispersal determine the degree of freedom that a given individual has to inbreed or to outbreed, as well as the costs and benefits of mate choice and sexual reproduction (section 12.4.3).

As was the case with mortality, the pattern of dispersal within the life history is at least as important as its overall rate. If both juveniles and adults are sedentary, most social interactions will be with inbred siblings. If both juveniles and adults are dispersive, populations will be outbred and probably genetically heterogeneous, and there will be no continuous social interactions among the same individuals. Dispersive young and sedentary adults likewise produce an outbred and heterogeneous population, but if the adults are long-lived they may interact with a fixed set of neighbours for a long time. If adults are dispersive and young are sedentary, the population will generally be outbred, but its structure otherwise depends upon the pattern of production of offspring. If they are produced in batches, then young neighbours are likely to be siblings. If they are produced singly, then juveniles spend their youth in association with a fixed group of other, unrelated juveniles. Each of these different combinations of adult and juvenile dispersal has markedly different consequences for the evolution of altruism, sexual recombination, the relation of mating to dispersal, and competition and crowding at various stages of the life history (Horn, 1981).

Empirical studies of dispersal in the context of sociobiology have just begun. A number of excellent examples are available in a collection edited by Swingland and Greenwood (1981).

12.6 Other topics

There is a wide range of other sociobiological topics having no direct connection with population biology, but beginning to profit from quantitative approaches similar to those used in population biology. Here are selected examples involving strategies and consequences of decisions faced by individual animals. Further details and complementary examples are in Krebs and Davies (1978).

A particularly successful model was inspired by Verner (1964) and developed by Orians (1969) to examine the effect of mate choice on mating systems. The model suggests that a female bird might attain the same net reproductive output by mating bigamously with a male on an unusually good territory and sharing with his other mates, or by mating with a male on a mediocre territory and having both territory and husband to herself. From this simple idea, Orians predicts the mating systems (polygyny, polyandry, promiscuity, or monogamy) that should be favoured by various ecological and behavioural conditions like spatial variability in habitat quality, and ability of the male to contribute to the care of offspring. The model has been broadened qualitatively by Emlen and Oring (1977), and quantitatively by Rubenstein (1980). The simple form of the model has been tested in detail by Pleszczynska (1978). Pleszczynska studied lark buntings, which nest in fields of sparse grass and alfalfa in the prairie region of the United States. The quality of territories defended by males depends on the amount of shade available to protect the nest and young from intense sunlight. She confirmed a direct effect on reproductive output by adding plastic leaves to the crowns of plants above some nests and noting increased fledgling success of those nests. Pleszczynska successfully predicted bigamy, monogamy and batchelorhood of males from measurements of the density of vegetation on their territories.

Several simple analytic models of mate choice have been proposed and compared by Janetos (1980) to explore the limitations of available time for choice, ability to locate mates, and ability to evaluate them. It is not surprising that the optimal strategy is to examine as many mates as possible and to choose the best of them. However, strategies involving a fixed threshold of acceptable quality for a mate are surprisingly bad, whether the threshold is high or low.

Where optimal strategical decisions for a given individual involve the tactical behaviour of other individuals, the use of game theory is appropriate. An early example is Fisher's (1930) discussion of optimal sex ratio, extended with an elegant combination of rigor, generality and brevity by MacArthur (1965). Because all sexual individuals have both a male and a female parent, males and females are equally effective as means of transmitting genes to future generations. It is therefore optimal for a given individual to devote energy to the production of offspring of the minority sex; that is, the sex in short supply. If all individuals in a population attempt to achieve this optimal behaviour, the evolutionarily stable expenditure of energy is 50 per cent to female offspring and 50 per cent to male. The optimal sex ratio should therefore be the inverse of the ratio of energetic costs of producing an individual of each sex (but see section 12.4.3 for an exception).

The language and approach of game theory have been applied interpretively to conflicts between animals by Maynard Smith and Price (1973) and Parker and Rubenstein (1981) among others, but they are also useful in cases of more subtle conflicts of interest, like those involved in parental care. For example, Maynard Smith (1977, 1978b) has shown that desertion of a family by either the male or the female parent is an evolutionarily stable strategy when guarding of the offspring by at least one parent is necessary but the presence of the second parent does not add much. A game theoretic approach is also applicable to parent–offspring conflict (Trivers, 1974). A particular example has been given in the context of dispersal of offspring (Hamilton and May, 1977; Comins, Hamilton and May, 1980, section 12.5.2), and Parker and Macnair (1979) explore the qualitative arguments of Trivers (1974) with genetic and game theoretic models. However, as Alexander (1974) suggests, parents are so often in a position to enforce their best interest on their offspring, that the conflict is often resolved in their favour before it is joined, and the formal approach of game theory may be unnecessary. Suspected cases of manipulation of parents by offspring are of particular interest; any human parent can suggest examples in abundance.

12.7 Notes on human sociobiology

In his *Sociobiology: The New Synthesis*, E. O. Wilson (1975) reviewed applications to humans of sociobiological ideas that had been developed from studies of other animals or in a purely theoretical context; he also suggested many novel applications. Reviewers loudly praised or vehemently criticized him; the reviews, Wilson's responses, and other articles emerging from the debate have been gathered together and reprinted by Caplan (1978). Wilson's (1978) further speculations, *On Human Nature*, will surely provide the material for more debate (Mackintosh, 1979; Rosenfeld, 1980). The central issue is the degree and nature of biological determinism in human behaviour, especially the extent of genetic influence on behaviour, which the most severe critics claim to be supported only by hazily inferential or anecdotal evidence. However, the genetic issue may be irrelevant to much of human sociobiology, because mixed models of cultural and genetic transmission of behaviour (Cavalli–Sforza and Feldman, 1981; Lumsden and Wilson, 1980, 1981) show the possibility of evolution of behavioural patterns that are prevalent and firmly ingrained but still malleable and potentially evanescent.

One clear difference between humans and other creatures is the strong capability and propensity that we have for learning, teaching, culture, and the linguistic ability to exercise them. Bonner (1980), in *The Evolution of Culture in Animals*, reviews instances of learning, teaching, and culture in non-human animals, discusses their adaptive significance for those animals, and speculates upon their significance in the evolution of humans and of human behaviour.

12.7.1   *Mixed models of genetic and cultural inheritance*

Cavalli–Sforza and Feldman (1981) and Lumsden and Wilson (1981) have independently elaborated novel models of the evolution of behavioural traits that are under the influence of both genetic predisposition and learning; they allow cultural transmission variously from parents, siblings, contemporary cohorts, and/or other members of the population. Cavalli–Sforza and Feldman's most important results are that the details of the learning process and the pattern of mating are of very great importance in the dynamics of the prevalence of different patterns of behaviour. They find, as do Lumsden and Wilson (1980, 1981), that small quantitative differences in cultural parameters can lead to dramatic differences in evolved behavioural patterns of model populations. Lumsden and Wilson (1981) fit the parameters of their models to data from cognitive psychology and successfully predict data from comparative ethnography. A quantitative prediction spanning several academic disciplines is a bold venture that will surely provoke ardent discussion. However, it can be tested fully only by new forms of research within the social sciences, for which Lumsden and Wilson provide intriguing shopping lists and recipes.

The models with mixed cultural and genetic transmission provide both the possibility of local behavioural uniformity and the likelihood of regional behavioural diversity. Thus these models are a first step toward disentangling the paradox that both 'human nature' and 'human diversity' are simultaneously acknowledged by many students of human behaviour.

12.7.2   *Adaptation and comparative studies*

One of the primary difficulties with many current applications of sociobiological ideas to humans is that they concentrate on adaptive interpretation of the supposedly universal traits of humans, rather than on differences between humans and other animals or differences among populations of humans who live in different cultural and ecological settings. Statements about the adaptive nature of a single bit of behaviour are intrinsically untestable, because the very existence of the behaviour is taken as evidence of its adaptive value (Lewontin, 1978; Cuppy, 1941). Comparative statements that emphasize a supposed evolutionary trend from lower to higher forms of behaviour (which usually means less to more 'human') are tautologies at a transcendent level. However, comparative statements are meaningful and testable if they emphasize adaptive differences between behaviours of different populations in different environments (Horn, 1979). Exemplary reviews of primate behaviour from such a comparative point of view are those of Jolly (1972), Clutton-Brock and Harvey (1977), and Clutton-Brock, Harvey and Rudder (1977). Comparative studies have a distinguished history; Darwin (1859) used them to argue the cause of natural selection, and E. O. Wilson (1971, 1975) used them to define and to crystalize the academic discipline of sociobiology.

13
Palaeontology Plus Ecology as Palaeobiology

STEPHEN JAY GOULD

13.1 Introduction

Palaeontology ought to be one of the world's most exciting subjects: dinosaur mania among children indicates a primal fascination. Yet palaeontology, as traditionally pursued by professionals, has been the poor sister, and often the laughing stock, of the sciences. It has been portrayed, not unfairly, as the dullest variety of empirical cataloguing practiced by the narrowest of specialists. In 1969, the editors of *Nature* wrote (anonymous, 1969, p. 903): 'Scientists in general might be excused for assuming that most geologists are palaeontologists and most palaeontologists have staked out a square mile as their life's work. A revamping of the geologist's image is badly needed.'

Traditional palaeontology was the handmaiden of stratigraphic geology. In 1947, a famous geologist defined the profession this way (Kay, 1947, p. 162): 'Paleontology is a means by which rocks are better classified in time and in environment of origin.' As an auxiliary to geologists, it became a profession of atheoretical, taxonomic specialists: a budding palaeontologist indentured himself to a biological group and a small section of geologic time; he then learned all the names, coined some new ones, and refined the stratigraphy of his chosen section. But a remarkable transformation has taken place during the last decade. In short, palaeontology has allied itself with evolutionary biology, and has experienced all the challenge and excitement of its most rapidly developing subdiscipline—theoretical ecology.

The change in emphasis has been dramatic. I began my subscription to the *Journal of Paleontology* in 1964. Of the 18 articles in my first issue, 11 had titles in the form: 'New species of . . . . from the . . . . (age) of . . . . (place).' Of the seven remaining, 4 were purely taxonomic. In 1975, the Palaeontological Society founded a new journal, *Paleobiology*, to receive more theoretical papers, thus preserving the documentary role of the *Journal of Paleontology*.

Nonetheless, of 12 articles in the first issue of the *Journal of Paleontology*, published after the founding of *Paleobiology*, only 3 list taxon, time and place. The others focus upon biological questions and cast descriptive sections in their light.

Traditional palaeontology was preeminently a geological discipline, but it did not ignore biology completely. It extolled the importance of 'evolution' but used the term to mean little more than the tracing of phylogeny and the identification of macroevolutionary trends in morphology. Moreover, it did not, on the face of it, deny the importance of ecology—'palaeoecology' was the fad subject of the 1940's and 1950's. But 'palaeoecology' bore little relationship to anything an ecologist would recognize; as a creation of the petroleum industry, it remained largely geological. The earth's strata are not arranged in a layer cake; rocks of the same age record different environments and some fossils, restricted to environments rather than time planes, do not serve traditional stratigraphy well. Thus, a limited autecology nearly exhausted the content of traditional palaeoecology: (1) environments (rock types) preferred by given taxa were recorded; (2) for fossils entombed in more than one rock type, correlations were sought between environment and morphology, largely in the interest of taxonomic refinement.

A glance at the most celebrated document of traditional 'palaeoecology', the 1000 page multi-authored Treatise of 1957 (Ladd, 1957), illustrates its largely atheoretical concern. It begins with an indifferent section of 100 pages on generalities (mostly nonbiological) and then proceeds via the conventional catalogue of times and taxa. Its body of 900 pages contains: (1) a set of case studies in environmental reconstruction ordered not by problem or concept, but by the older layer cake from Precambrian to Recent; (2) a set of essays on taxonomic groups, ordered by the conventional ladder of amoeba to man. Yet, we should not ascribe this atheoretical approach only to the biological naiveté of most traditional palaeontologists. Ecology, at the time, also emphasized highly specialized, inductive and largely atheoretical work. Palaeontologists had looked in (the 1957 Treatise has a companion volume on modern ecology—Hedgpeth, 1957), and they found little to inspire a change in direction.

The transformation of palaeontology during the last decade into a 'creative, chancy young man's game' has many roots (to borrow E. O. Wilson's description of modern evolutionary biology, enlightened by the approach of theoretical population ecology—Wilson, 1969). But I believe that the most important cause has been the direct impact of theoretical population ecology. Palaeontologists have looked in again, and this time they have found insight into their traditional problems of diversity and its history. We make no apology for this borrowing. Scientific change usually involves a fertilization from other disciplines (Kuhn, 1962); new directions rarely arise *sui generis* from the established professionals of a field. Moreover, we are beginning to repay the debt by providing our own theoretical insights for ecologists to consider.

I write this chapter to document the modern interaction of ecology and palaeontology. It is not a review article (for it leaves out far more than it includes); I am only seeking to illustrate in a highly selective way how palaeontology is incorporating ecological theory to produce a conceptual science worthy of the name palaeobiology.

13.2   Specific puzzles

13.2.1   *What are the primary controls on the diversity of fossil communities?*

Geological methods must be used to answer one of the most crucial general questions for any biological study of fossil communities: how well does a fossil assemblage reflect the living community. The biases of differential preservability, burial and subsequent alteration must be recognized and subtracted. In many cases they cannot be measured accurately, and palaeobiologists have sought the ideal situation of rapidly buried communities, entombed *in situ* within sediments that suffered little subsequent alteration.

We find (not surprisingly) enormous variation in numbers of species within fossil samples that represent their living communities adequately. The explanation of these differences occupied much of classical palaeoecology. Favoured modes of explanation went little beyond: (1) historicism, e.g., low diversity in areas only recently available for habitation; and (2) 'physicalism', the notion that species are under the direct control of physical variables, and that communities of low diversity reflect extreme values of a controlling variable—hypersalinity or low temperature, for example. The interaction of species and their evolutionary strategies (beyond morphological adaptation to a substrate) did not figure in classical explanations.

The framework of stability–diversity theory in modern ecology has inspired a new approach to the study of fossil communities. Organisms adapt to the relative stability of environments by altering their life-history strategies; intensity and frequency of disturbance may serve as the main environmental control upon diversity. Since the inevitability of environmental fluctuation (at least in the long run) is a cardinal principle of geology, this new theme has intrinsic appeal to palaeontologists. (It matters little that classical stability–diversity theory may have suffered a spectacular collapse within ecology (e.g., May, 1975a, and section 9.5). The surviving aspects have been emphasized by palaeobiologists; moreover, data from the fossil record have helped to spur the collapse.)

Using the old approach in his early papers, for example, Hallam (1969) invoked 'physicalism' to explain the low diversity of the Boreal in contrast to the Tethyan Realm among European Jurassic faunas. Seas of the Boreal Realm, he argued, maintained abnormal salinities. In later works (Hallam, 1975), he has tied low Boreal diversity to the frequent and severe disturbance of this realm; high diversity in Tethys reflects the stability of its environment. One may fairly ask what has been gained by this shift; does it represent anything beyond a more fashionable explanation for the same old data. I prefer it because it suggests, in the context of ecological theory, much more to test and a richer domain of biological insight (it may, of course, be wrong). With the physicalism of unusual salinity, one can scarcely go beyond two lines of biological evidence: are the fossils related to modern taxa that tolerate unusual salinities; and, do the fossils bear morphological signs (e.g., stunting) that often accompany life in these rigid environments? Modern ecological theory opens the whole field of adaptive strategies, and its criteria of quantitative population dynamics. New properties for study and test include survivorship curves, recruitment, measures of evenness among coexisting taxa and the attributes of r- and K-strategists preserved in fossils (and utilized, for example, by Snyder and Bretsky (1971) in arguing that paedomorphosis for rapid generation time in unpredictably fluctuating environments furnishes a better explanation than stunting by abnormal salinity for the depauperate 'dwarf fauna' of the Ordovician Maquoketa Formation).

In a pioneering paper, Bretsky and Lorenz (1969) noted that Paleozoic nearshore communities are generally less diverse, but geologically more persistent than offshore communities (data challenged by Thayer, 1973). They argued that nearshore environments suffer severe and unpredictable fluctuations. These environments support a limited number of physically controlled r-strategists. Stable (subtidal) offshore environments maintain diverse, biologically accommodated communities (*sensu* Sanders, 1968). But classical stability–diversity theory maintained that biologically accommodated communities should be stable in time. Why, then, do the nearshore, physically controlled communities persist for longer periods? Bretsky and Lorenz argue for an effect of scale. Offshore communities are more stable in the short run of ecological time; a nearshore community cannot survive in a desiccated estuary (though its elements might move to another area). But in the long run of geological time, ultimate control is exerted by the very rare major fluctuation. These rare events can wipe out a complex, biologically accommodated community, but the few species of nearshore communities are r-selected for surviving environmental fluctuation and have a better chance of 'weathering the storm'.

This paradoxical result of greater geological success for species usually judged less able to persist in single spots during ecological time has also been emphasized by Fortey (1980), who demonstrates the high generic longevity of trilobites adapted to certain stressed environments, and by Hoffman (1978) who emphasizes a general correlation between eurytopy and geological duration. (The paradox should disappear once we understand that effects at one level cannot be automatically extrapolated to superficial similarity at other levels or scales in the analysis of evolutionary processes—see Gould, 1980.) Yet, to emphasize the inadvisibility of simple extrapolation across levels of analysis, the greater geological longevity of stressed eurytopes does not imply that evolutionary trends towards their domination should exist within clades—for stenotopes may, on average, persist for shorter times, yet speciate at such a markedly greater rate that they increase steadily in relative frequency for species within clades. Vrba (1980) has invoked this argument as a basic premise for an important reconstruction of macroevolutionary theory. Hansen (1978) presents an interesting case, though he interprets it differently, involving a trend within volutid gastropods during the Tertiary for replacement of species with planktonic larvae by species with non-planktonic larvae.

The correlation of low diversity with frequent, unpredictable and severe disturbance has not been challenged in the current attack on stability–diversity theory in ecology. This is the correlation that palaeobiologists have exploited most often. Levinton (1970) may have

resolved the old dilemma of why some fossil beds are so jammed full
with remains of a single species while beds below and above do not
record it at all. These accumulations do not just represent another
example of current winnowing or differential transport—another
depressing example of the biological inadequacy of our fossil record.
They hold the opportunistic species of physically controlled communities caught in the acme of their transient success. Rollins and Donahue
(1975) explain patterns in diversity within Pennsylvanian cyclothems
by showing that initial faunas of low diversity are composed of opportunistic species. As sea level rises, these are replaced by a more
diverse community of biologically accommodated species. Low diversity
is not a function of history (early stages of transgression), but of
adaptive strategy in harsh, fluctuating environments. Kauffman (1972)
relates the low diversity of North Temperate Cretaceous molluscs
to the irregular fluctuations of their environment. Correlation between
low diversity and severe, unpredictable disturbance has also been
found by Ashton and Rowell (1975) for Cambrian tribolites and by
Watkins and Hurst (1977) for crinoids.

On the other hand, the necessity for a correlation between high
diversity and environmental stability has been successfully challenged
(e.g., by May, 1975a). Highly stable environments may eventually
beget low diversity, since a few excellent competitors may eventually
assert their dominance (as many natural monocultures testify). Maximal
diversity may arise under conditions of intermediate disturbance (too
much disturbance eliminates all but a few physically accommodated
species; too little permits the complete usurpation of space by best
competitors—see Connell and Slatyer, 1977). This reechoes one of
Horn's themes from chapter 11 (Horn, 1975a, b). Good evidence for
control by disturbance comes from work on modern communities by
palaeontologist R. Osman (1975). At Woods Hole, Massachusetts, he
found that maximal diversities characterize rocks of intermediate size
(large rocks are too big to be disturbed by turning; little rocks roll
over too often). But didn't Bretsky and Lorenz anticipate this revision
of ecological theory (admittedly at a different scale)? They challenged
the dogma that diverse communities are necessarily stable, but only
for the long perspective of geological time with its assurance of an
eventual, major disturbance. They did not realize that this lack of
stability might also extend into the microworld of ecological time as
well. Since the publication of Simpson's seminal treatises (1944, 1953),
palaeontologists have been trying to extrapolate the world of micro-

evolution to their time scales. Perhaps it is time for ecologists to examine the insights of palaeontologists for potential compression into their own worlds.

13.2.2 How does genetic variability relate to the diversity of communities and the probability of extinction?

Bretsky and Lorenz (1969) linked their pattern of nearshore low-diversity-long-term persistence vs. offshore high-diversity-earlier extinction to genetic speculation. They proposed that nearshore species are 'heteroselected' for genetic variation needed for flexibility of populations in frequent times of severe fluctuation. Offshore species are 'homoselected' for restricted variation since they adapt so narrowly to their stable habitats. They then claimed that the celebrated mass extinctions of the geological record differentially removed species from homoselected, offshore communities. These species did not have enough genetic variability to meet the adaptive requirements of rapidly changing environments. At times of crisis, in other words, the offshore species became victims of their own previous success.

This palaeobiological speculation has directly inspired some of the most fruitful work in recent years at the interface of ecology and genetics. Palaeontologists can rarely study genes directly (but see Prager et al., 1980), but many of them have learned the techniques of electrophoresis to probe modern organisms for insight into the palaeobiological dilemma of extinctions. Is there a relationship between stability of environment, adaptive strategy and the amount of genetic variability present in populations?

Schopf (a palaeontologist) and Gooch (1972) studied genetic variability in 8 deep-sea species (1000–2000 m) and reported their results with the title: 'A natural experiment to test the hypothesis that loss of genetic variability was responsible for mass extinctions of the fossil record.' In this highly stable environment, they found 20–50 per cent of loci polymorphic, thus casting great doubt on Bretsky and Lorenz' contention that reduced variability in stable environments favoured the extinction of offshore forms. Palaeontologist J. W. Valentine, with geneticist F. J. Ayala and several colleagues have extended this approach to reach empirical conclusions directly opposite to the speculation of Bretsky and Lorenz (Valentine et al., 1973; Ayala et al., 1973b; Valentine and Ayala, 1974). They studied species on a geographic gradient in trophic resource regimes from stable-tropical to unstable-high latitude; they found consistently that tropical species maintained more genetic variability in their populations than high latitude forms. Their first work compared the incommensurable (tropical clams, temperate horseshoe crabs and antarctic brachiopods) and permitted no firm conclusions. But Ayala *et al.*, (1975) have also studied closely related species of krill along geographic gradients. Again, variability increases steadily towards the tropics.

Bretsky and Lorenz, in retrospect, may have made the fatal error of confusing the physiological plasticity of individuals with the genetic variability of populations. Individuals in physically-controlled communities must be able to tolerate a wide range of environments, but they may achieve this by rigorous selection for genes conferring plasticity; variation among individuals is a different matter. Valentine believes that populations of tropical environments are more variable because the richness of stable habitats provides a wide range of microenvironments to which different subpopulations adapt. In any case, if this correlation between environmental stability and genetic variability holds, it will provide a powerful selectionist argument against the hypothesis of neutralism as a source of most genetic variation. The selectionist–neutralism debate may now be the cardinal controversy in evolutionary biology. A large piece of its resolution may have been inspired directly by a palaeobiological speculation. Valentine and Ayala (1974, p. 70) are justified in praising Bretsky and Lorenz 'for producing an elegant and testable hypothesis'—even though it was probably wrong. It is a good measure of the excitement of ecology that scrupulously correct factual documentation is no longer the only activity that wins praise among palaeontologists.

13.2.3   *Succession as a model for microtemporal change*

Palaeontologists may love to speculate about the general course of life, but most of their work involves historical sequences in local sections encompassing a relatively small amount of time. Changes in taxonomic composition up a section have traditionally been interpreted as sets of evolutionary events or evidences of migration mediated by changing environments. Several recent studies have suggested that such common patterns often represent autogenic successions. Since periodic wipe-out is the geological fate of any community at a given spot, the opportunity for rediversification through succession occurs with great frequency in the fossil record. Bretsky and Bretsky (1975) have interpreted some late Ordovician sequences as successions initiated by the colonization of offshore barren muds by two or three opportunistic deposit-feeding species. Walker and Alberstadt (1975) detect succession sequences in several fossil reefs (Ordovician to Cretaceous in age).

A problem in the Walker and Alberstadt (1975) paper illustrates the potential of ecological theory. They are puzzled by the sharp decrease in species diversity at the final stages of apparently autogenic successions. Since they accept the dogma of classical stability–diversity theory, they are stumped by this 'uppermost domination zone' of drastically reduced diversity (far greater than the orthodox slight reduction in climax communities). They are led to the *ad hoc* suggestion of allogenic control by changed water turbulence to produce 'a physically accommodated community from what was a near climax stage biologically accommodated one' (1975, p. 243). But they might avoid this *deus ex machina* by recognizing May's demonstration (1975a, c) that highly stable environments might eventually yield communities of low diversity. In any case, palaeontology can provide the only record of complete *in-situ* successions. The framework of classical succession theory (probably the most well known and widely discussed notion of ecology) rests largely upon inferences from separated areas in different stages of a single, hypothetical process (much like inferring phylogeny from the comparative anatomy of modern forms). We can provide direct evidence to supplement the revisionist discussion in chapter 11.

Yet palaeobiologists must not uncritically extrapolate to their domain the models that operate so well in ecological time. Several postulated 'successions' in the recent palaeontological literature probably span too long a time for proper application. (How can the first arrivals be called pioneers or opportunists if they range through sedimentary thicknesses measured in thousands, if not millions, of years? May we speak of true succession if most species disappear by total extinction rather than by local replacement?) Thus, the Bretsky and Bretsky paper (1975), cited above, reports sampling at 25-foot intervals; each of three successions they identify probably spanned millions of years and generations. Moreover, the basin of deposition was shallowing throughout their sequence and this directional change in environment may have been the main determinant of the faunal changes they observed.

Rollins *et al.* (1979) are more sensitive to scaling and the differences between succession within a community and replacement engendered by evolution and environmental change. They relate the systematic changes they observe in Pennsylvanian faunas of Ohio to the geological phenomena of transgression–regression cycles and attendant shifting in position of onshore–offshore stress gradients. Classical ecological succession may well apply, however, to local sequences of short duration (see Toomey and Cys, 1979, on succession in small algal-sponge bioherms from the Permian or New Mexico).

13.2.4 *Quantification of biases in the fossil record*

Perhaps the major impediment to an analysis of patterns in the history of life is the notorious imperfection of the fossil record. Palaeontologists have traditionally lamented about the woefully poor, imperfect, and sporadic preservation of fossils and have assumed that empirical patterns reflect the imperfections primarily, or that the patterns are an inextricable mixture of biological reality and missing data.

In this context, an attempt to quantify biases of preservation in the fossil record must be a major focus of palaeobiological work. If we cannot learn to recognize and remove such biases, prospects for a truly scientific palaeobiology are dim. In the initial and promising spate of articles on this subject, ecological studies have been useful in at least two ways. First, comparisons of imperfect fossil data with more complete ecological information of the modern world have demonstrated that the fossil data retain essential patterns. Campbell and Valentine (1977) demonstrated that geographic provinces recognized from the distribution patterns of fossil genera and families are comparable with modern provinces determined from the distribution of species. Schopf (1978) studied the fossilization potential of a modern intertidal fauna at Friday Harbor, Washington. He determined that about 40 per cent of all species could be retained in the fossil record and that such a level of preservation would accurately reflect at least some of the major habitats and modes of life—herbivores and filter feeders, in particular.

In a second approach, several palaeobiologists have applied some analytical techniques of ecology to problems in the fossil record. If, for example, rarefaction can lead to comparisons of faunal composition in samples of markedly different size (Sanders, 1968), then differential preservation in the fossil record may be a fruitful realm for application

—especially when many of our fossil samples are spotty. Raup (1975) applied rarefaction to several large samples of post-Paleozoic echinoids (some 8000 species in all) and showed that observed increase in numbers of echinoid families cannot solely be an artifact of increasing preservability of species through time (see also Tipper, 1979 on rarefaction in palaeontology). The species–area relationship has also been a fruitful guide. Sepkoski (1976), for example, performed a multiple regression analysis of faunal diversities upon estimates of rock volume and area of continental seas. He removed the correlation with rock volume as an artifact and found a good fit of residuals with estimates for areas of continental seas. Not only did this curve match the common pattern of negative allometry in species–area relationships, but patterns of residuals from it made sense in terms of our geological knowledge. Points for the Cambrian and Triassic fell below the line, the Cambrian because oceans were just then being populated with metazoan life, the Triassic as a period of recovery from the greatest mass extinction in earth history (end of Permian).

13.3 Basic questions in the history of life

The history of life has been anything but smooth, and its major episodes have provided the greatest problem of palaeontology since Cuvier's day. A list of theories for the Cambrian 'explosion' and Permian extinction alone would probably fill the London telephone directory. Most palaeontologists (not including myself) also believe that directional patterns can be discerned through all this tumult of pulsating extinctions and radiations, and the causes of generally increasing diversity or morphological complexity have been much debated. During the past five years, ecological theories have inspired new explanations for all these ancient dilemmas; some seem so satisfactory that I might even be led to proclaim certain issues as settled if I did not so well appreciate the short half life of previous 'definite' solutions.

13.3.1 *Pulsations in the history of life*

About 600 million years ago, almost all the phyla of marine invertebrates entered the fossil record for the first time during the short space of a very few million years. Yet, for the previous 2500 million years of earth history, little more than prokaryotic bacteria and algae populated the world, forming extensive stromatolites (algal mats) in their favoured habitats. Most explanations for this Cambrian 'explosion' have been crassly physicalist. The highly touted Berkner–Marshall hypothesis (1964), for example, holds that it marks an attainment of atmospheric oxygen levels sufficient to screen ultraviolet light (by an ozone shield) from shallow waters and provide for respiration—as if organisms were inert billiard balls, responding immediately and automatically to any external stimulus.

Stanley (1973a, 1976) has recently proposed an ingenious explanation by importing a key notion of contemporary ecological theory. Contrary to intuition, the introduction of a higher trophic level (a herbivore in a plant community, a carnivore among herbivores) tends to increase rather than decrease diversity at levels below it. 'Cropping' frees space otherwise completely usurped by the one (or very few) dominant competitors of the uncropped system. Precambrian stromatolites look, to Stanley, like a classical uncropped ecosystem of very low diversity. The 'hero' of the Cambrian explosion may have been the first single-celled herbivorous protist. Stanley's thesis does not tell us how this herbivore evolved (in part, the knotty problem of the origin of the eukaryotic cell); but he does provide a reasonable explanation for the rapidity of subsequent change once it began to crop the Precambrian monoculture. The cropping would be geologically instantaneous on a worldwide scale. The combination of available ecospace and the unexploited potential of a recently evolved eukaryotic cell may have determined the greatest burst of evolutionary activity this planet has ever known.

Sepkoski (1978, 1979) has made an ingenious application of another common principle in ecology (and many other disciplines) to the Cambrian explosion. He constructed plots of ordinal and familial diversity vs. time and found an excellent fit to the sigmoidal pattern with its lag phase, log phase, and plateau. He argues that the Cambrian explosion is merely the log phase of a sigmoidal process and may not have, contrary to the entire catalogue of previous theories (traditional and non-traditional), any specifically Cambrian cause. It may be the natural result of a process initiated by a Precambrian key innovation (perhaps the evolution of Stanley's cropper) that engendered the exponential filling of a largely empty environment up to the limits of its carrying capacity. The slow increase of late Precambrian times represents the lag phase, the Cambrian explosion the log phase, and subsequent stability in numbers at high taxonomic ranks a plateau at carrying capacity.

Two hundred and seventy five million years later, the debacle came as fully half the families of invertebrate animals became extinct in an equally short period of time at the boundary between Paleozoic and Mesozoic eras (Newell, 1973). This 'Permian crisis', unlike the Cambrian explosion, lies within the part of earth history well documented in the fossil record. Hence, it has received more attention from the profession, and may rightly be called palaeontology's outstanding dilemma (Gould, 1974). Favoured explanations again have tended to be physicalist—extraction of salt from the oceans and explosion of supernovae being among the most popular recent suggestions.

I believe that a solution—or at least a first order control—is now emerging thanks to the union of two theoretical positions that scarcely existed a decade ago: the revolutionary paradigm of plate tectonics in geology, and the equilibrium thinking of the MacArthur-Wilson (1967) approach to theoretical ecology, which was discussed in chapter 10. The late Permian represents the singular time (at least since the Cambrian explosion) during which the earth's continents coalesced to the single super-continent of Pangaea. The coincidence of this coalescence with the Permian extinction is not likely to be accidental, but we still must ask why the union of land masses should provoke such a catastrophe for shallow-water marine life.

Schopf (1974) has recently argued that the ecological literature on species-area curves may provide a key insight: the relationship between species diversity and habitat area often conforms to a simple power function with a slope considerably less than unity, as discussed in section 10.2 [see eq. (10.1)]. The fact of correlation is enough for Schopf's argument; I will bypass the issue of whether area *per se* is the controlling variable, or whether it merely acts as a surrogate for such determining factors as habitat diversity. The joining of all major continents produced a marked reduction in the area of shallow seas for two reasons: (1) basic geometry: shallow seas are peripheral to continents, and there is far less periphery around a single large continent than around several small continents of the same total area; (2) the mechanics of plate motion. When continents join, the margins of plates lock together since continents are too light to be subducted. Plate tectonics requires a balance between creation of new sea floor at mid-oceanic ridges and subduction of old sea floor at plate margins. If subduction stops, creation ceases as well. The oceanic ridges collapse as the supply of molten material (previously designated for new sea floor) dwindles. The ridges make up a substantial amount of undersea topography, and their collapse would cause a marked reduction of sea level. This reduction would drain the continental shelves and drastically reduce the area available to shallow water marine animals.

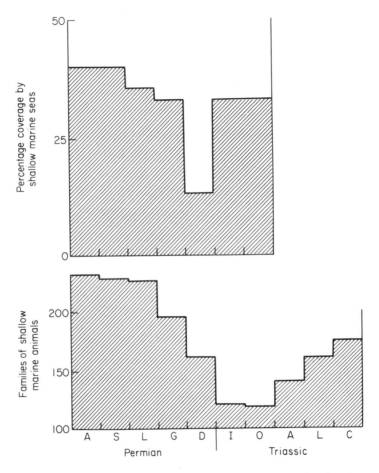

Fig. 13.1. Correlation of faunal extinction with the reduction in area of shallow water seas during the Permian crisis. (From Schopf, 1974.)

Schopf (1974) tested his hypothesis of areal control by using the best available evidence of geological maps to measure the reduction in shallow seas during the last three Permian stages (they were reduced, he concludes, from 40 per cent of possible early Permian distribution to 15 per cent for the latest Permian). He also calculated the reduction of taxa by extinction during these stages (Fig. 13.1). In a companion paper, ecologist D. Simberloff (1974) then developed a way of estimating species-area curves from the faunal data of families (for a variety of reasons—including the sins of past taxonomists and the imperfections of the geological record—we can form no reliable estimates for numbers of fossil species). Using Schopf's data for faunal and areal reductions, he then obtained a good fit to standard species-area curves. Schopf (1974) concludes that reduction of area itself served as the first-order control of the Permian extinction.

The second great extinction in the history of life, the late Cretaceous debacle that wiped out dinosaurs along with 25 per cent of the families of marine invertebrates, has also received an ingenious explanation based upon an integration of plate tectonic and ecological principles (Bakker, 1977). More recently, however, astronomical catastrophes have made a comeback, but this time with evidence rather than in the conventional mode of vain speculation (an anomalous increase in iridium right at the Cretaceous–Tertiary boundary; iridium, one of the platinum elements, is virtually absent from indigenous crustal sediments on earth, but accumulates through the influx of extraterrestrial bodies). These theories, involving impacts of asteroids (Alvarez et al., 1980) or comets (Hsü, 1980) may seem to provide a small role for ecology in explaining mass extinction. Yet, if confirmed, such explanations do emphasize the importance of density-independent extinction and may point to a fundamental difference between reasons for local extinction in ecological time (primarily density-dependent) and mass extinction in geological time—the primary determinant of pattern in higher taxonomic ranks through the history of life. Moreover, these catastrophic theories will not compel assent until they can provide an ecologically reasonable scenario for the differential patterns that characterize mass extinctions—a consideration notably lacking in papers by non-biologists published so far. Why do dinosaurs die and mammals and fresh water vertebrates remain largely unaffected? Why does the oceanic phytoplankton virtually disappear at an instant, while land plants scarcely suffer?

13.3.2 *Basic patterns in the history of diversity*

If we count the number of fossil taxa recorded in the palaeontological literature, we note a pattern of steady increase modified by the pulsations discussed in the last section. A rapid increase marks the Cambrian explosion. A slow decline sets in towards the late Paleozoic, culminating in the precipitous drop of the late Permian. Since Triassic times, the trend has been steadily upward. (The Cretaceous extinction did not markedly affect marine invertebrates.) Valentine (1970, 1973, and Valentine *et al.* (1978) have argued that this empirical pattern is a fair representation of true history and that, following the Cambro-Ordovician filling of the ecological barrel, diversity pretty well maps the configuration of continents—moderately high with moderately separated mid Paleozoic continents, declining as the continents join to form Pangaea, and steadily increasing with post Permian continental separation (Fig. 13.2). Opportunity for endemism among shallow water faunas becomes the chief control of global diversity (augmented by such secondary factors as increased climatic stability in a world of small, widely separated continents).

Raup (1972, 1976) has challenged Valentine's interpretation by arguing that the empirical curve might represent little more than an

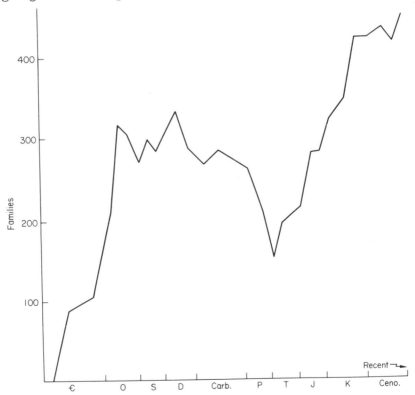

Fig. 13.2. Empirical curve for diversity through time for families of shallow water invertebrates. (From Valentine, 1970.)

artifact of sampling. After all, preservability (more rocks and less alteration) increases markedly as we approach the present day. Raup makes a rough estimate of the biases and demonstrates that the empirical curve can be obtained by sampling a real distribution with the following pattern: rapid rise in the Cambrian, increase to a mid Paleozoic high, slow decline to a late Paleozoic intermediate level, and constancy thereafter.

Hidden in this debate is one of the most interesting ecological questions we can ask within a palaeontologist's time scale. If diversity increases generally through time, why? One answer, emphasized by Valentine, holds little interest for an evolutionary theorist. If diversity simply maps continental positions, then its vector of increase is an 'accidental' result of historical trends in continental separation; different patterns of plate motion would have yielded a different curve of diversity through time. But Valentine (personal communication) also accepts an 'intrinsic' influence upon increasing diversity. He argues that, all things being equal, average niche breadth should tend to decrease with time—i.e., we should find increased species packing *within* a constant amount of geographical and ecological space. An alternate view holds that the ecological barrel filled up quickly following the Cambrian explosion. Limiting similarity is rapidly approached and there are no intrinsic trends in species packing. The same general equilibrium number of species applies to similar environments through time. Actual changes in diversity reflect the influence of external events in altering equilibria by changing environments—e.g., the decrease in equilibrium number caused by draining of the continental shelves in late Permian times.

This debate is merely the latest incarnation of the most fundamental question in palaeontology: does the history of life have an intrinsic direction (towards greater morphological complexity, increased diversity, etc.)? In more general terms, is steady-statism or directionalism a more appropriate metaphor for the history of life? This is an ancient argument, long predating any belief in organic evolution. It was, contrary to conventional historiography, the major issue separating Lyell from the catastrophists (Hooykaas, 1963; Rudwick, 1972; Gould, 1975). Lyell argued for a Newtonian timelessness in earth history—constant overturn of species through extinction and new creation, but no trends in complexity or diversity. The catastrophists, as progressionists, claimed that each new creation approached more clearly the perfected forms of our current earth.

The obvious test for an intrinsic increase in diversity requires a comparison of perfectly preserved faunas in similar environments covering the same area through time. R. Bambach (1977) has attempted to do this, although the vagaries of an imperfect fossil record do not permit great confidence in the measure of control (especially over areal extent of a community). Bambach's results are interestingly ambiguous. He divides communities into nearshore, offshore shallow and offshore deep, and times into early Paleozoic, mid Paleozoic, late Paleozoic, Mesozoic, and Cenozoic. For the opportunistic, low-diversity nearshore communities, Bambach finds no trends in species diversity through time. Both categories of offshore communities display the same pattern: low diversity in the early Paleozoic, no statistically significant differences for nearly 400 million years from mid Paleozoic through Mesozoic, but a rise in diversity for Cenozoic communities. The low diversity of early communities surprises neither side: it represents the filling of the ecological barrel following the Cambrian explosion. The long period of subsequent stability seems to favour the steady–statist metaphor; but the Cenozoic increase, if real, demands a special explanation. Bambach believes that it may reflect increasing replacement by narrow niched molluscs of the broader niched dominants of earlier communities (brachiopods and bryozoans, for example).

Raup *et al.* (1973) have followed a deductive approach in supporting the steady-statist position (Raup and Gould, 1974, have also tried to model the history of morphology with similar techniques). The success of equilibrium thinking in theoretical ecology (MacArthur and Wilson, 1967; Levins, 1968; May, 1975a) has supplied the model for this enterprise. Raup *et al.* simulate evolutionary trees under conditions of equal probability for branching and extinction of lineages, and a constant equilibrium number of coexisting lineages for all time periods. They classify the tree into a smaller number of higher taxa (clades) and draw their diversity diagrams through time (Fig. 13.3). The simulations reproduce many features of real clades, universally attributed to directed causes in the past (e.g., strong waning of one clade while another waxes as evidence of competition).

Gould *et al.* (1977) have made a statistical comparison between random clades constructed over a wide range of probabilities for branching and extinction and real clades calculated for the entire history of life at several taxonomic levels. Some interesting differences emerge (greater fluctuation in size of real clades, for example) but the predominant impression is one of conformity between patterns in the real and random worlds. In particular, the average centre of gravity for 350 post-Ordovician extinct clades of genera within families for 8 major invertebrate groups is 0·4993, a more than satisfactory match to the stochastic prediction of 0·5 (maximum diversity in the middle of a clade's duration). For 353 Cambrian and Ordovician clades, average centre of gravity is 0·482 (maximum diversity below the middle), a situation also simulated in pre-equilibrial worlds with initially high rates of origination and stochastic balance of origination and extinction following this short initial period. The Cambro–Ordovician world, marking the first filling of the earth's oceans with metazoan life, was a time of pre-equilibrial increase. Clades arising during these times

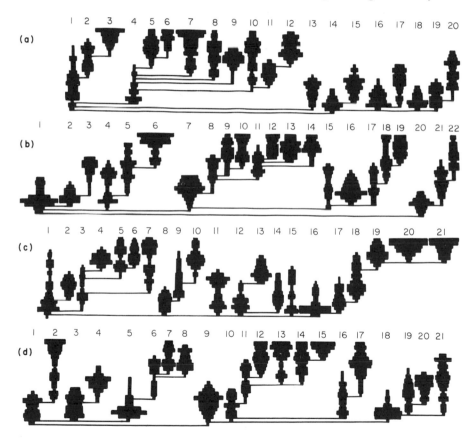

Fig. 13.3. The diversity and form of stochastic clades simulated within an equilibrium system by Raup *et al.* (1973). Only the seed of the random number generator (for decisions on extinction and branching) varies among the runs.

tend to have fat bottoms (initially rapid increase in diversity) followed by tapering demises during later, equilibrial times. Fat bottoms and tapering tops yield centres of gravity below 0·5.

13.4   Conclusion

13.4.1   *The ecological way of thought*

I have reviewed some specific contributions of ecological theory to the solution of long standing palaeontological dilemmas. Ultimately, however, its greatest impact may lie in a more subtle and pervasive direction: the reorientation of basic attitudes towards evolutionary questions. In discussing adaptation within phyletic sequences, it is legitimate to consider both 'immediate' significance in local situations and 'retrospective' significance for future evolutionary success (since we read phyletic stories after they have occurred). Darwin argued (largely in vain among palaeontologists) that evolutionary change was nothing more than adaptation to local circumstances. A 'trend' can only be a concatenation of these local adaptations; there are no intrinsic directions in evolution. Ecological time is rooted in this Darwinian present; ecological themes encourage palaeontologists to study adaptation for its immediate significance. Traditional palaeontology rarely worked at this level; it focused instead on the meaning of adaptation as a contribution to long-term evolutionary trends. And when it considered the immediate significance of adaptation at all, it did not venture beyond morphology in a 'physicalist' perspective— i.e., to what aspect of the physical environment is this structure fitted? The ecological themes of population dynamics, life history strategies and species interactions were simply not categories for consideration.

If we are now treating immediate significance seriously with those new themes, then almost every empirical study in palaeontology will be conducted differently in the future. I can only offer personal testimony. Two very similar species of land snails inhabit all environments and times of the Bermudian Pleistocene. They are usually found together in the same formation, but one or the other is sometimes absent. Fifteen years ago, I noted that one species in several localities from western Bermuda seemed to converge slightly upon the other. For years, I could not drop the blinders of traditional thought. I racked my brains for either phyletic meaning in the long run, or for the physical parameter that conditioned it—i.e., I searched for a way in which the sedimentary environment resembled that of samples dominated by the other species. I never considered interaction. Only a few years ago did I realize that these western samples included the only allopatric occurrences of my converging species and that the greater separation in sympatry probably represents character displacement (Schindel and Gould, 1977). In an excellent example of shifting perspectives, Stanley (1973b) has invoked differences in intensity of competition to explain the more rapid rates of evolution in mammals vs. clams. Simpson (1953) set this issue as a classic dilemma in macroevolution. Previous explanations had been confined to the traditional categories of morphology (e.g. mammals more complex).

To illustrate the blinders of macroevolutionary phylcticism, I submit the following: In a type of paedomorphosis, termed progenesis or paedogenesis (Gould, 1977), descendants retain the juvenile morphologies of ancestors because ontogeny is truncated by precocious sexual maturation. Progenetic organisms are usually very small and rapidly developing. Traditional wisdom (e.g. de Beer, 1958) has denied progenesis any importance in evolution because it is essentially 'degenerate', leading to simplification of structure and shortening of ontogeny—fine for some dead end parasites, but not for anything of phyletic significance. Note first of all, how this denial assumes that retrospective significance is the only thing that matters in evolution— a typical palaeontological chauvinism. Traditional thought could find no general immediate significance, but the ecological theme of life history strategies almost surely provides it (Gould, 1977, chapters 8-9): progenetic organisms are r-selected for rapid maturation in unpredictably fluctuating environments. Secondly, palaeontologists have had to face a dilemma of their own making. The origin of many higher taxa has been attributed to paedomorphosis by the sexual maturation of marine larvae. Clearly, the easiest mechanism for such paedomorphosis is progenesis (precocious maturation). But progenesis is supposed to have no evolutionary significance, and speculators have been driven to fanciful arguments for why paedomorphosis in these marine larvae might have been neotenic (delayed somatic development with no shortening of ontogeny). Slowly developing giant larvae become the fanciful models for new higher taxa. But: (1) we now have a good explanation from ecological theory for the immediate significance of progenesis; (2) immediate and retrospective significance are different categories. I see no reason why a small, rapidly maturing organism living in a precarious environment cannot be the stem species of a successful group.

13.4.2   Cross fertilization

Palaeontologists should be grateful to theoretical ecology for the reinvigoration of their field. But we surrender no sovereignty and look for reciprocal influence. I have already suggested that palaeontological data for truly temporal sequences might resolve some of the issues surrounding succession theory and the relationship of diversity to disturbance. I also believe that a bit of palaeontological traditionalism might serve as a good antidote for some excesses of simplification often encountered in ecological models. Too many models treat species as billiard-balls of no recognizable shape (though perhaps of different sizes). Yet adaptive morphology must be consulted for the solution of many problems in diversity. Take, for example, the remarkable species swarm of cichlid fishes in deep African lakes. A theoretical ecologist might look to stability of environments, breeding structures and life histories. But he would miss an important part of the story if he neglected the details of morphology. In an elegant paper, Liem (1973) demonstrates that explosive speciation was permitted by a shift in the insertion of two fourth levator externi muscles. This shift permits the pharyngial jaws to prepare food (rather than just transporting it as before). This frees the premaxillary and mandibular jaws to evolve exuberant specializations for the collection of food—scale scraping, eye biting, etc. As Deevey wrote in reviewing a founding work in modern palaeoecology (1965, p. 593): 'Older ideas of functional morphology are still valid; ecologists can learn much by simply looking at animals, before mentally decomposing them into fluxes of organic carbon.'

Palaeontology will retain its autonomy because the world allows no complete translation across its hierarchical levels of complexity. We should not surrender our uniqueness to the prestige and success of ecology; extrapolation from ecological to evolutionary time remains a dangerous game. When Tappen (1971) refers to patterns of diversity over many millions of years as a 'succession' (if only analogously), I begin to feel very uncomfortable. The concept of succession does not include replacement by evolution. The laws of scaling require different explanations at contrasting levels. An elephant cannot be built to look like a mouse and millions of years demand their own set of concepts.

Consider evolutionary 'trends' as one phenomenon that probably requires explanation at its own level. In the traditional view of phyletic gradualism, trends occur within lineages and are merely extensions of selection acting for the moment within populations. But Eldredge and Gould (1972) and Gould and Eldredge (1977) have argued that rapid events of speciation provide the primary input for macroevolutionary change; large populations rarely transform slowly and steadily. Each event of speciation represents a local adaptation, not a stage of a developing trend. Events of speciation provide an essentially stochastic input to evolutionary trends. The trends themselves represent differential preservation and success of a subset of speciations. Trends cannot be explained in the ecological time of speciation itself, but only in the evolutionary time of a higher-order 'selection' of speciation events (Eldredge and Gould, 1972, pp. 111–113; see also Stanley, 1975 and 1979, who refers to this mechanism as 'species selection'). The laws of ecology will not encompass trends, but they do set the speciations that serve as their building blocks. Ecological time will remain a fundamental level of palaeontological analysis. We give our thanks to modern ecology for teaching us the rudiments of perception in Darwin's own sphere of operation.

14

Population Ecology of Infectious Disease Agents

R. M. ANDERSON

14.1 Introduction

The earliest ecological studies of parasitic organisms were principally concerned with human disease agents. Recorded accounts of epidemic outbreaks, and speculations as to their possible origins, go back at least as far as the ancient Greeks [e.g. the Epidemics of Hippocrates (459–377 B.C.)]. Interest in epidemic phenomena led to one of the first theoretical studies in ecology, namely that of Bernoulli (1760) who used simple mathematical models to explore the dynamics of small-pox within human communities. Since this early beginning there has been considerable progress in the field of mathematical epidemiology (Bailey, 1975), although, surprisingly, this research has had little impact (until very recently) on the ecological study of animal and plant parasites (Anderson and May, 1979a; May and Anderson, 1979).

This chapter examines the dynamics of a wide spectrum of disease agents, ranging from viruses to helminths, within both vertebrate and invertebrate host populations. Attention is focused on two main areas, namely (1) the population mechanisms which enable parasites to stably coexist with their host species and (2), the regulatory impact of disease agents on host population growth.

14.2 Model structure

Population models of disease transmission have traditionally been couched in differential equation form, a pattern set by some of the earliest mathematical treatments (Ross, 1915; Kermack and McKendrick, 1927; Soper, 1929). It is generally assumed that birth, death and infection processes are continuous where generations overlap completely.

Two distinct types of mathematical framework have been employed. *First*, models of human infections are usually compartmental in structure, where the host population is assumed to be of constant size and interest is focused on the flow of hosts between compartments containing, for example, susceptibles, infecteds and immunes (Bailey, 1975). The disease agents studied are usually microparasites, the viruses, bacteria and protozoa; organisms which are characterized by small size, short generation times and the ability to multiply directly and rapidly within the host. Compartmental models do not describe changes in parasite population size, they simply mirror the dynamics of the number of infected hosts without reference to the abundance of organisms within these individuals.

These models seek to answer such questions as: can the infection be stably maintained within the population? Is the disease endemic or epidemic in character? How do the proportions of susceptibles, infecteds and immunes change through time after the infection is introduced into a virgin population? They are not generally used to address questions concerning the impact of infection on host population growth, although they can be adapted to do so (Anderson, 1979a; Anderson and May, 1979a).

The *second* type of model has arisen in the ecological literature, where attention has recently been given to the population dynamics of host–parasite associations with particular emphasis on the way macroparasites (the helminths and arthropods) depress the natural intrinsic growth rate of animal populations (Anderson and May, 1978; May and Anderson, 1978). The impact of a parasite on host reproduction and/or survival, the rate of production of transmission stages, and any resistence of the host (immunity) to further infection, all typically depend on the number of parasites present in a given host. A crude division of the host population into susceptible, infected and immune classes is therefore not ideal. A detailed description of the dynamics needs to deal with the full probability distribution of parasites within the host population (that is, with the number of hosts $n(i)$ harbouring i parasites, where $i = 0, 1, 2, \ldots$). This type of framework is somewhat broader since, in addition to the questions addressed by simple compartmental models, it seeks to answer problems concerning both the regulatory influence of the parasite on host population growth and the stability of the association between the two species.

In this chapter we will use both types of model, but first we briefly consider three basic concepts of disease dynamics which are central to later developments, namely: transmission; the basic reproductive rate; and threshold host densities for parasite maintenance.

14.3   Transmission between hosts

Disease transmission between hosts may be either direct, as a result of physical contact or the production of a free-living infective agent, or indirect via one or more obligatory intermediate hosts. Models of directly transmitted infections conventionally assume that the net rate at which susceptible hosts become infected is proportional to the density of susceptibles (X) times the density of infecteds (Y). Where β is some transmission coefficient and N represents total host population size, the net rate of transmission, $\beta X Y$, is formed from the total number of host contacts that susceptibles experience per unit of time $(\beta N X)$ times the proportion of those contacts which are infective (Y/N). These assumptions, although crude in the sense that no reference is made to the density of infective stages released into the habitat by infectious hosts, often provide a fairly accurate description of transmission, particularly if the life span of the free-living infective stage is short. A good example of this point is provided by the experimental work of Stiven (1964, 1967) who monitored the transmission of a protozoan pathogen *Hydramoeba hydroxena* within a population of the coelenterate host *Chlorohydra viridissima*. The time course of an epidemic of this disease, after the introduction of a few infecteds (numbering Y_0 at time $t=0$) into a population of X susceptibles, is accurately described by a model of the form

$$dY/dt = \beta X Y. \tag{14.1}$$

In Stiven's laboratory experiments, infected hosts rarely recovered from infection and host mortalities were negligible during the build-up of an epidemic (Stiven, 1964). The total host population size, N, thus remained constant and was defined by the sum of the number of susceptibles $[X(t)]$ and the number of infecteds $[Y(t)]$ throughout the epidemic.

The solution of eq. (14.1) is therefore

$$Y(t) = \frac{N Y_0 \exp(\beta N t)}{[(N - Y_0) + Y_0 \exp(\beta N t)]} \tag{14.2}$$

and as shown in Fig. 14.1a this model provides an excellent description of observed trends.

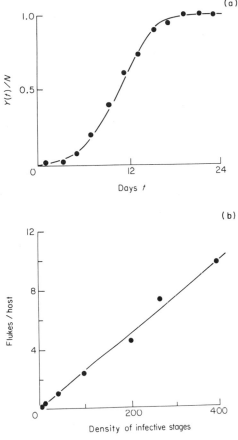

Fig. 14.1. (a) The temporal dynamics of an experimentally induced epidemic of the protozoan *Hydramoeba hydroxena* within a population of the coelenterate *Chlorohydra viridissima* [data from Stivens (1967)]. ●—observed proportions of hosts infected on day t, $Y(t)/N$. Solid line predictions of simple epidemic model defined in eq. (14.2) ($\beta = 0\cdot0043$). (b) The dynamics of infection of the fish host *Brachydanio rerio* by the cercarial infective stages of the ectoparasitic digenean *Transversotrema patialense* [data from Anderson *et al.* (1978)]. The relationship is between infective stage density and the number of parasites that establish on the surface of the host (host density and time of exposure held constant). ●—observed points. Solid line—best fit straight line constrained to go through the origin.

When direct transmission is achieved by means of free-living infective agents (produced by infected hosts) an accurate description of transmission dynamics should take into account the density of infective stages in the environment, particularly if the life span of these stages is not very short. It is frequently assumed that the rate of pick-up of infective stages is directly proportional to their density [$W(t)$ at time t] times the density of hosts in the habitat (N). Laboratory experiments confirm the validity of this assumption (Fig. 14.1b).

These assumptions of direct proportionality, however, are inappropriate for diseases which are transmitted indirectly (via one or more intermediate hosts), where infection results either from the consumption of one host by another, or due to the biting habits of one of the hosts. Predator–prey associations between hosts (which include the relationship between biting arthropods and their vertebrate 'prey') impose a maximum rate of transmission as a consequence of the functional response of the predator to prey density (see chapters 5 and 6). In the case of malaria, for example, and many other infections borne by biting arthropods, the intermediate host tends to make a fixed number of bites per week, independent of the number of vertebrate hosts available to feed on.

Other complications arise when the infection is passed from parent to unborn offspring. To distinguish it from the various forms of horizontal transmission discussed above, this process is labelled vertical transmission. It is clearly beneficial for the maintenance of parasites within low density host populations.

14.4 The basic reproductive rate of an infection (R)

One of the more important concepts to have emerged from theoretical studies of infectious diseases is that of the *basic reproductive rate* (MacDonald, 1957; Dietz, 1974; Anderson, 1980a). For compartmental models, based on the division of the host population into susceptibles, infecteds and immunes, this quantity (commonly denoted by the symbol R) represents the average number of secondary cases that one infected host gives rise to during its infectious lifespan if introduced into a large population of susceptibles. If a parasite is to persist within its host population the quantity R must clearly be equal to, or greater than, unity. When model structure takes account of the number or burden of parasites per host, R has a slightly different interpretation. In this case it represents the average number of offspring, produced by a mature parasite throughout its reproductive lifespan, which successfully complete their life cycle and attain reproductive maturity [note the similarity of this definition with Fisher's net reproductive rate R_0

(Fisher, 1930), see also eq. (2.16) in chapter 2]. Although the quantity R is widely referred to as a rate, it is a dimensionless quantity measuring the reproductive success of a disease in terms of the production of either infected hosts, or mature parasites.

To illustrate this concept we consider the dynamics of a directly transmitted viral microparasite, such as measles, within a host population of constant size N. This population consists of susceptibles, infecteds and immunes numbering respectively $X(t)$, $Y(t)$ and $Z(t)$ at time t. We assume that the net rate of input of susceptibles into the population (births) exactly balances the net rate of loss due to natural (bN) and disease induced (αY) host mortalities. We further assume that infected hosts are immediately infectious (no disease incubation period) and that, in line with earlier comments, the net rate of transmission is βXY. The average duration of infection is taken to be $1/v$ where v is the recovery rate and recovered hosts are assumed to be immune for the rest of their lives (as is the case for measles). These assumptions lead to the following differential equations,

$$dX/dt = bN + \alpha Y - \beta XY - bX \tag{14.3}$$

$$dY/dt = \beta XY - (\alpha + b + v)Y \tag{14.4}$$

$$dZ/dt = vY - bZ. \tag{14.5}$$

The model has two possible stable equilibrium solutions. If

$$\beta N/(\alpha + b + v) > 1 \tag{14.6}$$

the equilibrium proportion of infected hosts y^* adopts non-zero values, the disease persisting within the host population. If eq. (14.6) is not satisfied, the infection is unable to maintain itself and the equilibrium proportion of susceptibles (x^*) infecteds (y^*) and immunes (z^*) are respectively 1, 0 and 0. The basic reproductive rate of the infection is thus

$$R = \beta N/(\alpha + b + v). \tag{14.7}$$

This equation makes clear that the number of secondary cases produced by one infective throughout its infectious period is determined by the rate at which hosts join the infectious class (βN) times their expected lifespan in this class $[1/(b + \alpha + v)]$. The basic structure of the expression defining the quantity R is similar for all disease agents whether microparasites or macroparasites, and irrespective of

either their mode of transmission between hosts or the model framework used for the derivation of R (Anderson, 1980a). We shall see
evidence of this in the different models discussed in this chapter.

The value of R for an infection within a defined population can be
estimated from a knowledge of the biological characteristics of the
disease (i.e. host density and the rates of infection, recovery and host
mortality). In the case of directly transmitted diseases such as measles,
a simple relationship exists between R and the average age, A, at
which the host contracts the infection (Dietz, 1974). Specifically,

$$R = 1 + \frac{1}{bA} \tag{14.8}$$

where $1/b$ represents the expected life span of the host. The parameter
A can be estimated from data describing the proportion of individuals
in each age class of the population who have experienced the infection
(determined by serological surveys) (Muench, 1959; Dietz, 1974, 1976).
For example, Griffiths (1974) estimates that during the period 1956–69
in England and Wales the average age at which children experienced
an attack of measles was 5·16 years. With a life expectancy of roughly
70 years we may estimate R from eq. (14.8) as 14·6.

In the context of disease control we clearly need to reduce the
value of R to less than unity. This aim will be difficult to achieve if
the reproductive success of the disease is high (large R values). In the
case of our measles example, where an effective vaccine exists which
provides life-long immunity, the proportion p of the population which
must be protected at any one point of time to maintain the value of R
below unity is given by

$$p > \left[1 - \frac{1}{R} \right]. \tag{14.9}$$

It would therefore have been necessary to protect 93 per cent of the
population in England and Wales to eradicate measles in the 1950's.
The magnitude of this figure provides an explanation of why measles,
and similar viral diseases, are still endemic in Western Europe.

Before leaving the subject of measles, it is interesting to note that
some simple modifications of the model structure displayed by eqs.
(14.3)–(14.5) can result in considerable dynamical complications. For
example, the incorporation of an incubation period, such that hosts

may be infected but not infectious (roughly 12 days for measles), and the inclusion of annual periodicity in the transmission rate [$\beta(t)$] can produce complicated nonseasonal cycles in the prevalence (Y/N) of infection.

Between 1944 and 1964, measles in New York City (Yorke and London, 1973; London and Yorke, 1973) exhibited a regular biennial cycle, alternating between years of high and low incidence (Fig. 14.2). Recently, Yorke and London (1973) and Dietz (1974) [using models similar to that defined in eqs. (14.3)–(14.5)] have cogently argued that seasonality in the rate of transmission plus a brief disease incubation period are responsible for such observed patterns (see also Yorke *et al.*, 1979).

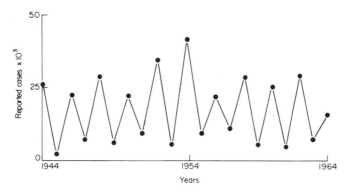

Fig. 14.2. The prevalence of measles in New York City between 1944 and 1964 (Yorke and London, 1973; London and Yorke, 1973).

14.5   Threshold host density for parasite maintenance (N_T)

By rearrangement of eq. (14.6) it can be seen that disease persistence is critically dependent on host population size, or density (N). A threshold density, N_T, exists below which the disease will be unable to establish if introduced into a susceptible population (Kermack and McKendrick, 1927). For the model defined by eqs. (14.3)–(14.5) this density is simply,

$$N_T = (\alpha + b + v)/\beta. \qquad (14.10)$$

Such threshold densities exist for virtually all diseases, whether caused by microparasites or macroparasites. The precise form of eq.

(14.10), however, is dependent on the mode of transmission between hosts (Anderson and May, 1979a; May and Anderson, 1979). An interesting exception is provided by the sexually transmitted infections, such as gonorrhoea, where the maintenance of disease is determined by the average degree of sexual promiscuity amongst the sexually active proportion of the population and is independent of total population density (Yorke *et al.*, 1978).

Where the disease agent's life cycle involves two or more hosts, such threshold densities are defined as a product (when transmission is achieved by means of free-living infective stages) or a ratio (when transmission involves a biting arthropod or predator–prey link) of the individual densities of each host species (Anderson and May, 1979a; May and Anderson, 1979) (see section 14.10).

The parameter structure of the equations which define these threshold densities provides many insights into the variety of strategies that parasites have evolved for persisting within populations of hosts. For example, many directly transmitted microparasites such as viruses and bacteria require high host densities in order to persist as a consequence of either high pathogenicity (large α), short durations of infection (large v) or low transmission efficiency (small β). Such parasites are therefore more commonly associated with animals that exhibit herd or shoaling behaviour, or breed and live in large colonies. Those directly transmitted diseases that do persist in low density host populations possess distinctive characteristics such as long-lived infective stages, failure to induce lasting immunity and the ability to persist within the host for long periods of time. Furthermore, they may often be transmitted vertically or by sexual contact.

Amongst the indirectly transmitted agents, many species traverse links in community food webs by virtue of predator–prey associations between final and intermediate hosts. Such associations (which include biting arthropods feeding on vertebrates), have played an important role in the evolution of complex life cycles. The high transmission efficiency (β) of these links reduces the threshold host density for maintenance and hence indirect life cycle infections predominate amongst parasites of hosts that exist at low density.

We now proceed to consider some models of specific infections, but throughout, the themes of reproductive success, threshold host densities and the ecology of transmission, will play a central role both in the development of our theories and their interpretation in the light of observed patterns.

14.6   Directly transmitted microparasites within laboratory populations of mice

Although there are relatively few studies of the influence of parasites upon the dynamics of laboratory populations, there is an exceptionally detailed study by Greenwood *et al.* (1936). These experiments, on laboratory populations of mice infected with various viral and bacterial diseases, have some simplifying features which make them particularly amenable to theoretical analysis. Specifically, the space available to the mice was adjusted to keep the population density constant as absolute levels changed. In addition adult mice were introduced at specified rates so that the basic population process was an immigration–death one which removed the time lags and other complications attendant upon recruitment by natural birth processes.

Wc now develop a simple compartmental model that captures the essentials of these experiments and discuss its fit to the data for two microparasites; one a bacterium, *Pasteurella muris*; the other a pox virus, ectromelia. Both parasites are transmitted directly between hosts by a very short-lived infective stage and induce a long lasting immunity to reinfection. Mice show some loss of immunity to reinfection by *P. muris*, but the immunity to ectromelia appears to be lifelong.

Using notation identical to that employed in eqs. (14.3) to (14.5), we define the total number of mice as $N = X + Y + Z$, where N is not assumed to be a constant but is determined by the dynamics of the infection. A is defined as the rate at which mice are introduced per day and thus where b is the natural mortality rate in the absence of the disease the mouse population will equilibrate at the immigration–death equilibrium $\hat{N} = A/b$. Since the infection is directly transmitted, and the infective stages are very short lived, we assume that the net rate of infection is $\beta X Y$. The mortality rate for infected mice is taken to be $(b + \alpha)$ with α representing the disease induced mortality. Mice recover from infection at a rate v and are initially immune, but this immunity can be lost at a rate γ (for permanent immunity, as for ectromelia, $\gamma = 0$). These assumptions lead to the following equations for the dynamics of the infection:

$$dX/dt = A - bX - \beta X Y + \gamma Z \qquad (14.11)$$

$$dY/dt = \beta X Y - (b + \alpha + v) Y \qquad (14.12)$$

$$dZ/dt = vY - (\gamma + b)Z. \tag{14.13}$$

Adding all three, the equation for the total population of mice is

$$dN/dt = A - bN - \alpha Y. \tag{14.14}$$

This system of equations has a stable equilibrium solution with the disease maintained in the mouse population if, and only if,

$$A/b > (\alpha + b + v)/\beta. \tag{14.15}$$

Failing this, the disease dies out and the population settles to its disease-free equilibrium value at $\hat{N} = A/b$. If eq. (14.15) is satisfied, the disease persists, and the total population is depressed below this infection-free level to the lower value

$$N^* = \frac{A + D \ (\alpha + b + v)/\beta}{b + D}. \tag{14.16}$$

Here D is defined for notational convenience as

$$D = \alpha/[1 + v/(b + \gamma)]. \tag{14.17}$$

Returning to our earlier themes note that, for an epidemic to occur after the introduction of a few infecteds into a population of susceptibles, the critical host density N_T is still as defined in eq. (14.10). The basic reproductive rate R, however, is now defined as

$$R = \beta N/(\alpha + b + v), \tag{14.18}$$

where N represents the potential mouse density as determined by the rate of input of susceptibles A times their expected life span, $1/b$ ($N = A/b$).

In their experiments on the maintenance of pasteurellosis ($P.$ $muris$) in mouse populations, Greenwood et al. (1936) introduced susceptible mice at rates ranging from $A = 0.33$ to $A = 6$ mice per day. Using data from Greenwood et al. (1936), Anderson and May (1979a) obtained the experimental results shown in Fig. 14.3a for the equilibrium mouse population N^* as a function of A. These data accord well with the linear relation between N^* and A predicted by eq. (14.16). In Fig. 14.3a, the dashed line depicts the equilibrium mouse population in the absence of the disease, $N = A/b$. The intercept of this line with the linear fit to the data for N^* in the presence of the disease yields the critical immigration rate A_T, below which R (eq. 14.18) becomes less than unity and the disease cannot persist. Anderson and May (1979a) estimated $A_T \simeq 0.11$ mice per day, corresponding to an equilibrium population of about 19 mice. Greenwood et al. (1936) suggested P. muris was always maintained in mice populations, but their lowest introduction rate was $A = 0.33$.

Anderson and May (1979a) crudely estimated the quantities b, α and v from life tables for uninfected populations, and from case mortality and recovery rates. Estimates of the parameters β and γ were obtained from the linear relationship portrayed in Fig. 14.3c and parameter-free predictions were calculated of the temporal development of the infection for a defined initial number of mice $N(0)$ and two introduction rates, $A = 6$ and $A = 0.33$. The fits between theory and data are shown in Figs. 14.3b and 14.3c and, bearing in mind the complete absence of adjustable parameters, both fits are extremely encouraging. They strongly suggest that simple deterministic models can be useful even when the host population is small.

How much does the disease depress the mouse population below the level that would pertain in its absence? This general question is answered in Figs. 14.3d and 14.3e, which show N^* as a function of disease pathogenicity for $A = 6$ and $A = 0.33$ respectively. Two interesting points emerge from this figure. First, the maximum depression of the host population is achieved by a disease of intermediate pathogenicity. Too small an α has little effect on N^*, while too large an α violates eq. (14.15) and makes R less than unity. This theoretical prediction has important practical implications for the use of pathogens as biological control agents of pest species (Anderson, 1979b).

Second, note that the higher the immigration rate A, the greater the degree of depression of the host population relative to the disease free equilibrium ($\hat{N} = A/b$) (see also Fig. 14.3a). This prediction supports the view that diseases caused by directly transmitted viruses and bacteria are more likely to persist within, and cause severe reduction of, host populations with high birth (or immigration) rates. This phenomenon derives essentially from the high inflow of susceptibles helping to maintain a disease which induces lasting immunity to reinfection in those hosts that recover from infection.

Greenwood et al. (1936) and later Fenner (1948, 1949), also studied the effects of the mouse pox, ectromelia. An analysis, akin to that just outlined, leads to similarly encouraging agreement between theory and the experimental data for ectromelia in laboratory populations of mice (Anderson and May, 1979a). Some of these results are presented in Fig. 14.3f.

The theory and facts of these experiments are in accord in showing how infectious diseases can stably regulate their host population below disease-free levels. Similar theoretical results emerge if we broaden the model structure of eqs. (14.11)-(14.13) by replacing the immigration rate A of eq. (14.13) with a true birth process of the form $a(X + Y + Z)$ where a is the per capita birth rate (Anderson, 1979a; Anderson and May, 1979a). The equation for the total host population then becomes

$$dN/dt = rN - \alpha Y \qquad (14.19)$$

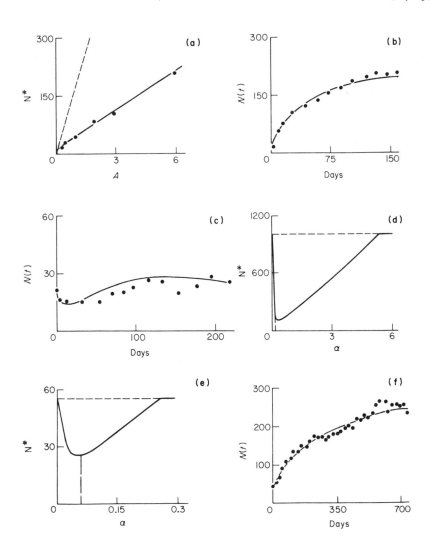

where the intrinsic growth rate of the disease free host population is
defined as $r = a - b$.

One of two circumstances now arises. If

$$\alpha > r \left[1 + \frac{v}{b + \gamma} \right] \tag{14.20}$$

the parasite regulates the host population to a stable value N^* where

$$N^* = \frac{\alpha(\alpha + b + v)}{\beta \left[\alpha - r \left(1 + v / \{ b + \gamma \} \right) \right]}. \tag{14.21}$$

Of this equilibrium population, the fraction infected, y^*, is given
from eq. (14.19) as

$$y^* = r / \alpha. \tag{14.22}$$

Conversely, if eq. (14.20) is not satisfied, the system eventually settles
to a state in which the total population grows exponentially at a
rate p given by

$$p = [B^2 - (b + \gamma) (\alpha - r) + rv]^{\frac{1}{2}} - B \tag{14.23}$$

with $B = \frac{1}{2} (\alpha + b + v + \gamma - r)$. This population growth rate is necessarily

Fig. 14.3. Population dynamics of *Pasteurella muris* in colonies of
laboratory mice [data from Greenwood *et al.* (1936), analysis described
in Anderson and May (1979) and text].
(a) Relationship between the equilibrium population of mice N^* and
the daily input of susceptible mice A. Solid dots are observed levels; the
solid line is the best linear fit (eq. 14.16). The dashed line shows the
estimated relationship between N^* and A in the absence of the disease
(the slope is $1/b$, where $b = 0.006$/day).
(b) and (c) Growth of mouse colonies harbouring the disease, from an initial population of 20 mice, for $A = 6.0$ and 0.33 respectively. ● are the
experimental data while the solid lines are the theoretical predictions
of the model defined by eqs. (14.11) to (14.13).
(d) and (e) Relationship between the equilibrium population of mice
N^* and the disease induced mortality rate α, as predicted by eqs. (14.11)
to (14.13), for $A = 6.0$ and 0.33 respectively.
(f) Population dynamics of ectromelia virus in laboratory colonies of
mice [data from Greenwood *et al.* (1936) and Fenner (1948, 1949)].
Growth of a mouse colony harbouring the viral infection from an
initial population of 45 mice where $A = 3.0$ [dots and solid curve as in
graphs (b) and (c)].

less than the disease-free one, $p < r$. In this crude model there are no density-dependent regulatory effects other than the disease itself and the population 'runs away' maintaining the infection within it at the diminished rate p, if α is too small to satisfy eq. (14.20). Conversely, if eq. (14.20) is fulfilled, the population settles to the equilibrium level N^* given by eq. (14.21). In either case, if N is initially less than the threshold value N_T defined in eq. (14.10) (in other words R is less than unity), then initially Y will decrease and X will increase exponentially at the rate r. However, once X exceeds N_T on this trajectory of exponential increase, then Y will increase, and the system either will converge on the N^* of eq. (14.21), or will grow at the slower rate p of eq. (14.23).

Equation (14.20) is a key one and it can be modified to take into account the known biology of a wide range of directly transmitted microparasitic infections. Anderson and May (1979a) discuss a variety of such refinements including the effects of incubation periods, vertical transmission and infections that reduce host reproduction. We may summarize certain of their conclusions as follows. (1) For a disease to regulate the host population, the case mortality rate α must be high relative to the intrinsic growth rate r of the disease-free host population. The ability to achieve this degree of regulation is decreased by lasting immunity (γ small) and high rates of recovery from infection (v large, corresponding to infections of short duration). (2) Diseases with long incubation periods, where hosts are infected but not infectious, have less impact on population growth. (3) Diseases which affect the reproductive capacities of infected hosts are more liable to suppress population growth. (4) Vertical transmission lowers both the magnitude of the threshold density, N_T, needed for successful introduction of the disease and the equilibrium population of the host in those cases where it is regulated by the disease.

14.7   Infectious diseases and population cycles of arthropods

A major complication in models of directly transmitted infections arises when the free-living transmission stages of the pathogen are longlived. This happens for many parasitic infections of insects and it can make the dynamical behaviour of the host–parasite system qualitatively more complex. We can no longer assume that the net rate of transmission is simply proportional to the density of susceptibles

(X) and infecteds (Y). As mentioned earlier (see Fig. 14.2), we must now take into account the density of infected stages in the habitat of the host (W) since the transmission term will be proportional to the rate of encounters between susceptible hosts and infective stages of the parasite, $\hat{\beta}WX$.

Such free-living infective stages include the spores of many bacteria, protozoans and fungi, and the capsules, polyhedra or free particles of viruses (Smith, 1976). Many baculoviruses of temperate forest insects (principally of lepidopteran, hymenopteran and dipteran species) tend to have very long-lived infective stages, partly because the soil environment of forests affords relative protection from the ultra-violet components of sunlight (Anderson and May, 1981; Jacques, 1977).

Some simplification of model structure stems from the fact that invertebrate hosts appear to be unable to develop an effective degree of acquired immunity to parasitic infection (Maramorosch and Shope, 1975). Thus, although invertebrate species are often able to mount cellular and/or humoral responses to parasitic invasion, the recovered individuals pass directly back into a pool of hosts susceptible to further infection (Bang, 1973; Bayne and Kine, 1970). This makes the dynamics somewhat simpler than is the case for most vertebrate species, where a distinct category of immune hosts needs to be accounted for.

Using notation similar to that employed in eqs. (14.12), (14.13) and (14.19), and defining $W(t)$ as the density of infective stages at time t, our new model is of the form:

$$dX/dt = aN - bX - \hat{\beta}XW + vY \qquad (14.24)$$

$$dY/dt = \hat{\beta}XW - (\alpha + b + v)Y \qquad (14.25)$$

$$dW/dt = \lambda Y - (\mu + \hat{\beta}N)W \qquad (14.26)$$

Here, infective stages are produced at a rate λ by infected hosts, and are lost by death (at a rate μ) or by absorption in hosts (at a rate $\hat{\beta}N$). The total size of the host population is now defined as $N = X + Y$.

Provided infected hosts produce transmission stages of the parasite at a sufficiently fast rate, the disease will regulate its host population so long as $\alpha > (a - b)$ (compare with eq. (14.20) of previous model defined in eqs. (14.12), (14.13) and (14.19)). More precisely λ must satisfy,

$$\lambda > \alpha(\alpha + b + v)/(\alpha - r). \qquad (14.27)$$

If this condition is not met, the disease will persist provided $R > 1$, where R is now defined as

$$R = \frac{\hat{\beta}\lambda N}{(\alpha+b+v)(\mu+\hat{\beta}N)} \qquad (14.28)$$

and the threshold host density N_T becomes

$$N_T = \frac{(\alpha+b+v)\mu}{\hat{\beta}(\lambda-(\alpha+b+v))}. \qquad (14.29)$$

These equations are to be compared with eqs. (14.7) and (14.10) derived from the model in which the rate of transmission was assumed proportional to the densities of susceptibles and infecteds.

The interest in this model, however, lies in the regulated state which may be either a stable point *or* a stable cycle (Anderson and May, 1980). The cyclic solutions tend to arise for infections of high pathogenicity (large α) that produce large numbers of long lived infective stages (λ very big and μ small). The period of these cycles may be long if the natural intrinsic growth rate of the host, r, is small. Fig. 14.4 illustrates the relation between the period T of the host's

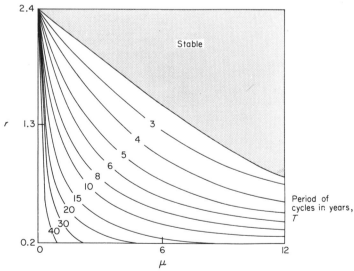

Fig. 14.4. The solution to eqs. 14.24, 14.25, 14.26 is indicated for various values of μ and $r(r = a-b)$, with the other relevant rate parameters fixed at $\alpha = 9 \cdot 0$, $b = 3 \cdot 3$, $\gamma = 0$ (all units in year⁻¹). For combinations of μ and r in the shaded region, the solution is a stable equilibrium point. The contour lines correspond to stable limit cycle solutions, with period, T (in years) as labelled (Anderson and May, 1980).

population cycles and the parameters μ and r, for large α (Anderson and May, 1980).

Many microsporidian protozoan and baculovirus infections of insects appear to possess the combination of relatively large α and small μ that produce cyclic changes in host abundance. For example, the polyhedra of many nuclear-polyhedrosis viruses of insects can survive in the soil of temperate forests for a number of years (Jacques, 1969; Thomas et al., 1972; Thompson and Scott, 1979). Moreover, insertion of a plausible range of values of μ and r in Fig. 14.4 suggests cycles with periods in the general range 3–30 years (Anderson and May, 1981). This range is to be compared with the observed periods of 5–12 years for cycles in populations of forest insect pests documented in Varley et al. (1973), many of which harbour virus and protozoan

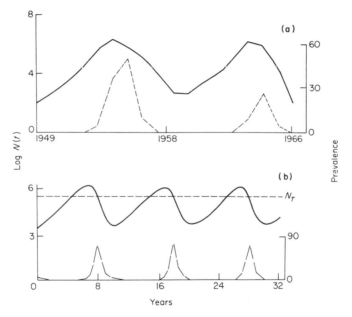

Fig. 14.5. (a) The figure shows observed changes in the abundance (solid line; plotted logarithmically) of the larch bud moth, *Zeiraphera diniana*, in the European Alps, and in the prevalence (dashed line, expressed as a percentage and plotted linearly) of infection with a granulosis virus (Auer, 1968).

(b) The asymptotic solution of eqs. 14.24, 14.25, 14.26 for host abundance, $N(t)$, and for the prevalence of infection, $Y(t)/N(t)$, are plotted as functions of time; the parameters in eqs. 14.24, 14.25, 14.26 are assigned values appropriate to the larch bud moth and the granulosis virus that infects it (Anderson and May, 1980, 1981).

infections. It appears likely that some of these population cycles are driven by the interactions between host insect and pathogen.

In support of this suggestion, Fig. 14.5 shows a comparison between theory and observation for the population dynamics of the larch bud moth, *Zeiraphera diniana*, in the European Alps. Fig. 14.5a shows the data for the abundance of the insect; and for the prevalence of infection with a granulosis virus, over a 20-year period (Auer, 1968; Baltensweiler, 1964). Fig. 14.5b shows the same quantities calculated from eqs. (14.24), (14.25) and (14.26) with approximate values of the parameters a, b, α, v, λ and μ estimated independent of this population data (Anderson and May, 1980, 1981). Estimation of the transmission parameter β is impossible, but fortunately it only enters in determining the relative magnitude or scaling of $N(t)$. The agreement between Figs. 14.5a and 14.5b, with respect both to the period and to the shape and magnitude of the oscillations in bud moth populations and prevalence of virus infection is encouraging.

Several general features of population cycles driven by host-pathogen associations are illustrated in Fig. 14.5. First, the peak in prevalence of the infection within the host population occurs shortly after the peak in host abundance. Second, the host population falls below the threshold value N_T during part of its cycle; the infection survives by virtue of its relatively long-lived transmission stages. Third, when N is below N_T the prevalence declines effectively to zero so that the disease appears to have disappeared from its host population. It is a mistake, however, to think that its apparent disappearance, or 'epidemic' reappearance is inconsistent with the pathogen driving the host cycles. This point is clearly illustrated in Fig. 14.5b.

Other mechanisms are capable of producing cycles in host parasite associations. As we saw earlier for measles in human populations (Fig. 14.2), seasonal changes in the transmission rate in eqs. (14.24) to (14.26) can produce annual cycles in the host population, or even, under some circumstances, cycles with periods of 2 years or other higher harmonics (Dietz, 1976; Yorke and London, 1973; Anderson and May, 1979a; 1981). Predation or other factors extrinsic to the host parasite interaction can produce cycles in the host population (see e.g. chapters 2, 5, 6 and 7). Furthermore, these different cycle producing mechanisms can interact to produce very complex oscillatory behaviour in host abundance (Anderson and May, 1981). The simple host parasite mechanisms represented in eqs. (14.24) to (14.26), however, are sufficient to account for many long-term population cycles of forest insects. This model is of more than academic interest, as it enables one to calculate the rate at which a virus or other pathogen must be artificially introduced if it is to be effective in the control, or extinction, of a population of insect pests (Anderson and May, 1981).

14.8 Human hookworm infections

Directly transmitted nematode parasites such as the hookworms *Ancylostoma duodenale* and *Necator americanus*, and the roundworm *Ascaris lumbricoides* are amongst the most prevalent of all human infections in the world today. World Health Organization statistics in 1975 record that more than one third of the world's population was infected with one or more of these species (Peters, 1978).

The stability of their populations is striking in many tropical, subtropical and temperate regions of the world; the prevalence and intensity of infection often remains remarkably constant over long periods of time (Nawalinski *et al.*, 1978a, 1978b; Muller, 1975). They are notoriously difficult to eradicate; the cessation of control programmes based on chemotherapy invariably results in the parasite populations rapidly returning to their pre-control levels.

Since hookworms cause morbidity rather than mortality, our model will focus on the dynamics of parasite population growth within a host population of constant size, N (the impact of infection on host population growth will not concern us). The assumption that the host population is effectively constant is a reasonable one partly because the dynamics of the parasite operate on a much faster time scale than that of the human host. In many regions of endemic hookworm, for example, human life expectancy is roughly 50 years while that of the adult parasite is in the region of 3.5 years (Hoagland and Schad, 1978).

Before we consider model structure and behaviour some brief comments on the developmental cycle of human hookworms are necessary.

The life history of the parasite begins with the production of eggs by mature female worms in the small intestine of the host. These pass to the exterior in the faeces and hatch to release a first stage or L_1 larva which undergoes two further moults to produce an L_3 infective larva. The L_3 stage is responsible for host location and infection, gaining entry by either direct penetration or ingestion. Once inside the host, a phase of migration is followed by growth to sexual maturity

in the small intestine. Two populations in this life cycle play a dominant
role in determining transmission success; namely, the sexually mature
parasitic worms and the free-living infective larvae. The former are
responsible for the single phase of reproduction while the latter determine the rate at which new hosts are colonized. Infection is essentially
a form of reproduction since it places the parasite in a habitat in
which it is able to reproduce.

The pathogenicity of hookworms to the host, the rate of production of eggs and the mortality rate of mature parasites, all depend
on the number of parasites present in a given host. Our model must
therefore take into account the full probability distribution of parasite
numbers per host and hence we now depart from the susceptible, infected and immune framework.

We restrict our attention to the sexually mature parasites and
infective larvae, numbering respectively $P(t)$ and $L(t)$ at time t. The
biology of parasite development via the remaining segments of the
life cycle is reflected in the time delays and mortalities associated
with transmission between these populations. The assumptions incorporated in the hookworm model are discussed at length by Anderson
(1980b) and our treatment here is necessarily brief.

We define b to be the per capita rate of human mortality ($1/b \approx 50$
years); $\hat{\beta}$ a transmission coefficient (the net rate of transmission is
assumed proportional to the number of hosts, N, and to the number
of infective larvae L (see Fig. 14.1b)), λ the per capita rate of egg
production by mature female worms (the net rate of egg production
assuming a 1 : 1 sex ratio of male to female worms, is $\frac{1}{2}\lambda\phi P$ where ϕ
is the probability that a given worm is mated); and μ_3 the per capita
death rate of infective larvae. Adult worm mortality is assumed to be
density-dependent where the death rate μ is linearly related to parasite
burden, i, such that $\mu(i) = \mu_1 + \hat{\alpha}i$ (Krupp, 1961; Anderson, 1980b).
We further define T_1 as the time delay from host infection to the
achievement of sexual maturity ($T_1 \simeq 40$ days) and D_1 as the proportion of successful infections which reach maturity. Similarly T_2 is
the time delay from the release of an egg to the development of an
infective larva [$T_2 \simeq 5$ days under optimum conditions (Augustine,
1923a and b)] and D_2 is the proportion of these eggs which develop
and survive to the infective larval stage.

Some simplification in model structure can be achieved by making a
phenomenological assumption concerning the statistical distribution of
worm numbers per host. On the basis of empirical evidence we assume

that the parasites are distributed in a negative binomial manner (Bliss and Fisher, 1953), with a characteristic degree of aggregation or clumping measured inversely by the parameter k. This distribution, in conjunction with the sexual habits of the worm, determines the likelihood of successful mating (the parameter ϕ). Human hookworms are dioecious and are thought to be polygamous. Figure 14.6 displays the relationship between the mating function ϕ and the mean worm burden per host M for various sexual habits and worm distributions. Irrespective of whether the worms are monogamous or polygamous, parasite clumping is clearly advantageous, provided, of course, that

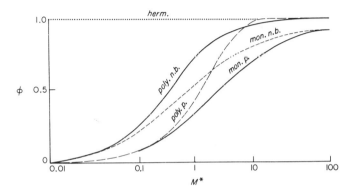

Fig. 14.6. The relationship between the probability ϕ that a given female worm is mated (and thus by assumption producing eggs) and the mean worm burden per host M for various assumptions about sexual habits: *herm* = hermaphroditic and self-fertilizing; *poly. n.b.* = polygamous worms distributed in a negative binomial manner ($k = 0.34$); *poly. p.* = polygamous worms distributed in a Poisson manner; *mono. n.b.* = monogamous worms distributed in a negative binomial manner; *mono. p.* = monogamous worms distributed in a Poisson manner [see May (1977b) for formal details of the derivation of ϕ].

male and female worms are distributed together and not in an independently clumped fashion (May, 1977b). Polygamous habits are obviously beneficial to reproductive success when worm density is low.

These assumptions lead to the following differential equations for M, the mean worm burden per host ($M = P/N$), and L, the density of infective larvae,

$$dM/dt = \hat{\beta}L(t - T_1)D_1 - (b + \mu_1)M - \frac{\hat{\alpha}(k+1)M^2}{k} \qquad (14.30)$$

$$dL/dt = \tfrac{1}{2}\lambda\phi M(t - T_2)D_2 N - (\mu_3 + \hat{\beta}N)L. \qquad (14.31)$$

A further simplification in model structure can be achieved by noting that (a) the two time delays T_1 and T_2 are short ($T_1 = 40$ days, $T_2 = 5$ days) and (b) the expected life span of the mature parasites ($1/\mu_1 = 3.5$ years) is very long in comparison to that of the infective larvae ($1/\mu_3 = 5$ days). We can therefore decouple eqs. (14.30) and (14.31) by assuming the 'short lived' infective stages are adjusted essentially instantaneously to the equilibrium level ($dL/dt = 0$) for any given value of M. This gives a single dynamical equation for the variable M:

$$dM/dt = M\left[\frac{\frac{1}{2}\lambda\phi\hat{\beta}D_1D_2N}{(\mu_3 + \beta N)} - (b + \mu_1) - \hat{\alpha}\frac{(k+1)}{k}M\right]. \quad (14.32)$$

An analytical solution of eq. (14.30) for arbitrary ϕ is not feasible but we can determine the qualitative behaviour by means of phase plane analysis. Two general patterns emerge. If the basic reproductive rate of the parasite R is greater than unity (where R is now defined as the average number of offspring produced by a female worm during her reproductive life span, which achieve sexual maturity), the infection is maintained within the host population. More precisely R is defined as

$$R = \frac{\frac{1}{2}\lambda\phi\hat{\beta}D_1D_2N}{(\mu_3 + \beta N)(b + \mu_1)}. \quad (14.33)$$

In this event there are two alternative stable states (one at $M = 0$ and one at a finite value M^*), separated by an unstable state M_u. If R is less than unity, the parasite is unable to persist and a single stable state exists at $M = 0$. The condition $R = 1$ is often termed the *transmission threshold* in the context of helminth epidemiology (Macdonald, 1965), while the unstable state M_u is sometimes referred to as the *breakpoint*. These two concepts are of major practical significance to epidemiologists concerned with parasite eradication. Ideally, we need to either reduce R to less than unity such that the population parameters of the parasite fall below the transmission threshold, or reduce the mean worm burden M^* below the unstable breakpoint M_u such that the dynamical trajectories are 'attracted' to the stable state of parasite extinction ($M = 0$). These notions are displayed graphically in Fig. 14.7, where the equilibrium worm burden M^* is plotted for various values of R.

The critical question surrounding the existence of an unstable breakpoint concerns its precise location (the value of M_u) in relation

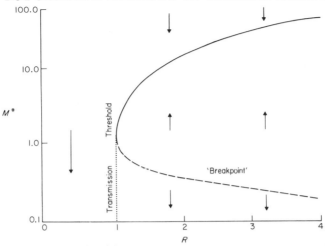

Fig. 14.7. The relationship between the equilibrium mean worm burden per host, M^*, and the basic reproductive rate R for endemic hookworm in a rural community in India (from Anderson, 1980b). The dashed line indicates the unstable equilibrium or 'breakpoint', M_u, and the arrows denote the dynamical trajectories of the system following a perturbation from the two stable equilibrium states, M^* and zero. Below the transmission threshold, $R = 1$, the infection cannot be maintained ($M^* = 0$). As the degree of worm clumping increases (k becomes smaller) the breakpoint comes to lie closer to the horizontal axis [hookworms are dioecious and polygamous, their distribution within human populations is invariably highly clumped].

to the two stable states M^* and $M = 0$. For example, if the value of M_u is close to zero, the stable state of endemic disease M^* is effectively globally stable since the region of attraction to $M = 0$ is very small (Fig. 14.7). The magnitude of this region is to a large extent determined by the degree of worm clumping, high levels of aggregation resulting in the value of M_u being close to zero. Unfortunately this often appears to be the case for endemic hookworm. For example, in a recent study of the dynamics of *N. americanus* infections in a rural community in India, Anderson (1980b) estimated that the values of M^*, M_u and R were respectively 51, 0·3 and 2·7. Similar conclusions have recently emerged from studies of the dynamics of schistosomiasis, an indirectly transmitted helminth disease (Bradley and May, 1978; May, 1977b).

It can be seen from Fig. 14.6 that when the mean worm burden is large the value of the mating function ϕ is effectively unity. In such cases, we can obtain an analytical solution of eq. (14.30) which enables us to derive models to predict the intensity (mean worm burden) and prevalence (proportion of population infected) in different age classes of a community. Where $M(a)$ and $p(a)$ are respectively the intensity and prevalence at age a, we obtain,

$$M(a) = K\left[\exp[-(R-\mathrm{I})(\mu_1+b)a] + \frac{\hat{\alpha}(k+\mathrm{I})}{k(R-\mathrm{I})(\mu_1+b)}\right]^{-1} \qquad (14.34)$$

and

$$p(a) = \left[\mathrm{I} - \left(\mathrm{I} + \frac{M(a)}{k}\right)^{-k}\right]. \qquad (14.35)$$

[In eq. (14.34) the constant K is defined by initial conditions.] Equation (14.34) describes a logistic curve which rises to the asymptotic equilibrium mean worm burden M^*. The age prevalence curve is obtained from the zero probability term of the negative binomial distribution with mean $M(a)$ and clumping parameter k.

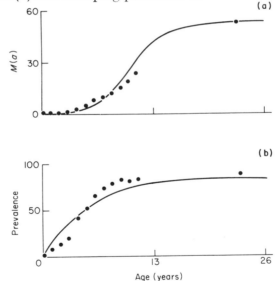

Fig. 14.8. Epidemiological data for *Necator americanus* infections in a population of children from a rural community near Calcutta in India (data from Nawalinski *et al.*, 1978a, 1978b).

(**a**) The average worm burden per host in different age classes of the population (age–intensity curve). ●—observed data; solid line—model predictions (eq. 14.34) fitted to observed data by non-linear least squares method ($R = 2\cdot7$).

(**b**) The prevalence of infection in different age classes of the population (age-prevalence curve). ●—observed data; solid line—model predictions (eq. 14.35).

Despite the crudity of these age intensity and age prevalence models, eqs. (14.34) and (14.35), they predict trends which are in broad agreement with observed data; an example of which is shown in Fig. 14.8.

The preceding analysis now enables us to return to the question of why populations of hookworms are remarkably stable through time and robust to control intervention. This observed stability is basically due to: (1) density dependent constraints on worm population growth within an individual host [probably acting on worm mortality *and* fecundity (Krupp, 1963; Hill, 1926)]; (2) high degrees of worm aggregation within the host population ($k = 0.01 \rightarrow 0.6$) adding to the efficiency of density dependent constraints; (3) R values which are a number of orders of magnitude greater than unity (in part due to polygamy and worm aggregation enhancing mating frequency but also to the enormous reproductive output of mature female worms (for *N. americanus* $\lambda = 15,000$/day)); and (4) equilibrium worm burdens which are *effectively* globally stable to perturbation.

These conclusions have many implications for the design of control programmes (Anderson, 1980b).

14.9 The impact of helminths and protozoa on host population growth

The model framework used to study the dynamics of hookworm infections can easily be extended to consider more general ecological (as opposed to epidemiological) questions concerning the impact of directly transmitted parasites on the dynamics of host population growth. Such models, with their detailed description of the distribution of parasite numbers per host, are appropriate for the study of helminths and protozoa where direct counts of the parasite burden per host are feasible.

The most detailed study of this type draws on a wide collection of data and describes the dynamics in terms of three differential equations for the number of hosts N, parasites P and free-living infective stages W (Anderson and May, 1978; May and Anderson, 1978):

$$dN/dt = (a-b)N - \alpha P \tag{14.36}$$

$$dP/dt = \beta W N - (\mu + b + \alpha)P - \alpha(k+1)P^2/(kN) \tag{14.37}$$

$$dW/dt = \lambda P - cW - \hat{\beta}WN. \tag{14.38}$$

Here, the host birth and death rates are as defined in eq. (14.19) and the transmission parameter $\hat{\beta}$ as in eq. (14.24). The parasite-induced host death rate (or, equivalently, depression of the birth rate) is taken to be linearly proportional to the parasite burden (i) in a given host, at a rate α per parasite (Anderson and May, 1978; Anderson, 1978). The parasites are distributed as a negative binomial with parameter k [observed distributions are well described by this probability distribution (Crofton, 1971; Anderson, 1978)]; μ is the natural mortality rate of adult parasites; λ is the rate of production of infective stages by mature adult parasites (where the mating function ϕ is assumed for simplicity to be unity, an assumption appropriate for hermaphroditic self-fertilizing helminths and asexually reproducing protozoa) and c is the death rate of these infective stages.

For many host–parasite associations the infective stages have a much shorter life span than that of the host and the adult parasites. We can therefore decouple the set of differential equations, in a similar manner to that shown for the hookworm model [eqs. (14.30) and (14.31)], such that

$$dN/dt = rN - \alpha P \tag{14.39}$$

$$dP/dt = \frac{\lambda NP}{H_0 + N} - (\mu + b + \alpha)P - \frac{\alpha(k+1)P^2}{kN} \tag{14.40}$$

(where $r = a - b$ and $H_0 = c/\hat{\beta}$).

This model possesses three patterns of dynamical behaviour. Provided the basic reproductive rate, R, is greater than unity the parasite persists within the host population, where R is defined as

$$R = \frac{\lambda\hat{\beta}N}{(\mu + b + \alpha)(c + \hat{\beta}N)}. \tag{14.41}$$

In this event the parasite regulates the host population to a stable equilibrium provided

$$\lambda - (\mu + b + \alpha) > r(k+1)/k. \tag{14.42}$$

In other words the parasite's effective rate of reproduction $[\lambda - (\mu + b + \alpha)]$ must be greater than the host reproductive rate r, weighted by a factor $(k+1)/k$ to allow for the clumped distribution of parasites. When regulation is achieved the average parasite burden M^* settles to,

$$M^* = r/\alpha. \tag{14.43}$$

If eq. (14.42) is not satisfied, but R is greater than unity, the host population is unregulated and grows exponentially at a rate

$$p = r - [\lambda - (\mu + b + \alpha)] [k/(k + 1)] \tag{14.44}$$

which is clearly less than the disease free rate, r. In this case the parasite is maintained and the mean parasite burden, in the exponentially growing host population, settles to the value

$$M \to (r - p)/\alpha. \tag{14.45}$$

In either event if the host population is initially below the threshold value

$$N_T = \frac{c(\mu + b + \alpha)}{\beta[\lambda - (\mu + b + \alpha)]}, \tag{14.46}$$

the parasite cannot become established. However, provided $R > 1$, the host population will grow exponentially until N_T is exceeded and the parasite will then persist, either regulating or at least slowing host population growth.

Finally if $R < 1$ the infection can never become established.

The similarities between these predictions and those derived from the susceptible-infected-immune microparasite model [eqs. (14.12), (14.13) and (14.19)] are striking. The concepts of reproductive success and threshold host densities emerge from both types of models. Moreover, these models clearly suggest that parasites of all kinds are able to stably regulate host population growth in the absence of or in conjunction with (Anderson, 1980c) other regulatory constraints. Some laboratory examples of this regulatory potential are displayed in Fig. 14.9.

The model defined by eqs. (14.37) and (14.40) can be modified to take account of a wide variety of biological complications specific to individual host–parasite interactions. Space forbids a detailed discussion of these but the major issues may be summarized as follows. The stable coexistence of host and parasite and the regulatory potential of the infection is facilitated by: (1) aggregated distributions of parasite numbers per host; (2) density-dependent parasite mortality or reproduction; (3) rates of parasite induced host mortality that

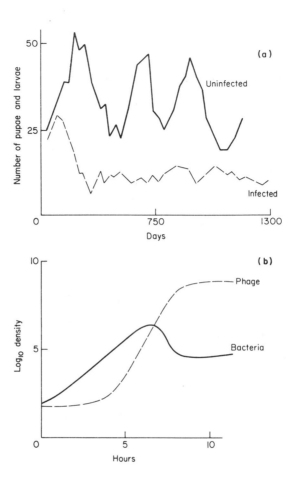

Fig. 14.9. Laboratory examples of the regulation of host population growth by parasitic species.

(**a**) The growth of a population of *Tribolium castaneum* infected with the protozoan parasite *Adelina triboli* [data from Park (1948)]. Dashed line depicts infected population while solid line represents the control uninfected population.

(**b**) The impact of a bacteriophage (V_i—phage A) on the population growth of its host, *Salmonella typhi* (V_i—type A) (data from Anderson, 1957). The dashed line depicts the density of phage particles and the solid line the density of bacteria. Note that the bacterial population grows exponentially, until the density exceeds a critical level, at which point the phage regulates bacterial growth [see Anderson and May (1981)].

increase faster than linearly with parasite burden; and (4) parasite–
induced host mortality rates that are more severe at high host densities. Conversely, a number of processes have a destabilizing effect:
namely, (1) parasite-induced reduction in the host birth rate; (2) direct
parasite population growth within individual hosts; (3) developmental
time delays; and (4) random or underdispersed (regular) distributions
of parasite numbers per host. The long term persistence of associations
which exhibit such destabilizing influences suggests that evolutionary
pressures have counteracted these with strong stabilizing mechanisms
(May and Anderson, 1978).

One biological complication of special interest is the now widely
recognized fact that the impact of an infection is often related to the
nutritional state of the host. Broadly speaking, malnourished hosts
have lowered immunological competence and are less able to with
stand the onslaught of infection (Steinhaus, 1958; Gordon, 1963;
Scrimshaw et al., 1968). The effective pathogenicity of a parasite
therefore tends to increase as host density rises to a level where competition for available food is severe. Given certain reasonable assumptions about the exact relation between pathogenicity (α) and host
density (N), two stable states may occur for a given set of rate parameters (Anderson, 1979a). The outcome of such a model [similar to
eqs. (14.39) to (14.40) but with the pathogenicity α expressed as a
function of N] is portrayed in Fig. 14.10. In contrast to the hookworm
model, both states reflect stable endemic disease: one equilibrium is
characterized by high host density and low parasite burdens; the other
by low host density (severely depressed by the disease) and high average parasite loads. The two states are separated by a breakpoint or
unstable equilibrium.

The discontinuous switch from low to high levels of infection
following a disturbance severe enough to cross the breakpoint will
show up as an apparent 'epidemic' outbreak of disease typically producing many host deaths. Interestingly, many documented accounts
of disease outbreaks are for host populations at high densities where
stress induced by overcrowding or malnutrition is present. Good
examples are provided by the observed association between populations of the Red Grouse Lagopus scoticus and the directly transmitted
nematode Trichostrongylus tenuis in northern England and Scotland
(Lovat, 1911; Lack, 1954) and outbreaks of fowl cholera, caused by
the bacterium Pasteurella multocida among populations of wild duck
in North America (Petrides and Bryant, 1951).

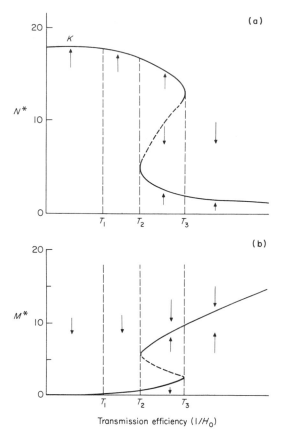

Fig. 14.10. Transmission threshold ($R = 1$) and alternative stable states arising in a model for directly transmitted helminth infections, where it is assumed that the pathogenicity of the disease is related to the nutritional state of the host (Anderson, 1979a).

(a) The graph shows the equilibrium host population size (N^*) as a function of a parameter, T, representing transmission efficiency $T = 1/H_0$, where H_0 is defined in eq. 14.40). The infection cannot persist below a threshold value T_1 (equivalent to $R = 1$): between T_1 and T_2 there is a unique stable state of low disease endemicity (M^* small in graph (b) and hence little depression of H^* from its disease-free level K); between T_2 and T_3 two stable levels of host population size may occur, one high, and the other low, separated by a 'breakpoint' (the unstable state denoted by the dashed line); above T_3 there is a unique equilibrium level, corresponding to high average parasite burdens per host and severe depression of the host population. The arrows indicate the stable state to which the system will return from a given initial value.

(b) Similar to graph (a) except denoting the relationship between transmission efficiency and the mean worm burden M^*.

14.10   Indirectly transmitted parasites

Many parasitic species are transmitted indirectly via one or more intermediate hosts. This happens for both microparasites (for example, the arthropod-borne viruses such as yellow fever and the protozoan malaria species of the genus *Plasmodium*) and macroparasites (for example, schistosomes, the filarial worms causing onchocerciasis, and other roundworms, flukes and tapeworms that utilize dipteran, molluscan and other intermediate hosts). Malaria and schistosomiasis in human populations are the two infections whose transmission cycles have been most fully studied (Ross, 1915; MacDonald, 1957; MacDonald, 1965; Dietz, Molineaux and Thomas, 1974; Cohen, 1979; Aron and May, 1981).

If we adopt the approach of eqs. (14.3)–(14.5), namely dividing the host population into susceptible, infected and immune categories, we will have a system of six differential equations: three for the intermediate host populations X', Y', Z' and three for the final host (alternatively referred to as the primary or definitive host) X, Y, Z. Most existing models, however, assume the total populations of both final host ($N = X + Y + Z$) and the intermediate host ($N' = X' + Y' + Z'$) are constant and not dynamically involved with the infection. This reduces the system to four equations. This is in essence the source of the classic malaria equations of Ross (1911) and MacDonald (1957), and the arbovirus equations of Dietz (1974).

These models have been subjected to various kinds of more refined treatment including age-structure (Dietz, 1974), the inclusion of incubation periods (MacDonald, 1957), the use of several immunological categories of hosts (Dietz *et al.*, 1974), and more careful treatments of the interplay between transmission rates and immune responses (Aron and May, 1981).

As a concrete example, we briefly consider a grossly oversimplified model for malaria in which incubation periods and complexities surrounding the nature of acquired immunity in man are ignored. First, note that an immune category of hosts does not exist for the mosquito population since, as noted in the previous section, current evidence suggests that invertebrate species are unable to develop an effective degree of acquired immunity. Moreover, in the case of malaria, infected mosquitoes do not appear to be able to recover from infection. Second, for malaria and many other infections borne by biting arthropods, the intermediate host tends to make a fixed number of bites per week, independent of the number of final hosts available to feed on. Thus the transmission rate from infected mosquitoes to man (and from infected people back to susceptible mosquitoes) is proportional to the biting rate B times the probability that a given human is susceptible (or infected), namely $B\,X/N$ (or $B\,Y/N$). Transmission is not simply proportional to the number of susceptible or infected people as in previous models [eqs. (14.3)–(14.5) and eqs. (14.11)–(14.13)].

Using notation similar to that defined in eqs. (14.3)–(14.5) our simplified model of malaria dynamics is

$$dY'/dt = B\,\frac{Y}{N}\,X' - (b' + \alpha')Y', \qquad (14.47)$$

$$dY/dt = B\,\frac{X}{N}\,Y' - (b + \alpha + v)Y, \qquad (14.48)$$

$$dZ/dt = vY - (b + \alpha + \gamma)Z. \qquad (14.49)$$

This type of model has two patterns of dynamical behaviour (Mac-Donald, 1957; Dietz, 1974; Bailey, 1975). If the basic reproductive rate of infection R is greater than unity, the infection will persist endemically at a stable equilibrium. Reproductive success is now defined as,

$$R = \frac{B^2\,N'/N}{(b' + \alpha')\,(b + \alpha + v)}. \qquad (14.50)$$

Interestingly, the term 'basic reproductive rate' was first used by MacDonald (1952, 1957) in the context of malaria transmission. If $R < 1$, the infection is unable to maintain itself and the system settles to the stable state, $Y = Y' = 0$, of parasite extinction.

For the maintenance of malaria, it can be seen from eq. (14.50) that

$$\frac{N'}{N} > \frac{(b' + \alpha')\,(b + \alpha + v)}{B^2} \qquad (14.51)$$

The simple threshold condition derived for directly transmitted infections [eq. (14.10)] is now replaced by a more complex expression defining the density of intermediate hosts relative to that of the final host. Malaria can therefore be maintained within low density human populations provided mosquito density is sufficiently high.

In the case of indirectly transmitted infections such as schistosomiasis, which pass from final to intermediate host (and from intermediate back to final host) by means of a free-living infective agent, this threshold condition is somewhat different. It now appears as the product, NN', of final and intermediate host densities (Nassell and Hirsch, 1973; May, 1977b; May and Anderson, 1979). These infections can again persist when one or other of the population densities, N or N', is low.

This brief discussion of indirectly transmitted infections glosses over many biological complications, the dynamical consequences of which are rather poorly understood at present. There is a desperate paucity of biological data concerning the many rate-determining processes which control the flow of these parasites through their complex life cycles. Many issues, concerning both the methods of data collection and parameter estimation by statistical analysis, remain unresolved (see for example: Warren, 1973; Dietz *et al.*, 1974; Barbour, 1978; Cohen and Singer, 1979).

The available epidemiological data are usually in the form of age-prevalence curves, denoting the proportion of intermediate or final hosts infected in various age classes of their respective populations. Figure 14.11, for example, records some of the best prevalence data available for schistosomiasis, a disease of man caused by helminth species of the genus *Schistosoma*, which are transmitted via molluscan intermediate hosts. These age prevalence patterns are more complex than those we saw earlier for hookworms (Fig. 14.8). The prevalence tends to decline in the older age groups of hosts (both in man and the mollusc), although the generative mechanisms of these patterns are as yet unclear.

Transmission rates may be estimated from the type of data present in Fig. 14.11, by the use of age prevalence models based on a susceptible, infected and immune framework [see Cohen (1976, 1977) for a detailed discussion of these methods]. These models are crude, but refinements must await the resolution of certain biological problems concerning, for example, the significance of age-dependent differences in host susceptibility to infections [both with respect to human exposure to infection and the significance of acquired immunity (Warren, 1973; Aron and May, 1981)].

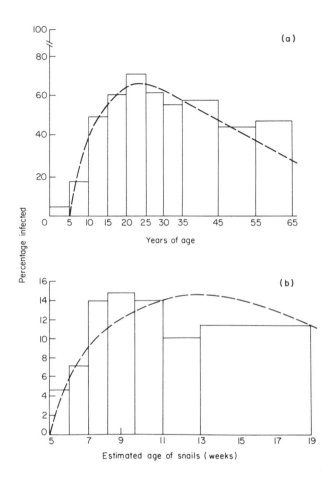

Fig. 14.11. Age-prevalence curves for schistosome infections in the human host and the snail intermediate host.

(a) Observed age prevalence (histogram) of human infection with *Schistosoma japonicum* in a coastal region of Palo in the Philippines (from Hairston, 1965). The dashed line is a theoretical age prevalence curve derived from a model described by Cohen (1976) based .on a susceptible, infected and immune framework.

(b) Observed age prevalence (histogram) of infection in a snail population (*Biomphalaria glabrata*) in St. Lucia (from Cohen, 1973). The dashed line denotes the prediction of an age prevalence model described by Cohen (1976) which assumes differential mortality between infected and uninfected snails plus loss of infection.

14.11   Evolutionary trends

Any discussion of the dynamics of host–parasite associations must ultimately take account of the evolutionary pressures on both host and parasite. Population dynamics is always confounded by population genetics.

The very nature of the parasitic mode of life implies that the action of the pathogen in reducing host survival or reproduction will select for hosts with reduced susceptibility to the disease (Haldane, 1949). Recent work on the genetic control in mice of both susceptibility to infectious diseases and the ability to develop acquired immunity has provided exciting evidence of the way selection operates (Blackwell *et al.*, 1980). Immunological responses to pathogens appear to be controlled by specific genes. In the two most intensively studied host species, namely mice and man, these genes are located at or near the major histocompatibility locus (H–2 in mice and HLA in man) (Blackwell *et al.*, 1980; Dausset and Colombani, 1972; Oliver, 1978). In wild populations of mice, for instance, it has been suggested that the levels of heterozygosity (over 95 per cent) found at the H–2 locus are the result of selection by pathogens acting on varying gene pools (Berry, 1980).

We might therefore expect the pathogenicity of a parasite to decrease through evolutionary time. Many examples of this phenomenon exist and conspicuous examples are provided by the phenomenon of man's aberrant red blood cells in regions where malaria is endemic (sickle cell anemia), and the history of myxomatosis (a virus) in rabbit populations in Australia and Europe (Bradley, 1977; Fenner and Ratcliffe, 1965). It is important to note, however, that selective pressures on the host will not always act to decrease susceptibility (and hence reduce the observed pathogenicity of a parasite). Reduced susceptibility, for example, may often be linked with deleterious traits which reduce reproductive fitness in response to other selective pressures. Observed susceptibility will be determined by the action of many forces of which the pathogen is but one.

We must also take into consideration the evolution of the parasite as well as that of the host. Contrary to popular belief, a decrease in pathogenicity is not always beneficial to the parasite (Dogiel, 1964; Smyth, 1977). The ability to multiply rapidly within the host, or

produce large numbers of transmission stages, may often be advantageous to overall reproductive success even if the host is eventually
killed by such action. This will certainly be the case if the immunological defences of the host are able to cope with, and eventually destroy, a slowly multiplying pathogen but are overwhelmed by the
rapidly reproducing individual. Selective pressures on the pathogen
may therefore act in opposition to those influencing the host. An
argument frequently put forward in the ecological and parasitological
literatures is that a 'well balanced' host–parasite association is one
in which the parasite does little 'harm' to its host. It is argued that
relationships which are long standing on an evolutionary time scale
tend towards commensalistic associations since low pathogenicity is
advantageous for both host and parasite. This is difficult to accept,
however, in light of the observed characteristics of host–parasite
interactions in natural animal and plant communities, and the apparent
scarcity of commensalistic associations. Moreover, and perhaps more
importantly, the temptation for either partner to cheat in a 'well
balanced' association (viewed in the light of game theory) may be just
too great (Maynard Smith, 1976).

The time span over which evolutionary changes occur is obviously
of central importance in such considerations. Because the generation
times of most hosts are many orders of magnitude longer than those
of the parasite it is tempting to conclude that selection acts more
rapidly on the pathogen. However, as illustrated by the history of
myxomatosis in rabbit populations, the way pathogens act to
reduce the reproductive success of their hosts makes it likely that parasites force the pace of host evolution to keep in step with, or even
ahead of, their own.

More broadly, it is becoming clear that parasites of all kinds play
an important role in determining community structure. In this chapter
we have principally been concerned with the direct effects of disease
on host survival and reproduction. Parasites, however, often act in
more subtle ways by increasing host susceptibility to predation or by
reducing both inter- or intra-specific competitive fitness (Anderson,
1979a).

A wide variety of field studies, for example, indicate that predators
tend to select the more vulnerable prey individuals (Rudebeck, 1950;
Hurst, 1965; Hornocker, 1970). Vulnerability is often associated with
the age of the prey, or its nutritional status, but in many cases the
presence or absence of parasitic infection is of major significance

(Murie, 1944; Crisler, 1956; Borg, 1962; Fuller, 1962; Mech, 1966; Holmes, 1979).

The strain of parasitic infection will often result in a decrease in the ability of infected hosts successfully to compete for available resources. In the context of intra-specific competition, Jenkins *et al.* (1963, 1964) found that red grouse (*Lagopus scoticus*) heavily infected with the nematode, *Trichostrongylus tenuis*, were less able to compete for territories and were more susceptible to predation, than lightly infected or disease-free birds. One of the most elegant studies of the interaction between parasites and interspecific competitive ability is that of Park (1948), who demonstrated that the protozoan parasite *Adelina triboli* is able to reverse the outcome of competition between two species of flour beetle, *Tribolium castaneum* and *T. confusum*.

A very dramatic account of the role played by parasites in moulding community structure is provided by Warner (1968), who suggests that the extinction of nearly half of the endemic bird fauna of the Hawaiian Islands is the result of the action of introduced bird pathogens such as malaria and bird pox. Similar ideas have been put forward to explain the observed distribution patterns of many mammalian species (Karns, 1967; Barkehenn, 1969; Weigal, 1969). A classic example is provided by recent changes in the distribution of the moose, *Alces alces*, in North America as a result of the nematode pathogen *Pneumostrongylus tenuis* (Telfer, 1967; Anderson, 1971, 1972; Kelsall and Prescott, 1971; Dauphiné, 1975). Both Barbehenn (1969) and Cornell (1974) have put forward some compelling arguments to look beyond conventional competition theory for the answers to host distributional patterns (see chapters 8, 9 and 10). We must clearly pay much greater attention to the population and evolutionary ecology of infectious diseases in future research on community structure and dynamics.

15
Man Versus Pests

GORDON CONWAY

15.1 Introduction

Pest control constitutes an ancient war, waged by man for 4000 years or more against a great variety of small and remarkably persistent enemies. Surprisingly, although the war is old, its · dynamics seem poorly understood. Even the objectives, at least of man, are ill-defined. It is as though man, in the heat of the battle, has not had the time to analyse in any sophisticated fashion the conflict in which he finds himself. Battles have been won and lost but lessons have been learned slowly and painfully. It is only in recent years that people have begun to ask the fundamental questions of principle and to raise doubts about implicit beliefs and objectives.

The term pest connotes a value judgement. A pest is a living organism (insect, fungus, bacterium, weed, etc.), which causes damage or illness to man or his possessions or is otherwise, in some sense, 'not wanted'. Thus, at the outset, pest control is a problem for the social sciences and in particular for applied economics. Simply stated, the problem is to assess whether the damage or illness caused by a pest can be reduced in a manner which is profitable or satisfactory. Formal economic analysis requires that the damage or illness be quantified, usually in monetary terms, so that the potential benefits ensuing from control of the pest can be compared with the costs involved. Such costs may be economic, social or environmental.

The classical economist's tool for tackling this kind of problem is marginal analysis (Fig. 15.1). A control action is termed rational when its cost is less than or equal to the net increase in revenue it produces and is economically most efficient (i.e., most profitable) when the marginal cost of control equals the marginal revenue produced (Southwood and Norton, 1973). If we consider an agricultural crop, then the object of pest control is to maximize the quantity

$$Y[A(S)] \; P[A(S)] - C(S). \qquad (15.1)$$

356

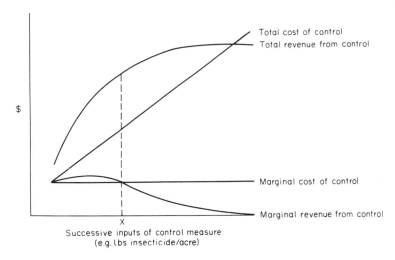

Fig. 15.1. Cost and revenue curves for pest control. The marginal curves give the increment in total cost or revenue for each successive input of control. Profit is greatest when marginal revenue equals marginal cost. (After Southwood and Norton, 1973.)

This expression, which subtracts the cost of control from the revenue, comprises four important functions.

(1) The quantity damage function $Y[A(S)]$, which relates yields to levels of attack $A(S)$.

(2) The quality damage function, $P[A(S)]$, which relates prices to levels of attack $A(S)$.

(3) The control function, $A(S)$, which relates levels of pest attack to the range of control strategies available S.

(4) The cost of control function, $C(S)$, which relates costs of control to the control strategies S.

Figure 15.2 illustrates how the functions are linked.

To this point analysis of the problem has been in purely economic terms. But, by definition, pests are also living organisms and as we dissect the constituent functions and investigate their properties biological, and in particular ecological, knowledge dominates the analysis. Populations of organisms—as the early chapters of this book make clear— respond, often in a complex and dynamic fashion, to outside interference. Only rarely will the level of pest attack be a simple function of control strategy. The kill produced by a pesticide may evoke a density dependent response, so that the pest population rebounds,

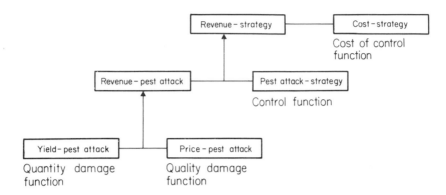

Fig. 15.2. Constituent functions of the relationship between revenue and cost of control.

sometimes to a new and higher level. Pesticides, for example, may kill off those predators and parasites which have an important function in regulating pest numbers. In the longer term there may be genetic responses. Pest populations may develop resistance to control, notably to chemical pesticides.

The damage function may also reflect complex ecological processes since the object of pest attack—usually a crop plant, domestic animal or man himself—is also a living organism. The outcome of attack is thus the product of the dynamic interaction of two or more populations. For example, the damage caused by a weed infestation is a function of competition between the weed and the crop population. Where crops are directly attacked by pests the outcome is complicated by the frequent ability of crops to compensate for low levels of attack. Similarly, where pests are vectors of disease, the ensuing incidence of illness in the animal or human population will depend on such features as the age structure and the level of immunity present.

In this chapter I shall discuss more fully how modern ecological and economic analysis illuminates the problem of pest control. I describe, in the next section, how the short and long term dynamics of pest populations determine the choice of appropriate strategies of control. I then show how knowledge of the damage function modifies this choice. Finally, I describe how optimal solutions to pest control problems can be theoretically attained and discuss the more difficult task of designing pest management systems which are resilient in both ecological and economic terms.

15.2   The control function

In traditional text books, pests are classified and dealt with in taxonomic order or according to the crop or other object of attack. A potentially more illuminating classification is according to the spectrum of
ecological strategies which are described by Southwood in chapter 3.
On this basis we can then speak of r-pests and K-pests with, in between,
'intermediate' pests.

15.2.1   *r-pests*

The word pest immediately conjures in the mind the image of a ravag-
ing swarm devastating all before it and moving on; indeed the word
derives from the Latin *pestis*, for plague. The classic pests which conform to this image, rats and locusts or the rusts of wheat for example, all
show the typical r-strategy features of high potential rates of increase
and strong dispersal and host finding ability.

An illuminating example of an r-pest is the plague caterpillar,
Tiracola plagiata, of South East Asia. This belongs to the moth family
Noctuidae which contains many other r-pests, including the army-
worms, cutworms and bollworms. The adult moths are strong fliers and
capable of laying batches of 1000 eggs or more. The life cycle from egg
to adult takes only 30 to 40 days. Conway (1971a) describes an outbreak of this species in an area of cleared primary rainforest in Northern
Borneo. A strip of forest about a quarter of a mile long had been cleared
some five years earlier, but had been abandoned and was in the process
of passing back through the long succession to primary forest. At the
time of the outbreak the vegetation consisted of a variety of secondary
forest trees standing six to fifteen feet high. When first seen the outbreak, a huge army of caterpillars comprising several million individuals, had progressed half way along the strip, removing the leaves
from virtually every plant. Since the land was abandoned, with no
value attached to the vegetation, the caterpillars were not pests. However, at the end of the strip was a young plantation of cocoa which had
also been planted on cleared forest land. The population was not
sprayed and two weeks later all the caterpillars pupated in the soil close
to the cocoa plantation. Only a few adult moths emerged from the
pupae, the majority dying from a viral or other infection, and the
cocoa was spared. But in other countries in South East Asia the plague
caterpillar has become a serious pest of cocoa.

The species is indigenous to the region, living within the boundaries of the primary rainforest, but adopting an r-strategy so as to exploit the scattered ephemeral clearings in the forest which, in the absence of man, are created by lightning strikes or by the overflowing of rivers. It is a species which is catholic in its food preferences, attacking many of the plants, themselves r-strategists, which characterize the early stages of secondary succession. This adaptation is clearly an excellent pre-adaptation to the role of agricultural pest. With the planting in the cleared forest of annual or perennial crops, species such as the plague caterpillar continue to pursue their life strategy but in a habitat that is no longer temporary.

This pattern of emergence of an r-pest, in this case from the tropical rain forest, has occurred many times in human history and in every part of the world, although in the industrialized, temperate regions the origins of the r-pests may be obscure. Agricultural and industrial development has greatly increased the availability of favourable habitats for r-strategists. For example, the natural habitat of the mosquito *Aedes aegypti*, which is a vector of yellow fever, is the primary forest where it breeds in the water contained in the forks and hollows of forest trees. Its invasion of urban areas and the accompanying spread of the disease it carries is a result of the numerous favourable sites provided by man, such as water containers, discarded tin cans, automobile tyres, etc., which furnish similar and often better breeding conditions than occur in nature. For this species there is evidence that there has been a shift from K- to r-strategy accompanying its move to the urban environment (Schlosser and Buffington, 1977). In comparable fashion mismanagement of the semi-arid regions of Africa and Asia, through over-grazing and repeated burning, has increased the availability of breeding sites for various locust species. The complicated and unique life cycle of the locust has evolved to exploit the ephemerally favourable habitats of semi-arid regions; human activity has both raised the probability of locust outbreaks occurring and extended their timespan.

In the temperate regions of the world perhaps the majority of pests are towards the r-end of the spectrum; they are species which have adopted this strategy in order to exploit the essentially short-lived plants which depend on each annual growing season. Often, as Southwood (1977b) points out, such r-pests spend part of the season at low population levels in one habitat, for example the sorghum midge on wild grasses, and then explode later onto the growing crops.

15.2.2 *K-pests*

K-pests, by contrast, have low rates of potential increase, greater competitive ability and more specialized food preferences.

A good example is another Southeast Asian pest, the bark borer, *Endoclita hosei*, which is a member of the Hepialidae, a family of large, slow flying moths with low fecundity and long life cycles. In nature the larvae bore in the trunk of a secondary forest tree, where they seem to cause little damage. But they also attack cocoa and there a single larva boring in a cocoa tree will cause its death. Thus even small populations can cause considerable damage.

Most organisms which show extreme *K*-strategies do not become pests; their specialized niche is not of interest to man. However, if man does expand the niche or provide a new niche of very similar composition then *K*-strategists can become important pests. On their natural primary hosts *K*-pests may cause little harm; the numbers are usually low and the damage is insufficient to cause the death of the host or seriously impair its reproductive capacity. *K*-pests thus act more in the manner of parasites than predators. Indeed many of the internal parasites of vertebrates such as the tapeworm or the schistosomes are characteristic *K*-pests.

Nevertheless even if the damage, at least in terms of biomass removed, is small it may be unacceptable to man when it affects him or his crops. Southwood (1977b) cites the example of the skylark which in eastern England often removes a small number of seedling sugar beet in the early summer. Until recently multigerm beet fruits were sown and the excess seedlings subsequently thinned by hoeing so that the skylark depradations were insignificant. Now fruit with a single seed is drilled to a stand and the skylark is a pest.

The codling moth provides another example of a pest towards the *K*-end of the spectrum. Its damage to fruits is biologically insignificant since few are attacked and normally they remain attractive to birds and mammals and the seeds are dispersed in the usual manner; but on cultivated fruit crops, such as apples, the damage becomes commercially important because of the high value we place on unblemished fruits.

The contrast between *r*- and *K*-pests is well illustrated by two important pests, the brown planthopper and the green leafhopper of rice in Japan (Table 15.1). For neither species are the indigenous parasites or predators (mostly spiders) significant in regulation. Each year the

Table 15.1. Comparison of two rice pests from Japan. (Data from Kuno and Hokyo, 1970.)

	Brown planthopper (*Nilaparvata lugens*) *r*-selected	Green leafhopper (*Nephotettix cincticeps*) *K*-selected
Population increase		
Generation time	similar	
Mean fecundity per female	429	47.6
Rate of growth in season	1855 ×	110 ×
Annual fluctuations		
Ratio highest to lowest numbers in peak generation	90·6	2·4
Regulation		
b-value (see Table 2.2,D)	$b \approx 0$	$b \approx 1$
Equilibrium density per hill	572	13·5
Spatial distribution	Patchy	Homogeneous
Mean crowding (Lloyd, 1967) (at density 20 per hill)	50	25
Vagility	Invasive, from China; density dependent migration	Local, within fields
Damage	Sapsucking causing loss of growth	Transmission of rice dwarf virus

brown planthopper invades from China and builds up rapidly in numbers, the eventual population size being primarily a function of climatic conditions. As the density increases the rice plants deteriorate and a growing proportion of the adults are of the large winged, migratory form. By contrast the green leafhopper persists throughout the year at a much lower, more stable level brought about by intraspecific competition. Both hoppers suck the sap of the rice plants, but the green leafhopper is never abundant enough to cause direct damage and is only a pest because it transmits a virus disease of rice.

15.2.3 *Intermediate pests*

Probably the great majority of pest species lie between these two extremes and can be classified as intermediate pests. As Southwood points out in chapter 3, there is a continuum, in nature, from *r*-strategist to *K*-strategist and a sharp demarcation between the categories is unrealistic. Intermediate pests may show any mixture of features, but probably their most important characteristic, from the viewpoint of control, is the high degree to which they are normally regulated by natural enemies (see Fig. 3.6). Insects may be controlled by parasites or predators, including other insects, nematodes, birds and mammals, and by insect pathogens; weeds by herbivorous insects and plant pathogens; and plant and animal pathogens themselves by other pathogens.

Commonly this degree of natural regulation is sufficient to keep the damage caused to crops or livestock or man below a level at which artificial control is required. But the regulation may break down and the insect or weed or pathogen 'escape' to become a pest. This may be a seasonal phenomenon or may only occur every few years, but often it is a permanent or semi-permanent change resulting from human intervention.

Many temperate aphids, lying towards the extreme r-end of the intermediate spectrum, may annually escape following the disruption to natural enemy regulation caused by the winter. For example, the bean aphid which overwinters on the spindle tree may in some years outstrip the build-up of its enemies (general predators such as ladybird beetles, hoverflies and lacewing larvae), given a good starting population and favourable weather in the spring. Often the enemies will then only re-establish control when the aphid population has already crashed. In other years conditions may favour the enemies and the pests will be held below the release point.

The longer-cycle, periodic release from natural enemy regulation is typified by many forest pests such as the spruce budworm. The detailed analyses of the dynamics of this pest (Peterman et al., 1979) suggest that the epidemics have followed long endemic periods (approximately 40 years) in which the absence of severe budworm damage has resulted in a maturing of the forest and a change in the tree species composition. The resulting increase in foliage and branch density then permit an explosive increase in budworms sufficient to allow their escape from their predators (Fig. 15.3).

The best known intermediate pests are those which have been imported to new regions of the world, but in the process have left their natural regulating agents in the country of origin. The examples of classical biological control, where the regulation has been reimposed by the deliberate introduction of the original natural enemies, fall in this category. One of the earliest and most quoted instances is of the cottony cushion scale introduced into California, probably on wattle from Australia in the 1860's, and controlled with spectacular success by the sub-

sequent introduction of two of its natural enemies, the vedalia beetle
and a parasitic wasp. Introduced pests, whether r or intermediate in
strategy, are particularly important in North America. Over 60 per cent
of the important insect pests of North America and of the weeds of
Canada are introduced, mostly from Europe.

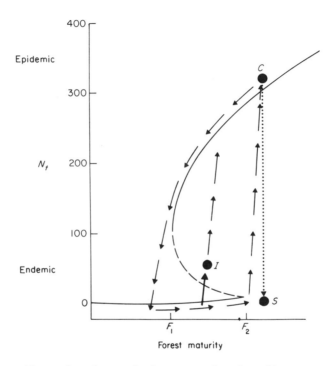

Fig. 15.3. Dynamics of spruce budworm as a function of forest maturity.
Epidemics occur either because the forest matures beyond point F_2 or
immigrants boost the population to point I. Spraying may only hold
the population at an unstable point S. (After Peterman *et al.*, 1979.)

The final category of intermediate pests comprises all those pests
which have become important because man-made changes have
eliminated or reduced the efficiency of their natural enemies. Included
are the 'secondary' or 'upset' pests which are the result of the inadver-
tent effect of pesticides on the natural enemies of hitherto insignificant
pest species. Some of the earliest and now most serious upset pests are
the spider mites, which attack various fruit trees. These mites first
became important as a consequence of the application of tar oil winter
washes to the trees. The tar oil destroyed the mosses and lichens which

gave protection to the hibernating predators (ladybird beetles and predacious mites) of the spider mites. Since the Second World War the problem has become worse because of the direct kills of the predators by modern insecticides.

15.2.4 Strategies of control

This conceptualization of pest strategies has important implications for control, since each different ecological strategy clearly invites an appropriate strategy of control. In practical terms a pest profile can be drawn up for the crop or host being attacked (Table 15.2) and used as a first step towards choosing a mix of control strategies.

Table 15.2. Crop pest profiles for cocoa in Northern Borneo and dessert apples in the U.K.

	Cocoa	Apples
r		
	Bagworms	Red spider mite
Intermediate	Nettle caterpillar	Winter moth
		Rosy apple aphid
	Cocoa looper	Apple grass aphid
	Branch borer	Codling moth
		Bullfinch
K	Bee bug	Tortrix
	Ring bark borer	Apple sucker
		Apple blossom weevil

Five major techniques of pest control are currently practised.

(1) *Pesticide control:* the use of chemical compounds to kill pests directly.

(2) *Biological control:* the use of natural enemies, viruses, bacteria or fungi, either by augmenting those already present or by introducing species from other regions and countries.

(3) *Cultural control:* the use of agricultural or other practices to change the habitat of the pest.

(4) *Plant and animal resistance:* the breeding of animals and crop plant varieties resistent to pests.

(5) *Sterile mating control:* the sterilisation of pest populations by various techniques to reduce the rate of reproduction.

Unfortunately the history of pest control has been dominated by the search for panaceas. With the advent of organochlorine and organophosphorus insecticides at the end of the Second World War, it was believed that most pest problems would be quickly solved. But within 15 years the limitations of pesticides had become widely apparent. Control programmes failed or sometimes created worse problems than they solved because of the effects of pesticides on natural enemies, or the development of resistance. Moreover evidence grew that many pesticides pose serious hazards to wildlife and in some cases to man himself. Biological control became the new panacea. When it, too, was seen to be limited in scope, attention shifted to sterile mating techniques and most recently has focussed on the development of animal and crop resistance. An attempt at a more rational approach began with the concept of integrating biological and chemical control. Latterly the term integrated pest management has been used to denote the mix of all appropriate techniques in a given situation (Apple and Smith, 1976). However, as Way (1973) points out, it has been largely an empirical approach; successful integrated control programmes have depended on a combination of insight and trial and error. There has been little attempt, so

Table 15.3. Principal control techniques appropriate for different pest strategies.

	r-pests	Intermediate pests	K-pests
Pesticides	Early widescale applications based on forecasting	Selective pesticides	Precisely targeted applications based on monitoring
Biological control		Introduction or enhancement of natural enemies	
Cultural control	Timing, cultivation sanitation and rotation	→ ←	Changes in agronomic practice, destruction of alternative hosts
Resistance	General, polygenic resistance →	←	Specific, monogenic resistance
Genetic control			Sterile mating technique

far, to define a theoretical basis for choosing appropriate strategies. Recognition of the r–K categorization, I believe, provides a step in this direction (Table 15.3).

r-pests are the most difficult to control. Their frequent invasions and the massive damaging outbreaks they are capable of producing require a fast and flexible response. Natural enemies cannot provide efficient control and pesticides remain the most appropriate means of attack. But because r-pests are resilient to disturbance, rebounding after even very heavy mortalities, pesticides have to be used more or less continuously. Resistant varieties of crops and animals may provide some protection but if it is to persist resistance has to be of the general, polygenic type providing, through broad physiological mechanisms, an overall reduction in the rate of increase of the attacking pest or pathogen (Browning *et al*, 1977). Many r-pests, such as the cereal rusts, regularly produce new races which will overcome narrow based resistance. Cultural control may also be of help where it serves to reduce the likelihood and size of the pest invasion. Early or late planting, for example, may permit the crop to escape; rotation prevents the build up of pests such as nematodes, while against r-selected weeds the production of weed-free seed and mechanical cultivation are appropriate.

By contrast K-pests are theoretically more vulnerable. They may be forced to low population levels and, in some situations, eradicated; for this reason they are the most suitable targets for the sterile mating technique. Pesticides may also give efficient control where small populations cause high losses, for example when fruit is blemished, but they need to be carefully timed and precisely applied. In general, because K-pests are specialists occupying narrow niches in the man-made world, they are most vulnerable to strategies aimed directly at reducing or eliminating the effective pest niche. Resistant animal breeds or crop varieties are a powerful approach to K-pests since permanent success can often be attained with simple, monogenic resistance. Cultural control may also be very effective; small changes in agronomic practices such as planting density or pruning may render a crop unattractive, or the K-pest may be greatly reduced by eliminating its alternative wild host.

Biological control has its greatest pay-off against intermediate pests and this must be the preferred strategy in every case. If pesticides have to be used they should be selective either in their mode of action or in the way they are applied so that only the target pests are affected, and not their natural enemies and those of other pests. It also, of course, follows that when r, K and intermediate pests coexist the pesticide applications against the r- or K-pests should not interfere with the biological control of the intermediate pests. Indeed the successful control of the intermediate pests on a crop or host should be seen as the initial taget, with which the strategies for control of the remaining r- and K-pests are then integrated.

15.2.5 *Pesticide resistance*

In much the same way that crops and animals can be artificially bred for resistance to pests, given sufficient time, pests can evolve resistance to virtually all of the control techniques devised by man (Corbett, 1978). The most serious contemporary problem, however, is the rapidly growing resistance of pests to pesticides. This is not a new phenomenon: in 1908 the San Jose scale was recorded as resistant to lime sulphur. But resistance has accelerated with the introduction of modern pesticides and more intensive use. Today there are 364 species of insect and mite resistant to one or more insecticides (Fig. 15.4) and resistance has occurred against every one of the insecticide groups, including the synthetic pyrethroids, juvenile hormone analogues and growth regulators,

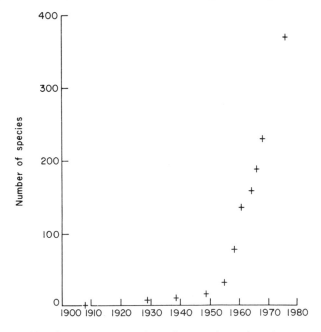

Fig. 15.4. Numbers of insect and acarine species resistant to one or more insecticides.

introduced in the last 5 years. There are fewer pathogens (67) and weed species (19) showing resistance, but certain fungicides and herbicides are very vulnerable.

Genes conferring resistance are often present in a pest population before the pesticide is applied and whether or not the population becomes resistant depends on the intensity of pesticide selection. In general, the more frequent the applications, the higher the kill produced and the more thorough the coverage of the population, then the more rapidly resistance develops. For these reasons resistance is more common among r-pests since their generally shorter generation times and high rates of population growth necessitate frequent pesticide treatment. Indeed the 23 insect and tick species which show resistance to all four of the major insecticide groups are all strongly r-selected. Resistance in mosquitoes has seriously hampered malaria control, while resistance in the cotton bollworm and tobacco budworm caused a near collapse of cotton production in the U.S.A. and great reductions in cotton acreage in Mexico.

It seems very unlikely that pesticides, and in particular insecticides, will be discovered against which pests cannot evolve resistance. The only practical way of minimizing the problem, and so extending the useful lifetime of existing or new pesticides, is to reduce the selection pressure. There is no shortage of tactical options for affecting the rate at which resistance develops. The dosage, distribution and timing of pesticide applications are important parameters, but unfortunately tactics for delaying resistance cannot be field-tested since resistance is essentially undetectable in its early stages. Moreover, resistance characteristically develops on a regional or even national scale, requiring a strategic rather than tactical response. These difficulties make the problem one which is particularly suited to mathematical modelling techniques.

Comins (1977a) has developed a model, based on a number of simplifying assumptions, which seems nevertheless sufficiently realistic as to provide guidelines that are generally applicable. The model assumes that the population is large and homogeneous with discrete generations. The resistance gene is also assumed to operate monofactorially and equally in both sexes. On these assumptions the typical pattern of development of the resistance gene occurs in three phases (Fig. 15.5). In the first phase, when the pesticide is not being applied, the resistance gene is likely to be a liability and hence will exist at a low equilibrium frequency determined by the opposing forces of mutation and adverse selection. In the second phase, once pesticide selection has commenced

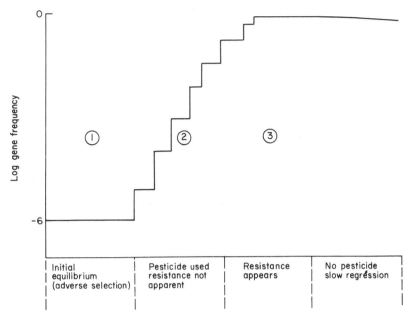

Fig. 15.5. The phases of development of pesticide resistance in a pest population. (After Comins, 1980.)

the resistance gene spreads rapidly through the population although still below the level at which control efficiency is noticeably reduced. Finally however, the crisis occurs; control measures fail, resistance is detected and after perhaps a period of increasing dosages, the pesticide is abandoned.

By concentrating on the second phase of resistance development it is possible to make a further important assumption, namely that the genes being selected exist at low frequencies. From this it follows that heterozygotes greatly outnumber homozygotes. Thus, in phase two, partially or fully dominant genes will be selected on the basis of the degree of resistance they confer in heterozygous individuals and this will be the most important factor determining the rate of development of resistance.

One of the most important questions that can be answered by this simple genetic model concerns the advisability of aiming for a high or low percentage mortality of the target pest population.

If we assume that 10 per cent of the target population escapes contact with the pesticide (there is some form of refuge due to incomplete pesticide coverage, for example) but otherwise every pest receives an equal standard dose, then the selection factor for heterozygotes (ratio of heterozygote to susceptible survival) is as shown in Fig. 15.6. At low pesticide dosages it is high but at high dosages it declines. This implies

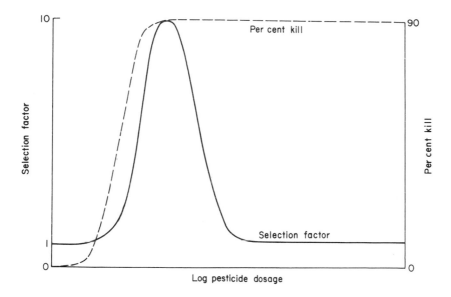

Fig. 15.6. Selection factor for heterozygotes (ratio of heterozygote to susceptible survival) as a function of pesticide dosage, where 90% of the pests receive the standard dosage. (After Comins, 1980).

that given a sufficiently high individual pesticide dose the resistant heterozygotes will be killed off, thereby permitting control of the pest population with negligible selection of the resistance gene.

This analysis suggests that, where individual dosage can be precisely controlled (for example when dipping cattle with non-residual acaricides against cattle ticks), resistance may be delayed by high kills.

In practice, however, the individual dose cannot be closely controlled under most field conditions. Applying more pesticide will simply tend to increase the proportion of the population subject to a low dose and this will simply accelerate the rate of selection. This situation can be conceptualized by a model which assumes that the pesticide, in effect, diffuses out from its point of application and individual pests or their descendants have a decreasing probability of entering the treated area (within a time that is short relative to the rate of resistance selection) the further they are from the centre. This model produces a graph of the form shown in Fig. 15.7 where the selection factor does not

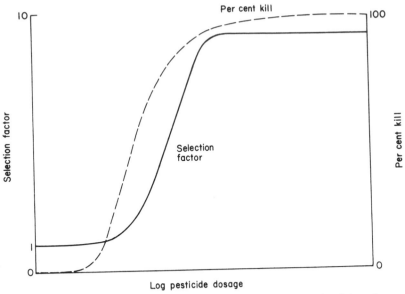

Fig. 15.7. Heterozygote selection where the dosage received is a function of the distance of the individual from the point of application. (After Comins, 1980.)

dip at high dosage. In this, which is probably the most common, field situation high kills will not delay resistance and indeed will markedly increase the selection rate for genes with high resistance factors.

The model also clearly demonstrates the importance of pest migration in delaying resistance. If the treated pest population in, say, a farmer's field is linked to a large untreated population in a wild habitat or in adjacent fields, then resistance in the treated population will develop much more quickly if the rate of migration from the untreated to the treated area falls below a certain critical threshold. Susceptible individuals migrating in provide 'susceptible' genes for the next generation and hence in effect 'dilute' the inheritance of resistance from the survivors in the treated area. Where the resistance gene is recessive the critical threshold is very sharply defined, the population suddenly jumping from a very low to a very high frequency of resistance (Fig. 15.8). However, even where the resistance genes are partially dominant the rate of development of resistance will accelerate as migration decreases.

In general any strategy which increases the effective migration rate or the effective relative size of the untreated population is to be recommended. Other considerations aside, it is good policy to leave untreated

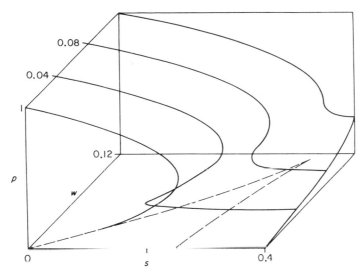

Fig. 15.8. Gene frequencies (p) for a recessive resistance gene in a pesticide-treated population; w is the initial gene frequency and $s = r/[(1 - r)(L - K)]$, where r is the migration rate and L, K are the survival rates of the resistant and susceptible homozygotes. (After Comins, 1977a.)

sub-populations of pests in neighbouring fields or on wild host plants. It also follows that for pests with discrete generations pesticides should be withheld in the period between immigration and reproduction, so producing the maximum degree of reproductive competition.

15.3 The damage function

It is common to find a linear relationship between the size of a pest population and the direct, immediate injury that it typically causes. For example, the leaf area of a crop consumed is roughly proportional to the number of pest individuals present in the crop, and the numbers of humans bitten by mosquitoes is proportional to the size of the man-biting mosquito population. The relationship becomes more complex, however, when we consider the yield of the crop or the ensuing incidence of illness.

15.3.1 *Crop compensation*

The basic form of the relationship between crop yield and pest population size is sigmoid (Fig. 15.9) but as Southwood and Norton (1973)

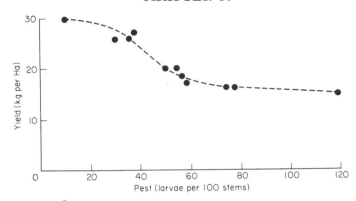

Fig. 15.9. The yield-pest relationship as shown by the African white rice borer, *Maliorpha separatella*, on rice. (After Brénière and Walker, 1970.)

point out, only a portion of this curve is usually exhibited. Where pests attack the foliage or roots and hence directly affect the growth of the plant, the plateau of high yields indicated by the left-hand portion of the curve is extended; severe losses are only caused by high populations. On the other hand, where pests attack the seeds or fruit, which are commonly the harvested product, there is a rapid decline in yields, as demonstrated in the right hand portion of the curve; losses can be severe from low populations. Most of the important pests of foliage and roots tend, therefore, to be r-strategists while many of the pests of fruit or seeds are K-strategists.

The plateau of high yields at low pest population densities, which is characteristic of foliage and root attacking pests, results from the ability of many crops to compensate for pest injury. Up to a certain level of attack a crop is able to replace the photosynthetic tissues removed by the pest population. The individual plant may respond by growing new leaves or shoots, or by extending the photosynthetic lives of existing leaves, or it may reallocate materials within the plant. Often when one plant is damaged, adjacent plants may produce extra growth to compensate. There are innumerable avenues for such compensation. In some cases it is so effective that pest injury early in the life of the crop actually produces higher yields than would have occurred in the absence of the pest.

It is thus important not to embark on control simply because of the presence in a crop of a population of pests known to cause injury. In order to make a rational decision both the density of the pest population and the current phenological condition of the crop have to be assessed and an estimate made of the likely level of damage that will eventually arise, taking into account what is known of the crop's ability to compensate for injury. This is not easy and all too often control is attempted when it is not strictly necessary. Profitability is reduced and, if pesticides are used, there is an unwarranted risk of developing pesticide resistance or of contaminating the environment.

In an attempt to minimize insecticide application in a situation of this kind Wilson *et al.* (1972) built a computer model describing the effects of the bollworm *Heliothis* on cotton in Western Australia. Their model was based on three observations. First, there is a definite maximum crop of bolls that can ripen. Second, the bolls require a minimum number of day-degrees to mature and hence have to be protected for this period in order for the maximum crop to be obtained. Third, the day-degree requirement to set the full crop of bolls depends on the size of the plant. The bigger the plant the more quickly the crop is set. As the model clearly demonstrated (Fig. 15.10), if treatment is delayed then the period of protection can be shorter. Less insecticide is used and the control is cheaper and more efficient.

Unfortunately there is very little information on the pest–yield relationship, even for the most common pests, and only rarely is there a

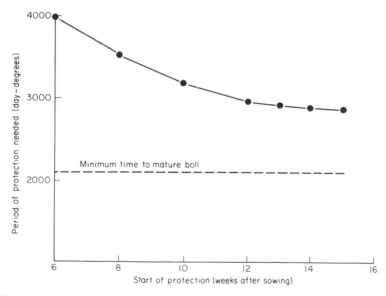

Fig. 15.10. Minimum period of time for which cotton in Western Australia needs to be protected from bollworm attack so as to ensure a maximum crop of bolls. (After Wilson *et al.*, 1972.)

good understanding of the compensatory mechanism, where present. If more were known, it would be possible not only to minimize wasteful control actions but also to enhance the degree of compensation which naturally occurs, using cultural techniques or plant breeding.

15.3.2 *Illness*

As discussed in chapter 14, where pests carry disease the relationship is a product of the dynamics of three populations: that of the pest or vector (e.g. a mosquito, *Anopheles* spp.); the host (e.g. man); and the disease organisms (e.g. the malaria parasite, *Plasmodium* spp.). Macdonald (1957) developed a simple and elegant mathematical model to describe this relationship for malaria. The basic model depends on six key parameters:

m = the density of anopheline mosquitoes in relation to man;

a = the average number of men bitten by one mosquito in one day;

b = the proportion of anophelines with sporozoites in their glands which are actually infective;

p = the probability of a mosquito surviving through one day;

n = the time taken for completion of the extrinsic cycle of the malaria parasite (neglected, for simplicity, in ch. 14); and,

r = the reciprocal of the period of infectivity in man.

When combined, these parameters give a measure of the rate of production of new infections arising from a single primary infection in a community. As explained more generally in chapter 14, this is called the basic reproductive rate of malaria and equals [see eq. (14.50)

$$\frac{m\ a^2 b\ p^n}{r\ \ln 1(/p)}. \tag{15.2}$$

In contrast to crop damage the relationship between illness and pest density (m) is simple and linear. The critical population parameter is the survival probability (p) of the pest which is contained in eq. (15.2) both raised to a power and as a logarithm. In consequence small changes in survival probability produce large changes in the malaria reproductive rate. As Fig. 15.11 shows, a decrease in the mosquito's mean expectation of life from 20 days ($p \sim 0.95$) to under 3 days ($p \sim 0.7$) is sufficient to push the reproductive rate below unity, and hence to cause extinction of disease transmission. This explains why pesticides such as DDT, which reduce survival probability, have been so heavily relied upon in

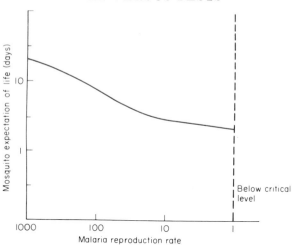

Fig. 15.11. Effect of change in the mosquito's expectation of life on the basic reproduction rate of malaria. (After Macdonald, 1961.)

malaria control campaigns. In practice, however, it has proved very difficult to achieve eradication (Harrison, 1978). The rapid rates of increase of mosquito populations, their powers of host finding, dispersal and adaptation to a wide variety of habitats, and their ability to evolve resistance, have limited the effective use of pesticides. Few nations have been able to provide the time, organization and funds which control of so powerful a pest requires.

Fortunately, the existence of three populations in the illness–pest relationship means that pest populations are no longer the sole target for control strategies. The disease pathogen populations can themselves be the subject of control, either through immunization or through chemotherapy. Macdonald *et al.* (1968) have suggested by means of a computer model how chemotherapy, based on drugs, and pesticides could be utilized in an integrated fashion for malaria control. The search is now on for a vaccine which will confer immunity against malaria. Other examples of the ecology of human diseases were discussed in chapter 14.

15.4 Ecology and economics

Knowledge of the dynamics of the pest and of the pest-damage relationship can provide a guide to those control strategies which, in ecological terms, are most likely to succeed. But the final choice of strategy depends on the objectives of the decision-maker and on the context within which the decisions are made.

15.4.1 *Optimal control*

At the beginning of the chapter I assumed that the objective of pest control was profit maximization, i.e. the farmer or other decision-maker wished to maximize the difference between the revenue and the cost of control. This is an objective which is probably pursued by most large scale farming operations. A good example occurs on the West Indian island of Trinidad, where sugar-cane is grown over several thousand acres by a single company. The cane is attacked by a frog-hopper, *Aeneolamia varia saccharina*, the adult of which causes a loss in the quantity of cane produced (Conway *et al.*, 1975). Each brood or generation is readily described by a simple mathematical function which can be used to determine the effects of pesticide control. In this case there is little evidence of compensation and field experiments suggest that the loss of sugar-cane yield is a simple linear function of the number of adult days.

Two kinds of insecticide are available: (1) a cheaper non-residual compound which only acts on the day it is applied, and (2) a more expensive residual compound which persists for 5 days. These can be applied at any time and in any permutation. If a limit is set of three applications per brood, ten dominant strategies are possible (five further strategies exist but these eliminate fewer adult days than other permutations of the same combination of insecticides). Table 15.4 gives the optimal spraying times for these strategies.

A set of net revenue lines can then be produced for different sizes of pest population, using information from the cost and damage function (Fig. 15.12). Each spraying strategy is rational when its revenue line is above that of the no-spraying strategy (1) and the most efficient strategy at each population level is the highest revenue line. The optimal control for the range of possible population sizes is given by the curve made up of the segments of revenue lines representing the most efficient strategies.

The froghopper, however, passes through four broods in a year. Clearly if there is no density dependent relationship between the broods then the problem of control remains relatively straightforward:

Table 15.4. Optimal application times for ten spraying strategies directed against a single froghopper brood. (From Conway et al. 1975.)

Strategy	Time from first adult emergence (days)																		Percentage of adult-days removed
---	1	2	3	4	5	6	7	8	9	10	11	12	13	14	15	16	17	18	---
1																			0·0
2											NR								19·3
3									NR										32·7
4														NR					37·6
5							NR			R	NR					NR			42·8
6								NR			R	R							49·0
7							NR			NR			R						57·2
8							R							R					60·9
9					NR		R			R					R				67·4
10				R												R			74·9

R = residual spray; NR = non-residual spray.

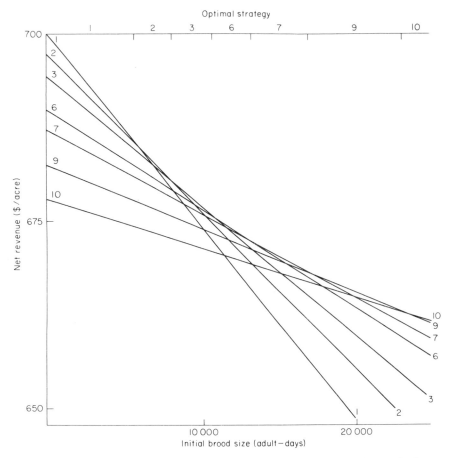

Fig. 15.12. Optimal spraying strategies (see Table 15.4) against a single brood of froghoppers. Net revenue is measured in Trinidad and Tobago dollars. (After Conway *et al.*, 1975.)

the size of all four broods is a function of the level of control of the first and the procedure, described above, for determining the best strategy is adequate. This would be true, for example, of the rice brown planthopper described in Table 15.1 which has a b value of approximately o. However, if density dependent regulation is present, as for example in the green leafhopper which has a b value of approximately equal to 1·0, the dynamics make the finding of an optimal control solution more complex.

Density dependence can be represented by an equation of the form described by eq. (3.3), that is

$$A_{t+1} = \lambda A_t^{1-b} \tag{15.3}$$

Here A_{t+1}, A_t are the numbers of adult days at time t and $t+1$; λ is the net rate of increase (as discussed in chapters 2 and 3); and b is a measure of density dependence (with the characteristic return time $T_R = 1/b$). In the case of the sugar cane froghopper we do not know the value of b but it is possible to derive an optimal control solution for different likely values of b, using the technique of dynamic programming (Shoemaker, 1973). Table 15.5 shows the results using the initial range of control strategies described in Table 15.4. As can be seen, with b equal to 0·5 (i.e. undercompensation) the spraying becomes concentrated in the second brood and with $b = 1·25$ (i.e. overcompensation) it becomes concentrated in the second and third broods.

Table 15.5. Optimal spraying strategies (see Table 15.4) for four froghopper broods with different degress of density dependence between the broods (After Conway *et al.*, 1975).

Initial size of 1st brood	Density dependence (value of b)											
	$b = 0$				$b = 0.5$				$b = 1.25$			
	Brood strategy				Brood strategy				Brood strategy			
	1st	2nd	3rd	4th	1st	2nd	3rd	4th	1st	2nd	3rd	4th
1000	2	2	1	1	1	9	2	1	1	10	10	1
5000	10	5	1	1	6	10	3	1	1	9	10	1
10000	10	10	1	1	10	10	3	1	2	9	10	1

15.4.2 *Resistance costs*

The remaining question is whether these solutions remain optimal if we take pesticide resistance into consideration. Comins (1977b) has extended his simple resistance model to tackle this problem. Underlying his approach is the notion that pesticide susceptibility in pest populations is an irreplaceable natural resource. It is essentially a free resource in the short term but, once lost, high costs are incurred in switching control to a usually more expensive substitute.

Comins makes two important approximations. The first is that the survival of the susceptible individuals is related to pesticide dosage by a simple 'random search' equation, so that

$$S_i = \exp(-C_i/A_i)$$

or $$C_i = A_i \ln S_i. \tag{15.4}$$

Here C_i is the cost of control of the ith pest generation; A_i is a 'cost parameter' specified separately for each generation; and S_i is the proportional survival of susceptibles from pesticide application for the ith generation. The value of A_i can be estimated from the results of field experiments. The second approximation is that the effective dominance of resistance stays the same regardless of pesticide dose. These assumptions do not appear to invalidate the results except, perhaps, for very low or very high pesticide application rates.

This model demonstrates very clearly that strategies which lead to a reduction in long term resistance costs are simply the optimal strategies which result when resistance is ignored, but the cost of applying insecticide is assumed to be somewhat larger than it actually is. Thus the cost parameter A_i for the control of each pest generation is increased by a constant amount Q,

$$A_i \rightarrow A_i' = A_i + Q, \tag{15.5}$$

where Q is the 'effective cost of resistance'.

The parameter Q represents the equivalent present cost of future resistance, evaluated in terms of the characteristic time for resistance to develop to the particular pesticide. Estimation of this time component implicitly takes into account the degree of dominance and initial gene frequency. In practice, values in the range of 4–12 years are probably reasonable (assuming that no cross resistance is anticipated). Values for Q can then be estimated using a standard discount formula and a range of values for the cost of switching to new pesticides once resistance has developed. Finally, the optimal value of Q can be determined graphically.

In the case of the sugar cane froghopper, Comins assumed a characteristic resistance time of four and a half years and that the cost of pesticide would be doubled once resistance to the present compound is established. The optimal strategies are the same as those when resistance is not considered but the effective prices are increased. As can be seen from Table 15.6, where density dependence is undercompensating the resistance cost becomes large when there is a large initial population, reflecting the much heavier pesticide treatment which is required for successful control. But the case of overcompensating density dependence is more complicated, since a large initial population produces a smaller second brood and a larger third brood.

Table 15.6. Effect of resistance on the cost of optimal spraying strategies for the froghopper. (After Comins, 1977b.)

Density dependence	Initial size of 1st brood (Adult-days)	Total cost ($)	
		No resistance	resistance
$b = 0.5$	2000	59	59
	5000	73	77
	10000	84	99
$b = 1.25$	2000	90	98
	5000	95	100
	10000	104	109
$b = 0$	2000	35	35
	5000	49	53
	10000	60	75

15.4.3 Resilient pest management

So far I have considered the situation of a single pest species on a crop, where accurate population estimates are possible and the farmer is aiming at a mathematically precise profit maximization. The real world is more complicated.

In most situations one knows only the probability of an approximate level of pest attack. Here the problem is best approached by the use of Bayesian decision matrix (Halter and Dean, 1972; Norton, 1976b). This approach is also useful where profit maximization is no longer the goal. Profit maximizers are assumed to be risk neutral; in other words they value each successive increment in income in the same way. But many, particularly smaller, farmers can be classified as risk averse. They value initial increases in income far higher than later increases. For example, a farmer who has just started farming and has little capital available will set his primary goal as the achievement of a certain minimum level of income. Such a situation will be particularly true of subsistence farmers in developing countries who will put a much higher value on achieving, each year, the necessary level of subsistence than on each increment of income above that level.

Most farmers also have to cope with an assortment of pests on a variety of crops. The control techniques which seem, on the previous analysis, to be most appropriate for a particular pest now have to be modified in the light of the other pest problems. Many of the pioneer integrated control programmes were expressely designed to cope with situations of this kind. For instance in the early 1960's, in northern Borneo, an integrated control programme was designed (see Table 15.2) for a complex of half a dozen pests on cocoa (Conway, 1971b). Most of these pests were intermediate pests, although close to the r-end of the spectrum. In their natural habitat of primary or secondary forest, they were regulated by a variety of natural enemies, but heavy spraying with organochlorine insecticides had apparently destroyed these enemies resulting in severe pest outbreaks on the cocoa. As a first step the use of organochlorine insecticides was stopped and within a few months several of the major pest species came under natural parasite control, reverting to levels which caused little damage to the cocoa. However, two pests remained and these had to be controlled in a manner which did not upset the balance, so recently re-established for the other species. One of these two pests was the tree borer *Endoclita hosei*, referred to earlier in the chapter as a typical K-pest. This was successfully controlled by combining injection with insecticide of any borings found and the destruction of the secondary forest tree species which was the primary host of the borer in the vicinity of the cocoa. The second pest, a bagworm (Psychidae), which showed a high reproductive rate characteristic of r-pests, was knocked down by a selective insecticide and then came under parasite control.

In such a situation, and indeed for the great majority of pest control problems, aiming for a precise mathematically optimum control is unrealistic. The farmer is more interested in having guidelines or decision rules, derived from a mixture of empirical studies and mathematical models, which provide a reasonably satisfactory level of control which is both durable and resilient to changing circumstances. At the outset much of the seemingly necessary biological information the farmer requires may be too difficult, too costly or too time consuming to obtain. In the long run crucial biological parameters may change, for example due to the evolution of resistance, and there may be major fluctuations in prices of products or the costs of control. The farmer requires a system of pest management which is able to cope with this considerable uncertainty.

Recently a study focussed on this problem has been carried out for the cattle tick *Boophilus microplus* in Australia (Sutherst *et al.*, 1979). The life cycle of this pest is relatively simple: eggs are laid on the pasture and the larvae are eventually picked up by passing cattle. The adult ticks cause serious damage to the cattle hides. Several broods a year are produced and the dynamics can be well represented by a Leslie matrix model (Leslie, 1945; Usher, 1972) of the form

$$n_{t+1} = An_t \tag{15.5}$$

where n_{t+1} and n_t are population vectors and A is a transition matrix representing fecundity, survival and development. The crucial dynamic feature of the life cycle, however, is the degree of density dependent competition between larvae for a place on the cows' hide, and the matrix model is modified to incorporate this using an equation of the form of eq. (3.3).

The model was used initially to obtain optimum strategies for the traditional means of control, which is dipping the cattle in an acaricide. The results of the simulations showed that the best strategy was to

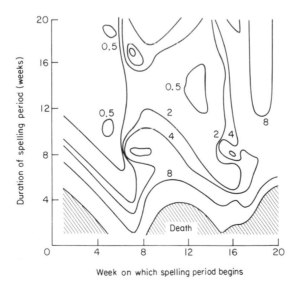

Fig. 15.13. Nomogram for single spelling and dipping of the cattle tick, showing the effect of spelling time and duration on beef losses (Australian $/head/annum). (After Sutherst *et al.*, 1979.)

dip five times at three-weekly intervals early in the season. This strategy appears surprisingly robust. Even quite large changes in the initial population density, egg and larval survival and the density dependence relationship require only the addition or subtraction of a single dipping. It is also robust to management errors, such as incomplete mustering of the cattle or mistiming of the dipping, and to changes in the price of beef.

However, cattle ticks are becoming increasingly resistant to acaricides and so the model was used to examine alternative methods of control, in particular the spelling of cattle. This entails the removal of cattle from an infested pasture for a period long enough that the tick population dies out. Various patterns of spelling, both alone and combined with dipping, were examined and a highly robust solution was found to be a single spelling period each year, with a dipping at its beginning and end (Fig. 15.13).

This strategy can be even further improved by the use of tick resistant breeds of cattle, on which there is much stronger density dependent competition between the tick larvae, and by carefully controlled high individual doses of acaricide which will, in this situation, effectively delay acaricide resistance. Such a strategy copes well with the existing known dynamics and promises to provide the flexibility which will permit adaptive changes to the unforeseen events of the future.

16

Bioeconomics

COLIN W. CLARK

16.1 Introduction

The term 'bioeconomics' is used in this chapter simply as a shortened term for 'biological resource economics'. The possible alternative meaning of the term (sometimes further reduced to 'bionomics'—see chapter 3) to describe the strategies of organisms or populations to maximize ecological fitness, will thus not be intended here.

The main problems of bioeconomics as a subject of study concern the way in which biological systems and economic systems interact with one another. These interactions are obviously of vital importance to human welfare, since virtually all of our food, much of our clothing, and many other fundamental human needs are met by biological resources. Because biological resources are so important, it is not surprising to find severe *conflicts of interest* dominating many of the problems of resource management. Any useful theory of resource economics (biological or other) must take cognizance of these conflicts of interest and their consequences. Broadly speaking, bioeconomics thus impinges on many fields besides biology and economics, including law, political science, ethics, and so on.

In classical economics, conflicts of interest between competitive producers is recognized as a 'good thing' for society—the 'invisible hand' of Adam Smith. It is important to keep in mind, however, that the classical notion of competition refers to *marketplace* competition: the most successful competitors are those that can place goods that people desire on the market at the lowest prices. Clearly such competition is often socially beneficial.

In resource economics, on the other hand, it turns out that competitive exploitation of resources is almost invariably a 'bad thing' for human welfare. Competitive resource use results in overexploitation, or depletion, of renewable resource stocks, and in extreme cases may lead to the actual extinction of populations or species. A successful bioeconomic theory must explain, both qualitatively and quantitatively, this phenomenon of competitive overexploitation, and furthermore should help identify management policies that will effectively counteract the phenomenon. Many of the results of such a theory do not seem at first glance to be self-evident. As will be seen later, for example, most of the management techniques that have traditionally been used for regulating commercial fisheries can be shown to lead, in a certain sense, to a *worsening* of the economic performance of the fishery!

It does not take much thought to become convinced that *time* is an essential dimension in any study of resource economics. Past and present use rates determine the current abundance of a resource stock, and thus specify the opportunities for future use. In the terminology of mathematical systems, it is the current *state* of the resource stock that determines future opportunities; the current state in turn is the result of the past history of exploitation. In economic terminology, resource economics thus becomes a special branch of the theory of capital and investment: a resource stock is (from the social viewpoint) a capital asset, capable of producing future goods and services. As with other capital assets, the size of a resource stock is subject to adjustment—for example, a fish population can either be decreased by harvesting in excess of the rate of natural replenishment, or it can be built up by harvesting at a suitably low rate. Presumably, therefore, there exists some 'optimal' stock level, and 'optimal' harvest rate. In fact, the theory of 'optimal' resource harvesting and stocking is one of the main themes of bioeconomics.

Of particular importance in resource economics is the concept of *time preference*. This phrase refers to the fact that people have a distinctly greater personal concern over the present and immediate future than over the distant future. Time preference is reflected in the economic system by means of the *discounting* of future revenues. If i denotes the annual rate of discount, the *present value* of a payment P due in n years' time is defined as $P/(1+i)^n$. Note that this is also the amount that must be invested today at compound interest with annual rate i, in order to accumulate a sum P after n years. Thus the discount rate i also represents the rate of interest on invested funds.

Clearly the present value of a future payment P approaches zero at an exponential rate as $n \to \infty$. At a 5 per cent annual discount rate, for example, a payment of $1000 due in 20 years' time has a present value of $376.90, but a present value of only $7.60 if due in 100 years.

As an illustration of the importance of time preference in bio-economics, let us consider the following simplistic problem of forestry management. Tree seedlings are to be planted today and harvested at maturity after 100 years. If the estimated value of a tree upon harvesting is $1000, how much can be expended on planting? If the discount rate is 5 per cent per annum (a very conservative rate for private investors, even ignoring inflation!), then the maximum that can be spent profitably is $7.60 per tree. In other words, if exactly $7.60 is spent for each tree planted (and if all trees planted reach maturity, and if there are no expenses incurred in managing and protecting the maturing forest, etc.), then the rate of return on this investment will equal 5 per cent per annum over the 100-year period.

Of course, the trees in question might be harvested sooner than 100 years. Suppose V_n represents the value of a tree if harvested after n years. As an exercise, the reader may wish to show how the value $n = n^*$ that maximizes the present value of a tree can be determined. How does n vary with i? (There is somewhat more to this problem than may be evident at first glance.)

Before proceeding, a remark on the effects of *inflation* is in order. If every price in the economy inflates at the same rate, then inflation can be ignored by carrying out all calculations in terms of constant–say 1970 —dollars. Then r represents the *real* interest rate in the economy, which equals the nominal ('bank') rate minus the rate of inflation. Although this simplification may be questionable, especially when inflation rates are high, it will be adopted throughout this chapter.

Another important aspect of many problems in resource economics is *uncertainty*, and this applies both to the biology and to the economics. In the case of a fishery, for example, the actual size of the fish population is seldom known with any precision: growth rates and other population parameters can only be roughly estimated; consequently the reaction of the fish stock to exploitation cannot be accurately predicted; the relative influences of fishing and natural fluctuations of abundance can seldom be separated; in many cases, even actual catches are not known accurately, either for lack or inaccuracy of reporting. On the economic side, neither the future demand for fish nor future changes in fishing technology and fishing costs can be predicted with any certainty.

Ideally, of course, resource management should take account of uncertainty, which can often be reduced (but seldom eliminated) by means of scientific and economic studies. In practice, unfortunately, many implications of uncertainty are often ignored, with the result that serious and avoidable errors are often made. Decision makers (often politicians) seldom understand—or wish to understand—uncertainty, and scientific advisers are thus expected to proffer simplistic advice in which risks and uncertainty are not clearly delineated. Time preference implies a further bias towards ignoring uncertainty, since in many resource situations benefits accrue largely in the present while risks fall on future generations. Thus we come full circle, with conflicts of interest, time preference and uncertainty intimately intertwined.

16.2 A bioeconomic model

We now describe a simple model pertaining to the economics of an exploited population. The model is not devised with the purpose of actual application to the management of any particular population, but rather is intended to clarify some of the ideas noted in the introduction, and to provide the basis for further discussion.

The biological basis of our model is the single-species logistic model introduced in ch. 2, adjusted to allow for harvesting:

$$\frac{dN}{dt} = rN(1 - N/K) - h \tag{16.1}$$

where $h \geq 0$ denotes the rate of harvesting. Actually, little difficulty is introduced by studying the slightly more general model

$$\frac{dN}{dt} = F(N) - h \tag{16.2}$$

where $F(N)$ is a function with characteristics similar to the above quadratic:

$$F(0) = F(K) = 0; \; F(x) > 0 \text{ for } 0 < x < K; \; F''(x) < 0. \tag{16.3}$$

This model has in fact often been used in the management of marine fisheries, where it is referred to as a 'general production model' (Schaefer, 1954; Pella and Tomlinson, 1969). In the fishery setting, it is customary also to assume the relationship

$$h = q E N \tag{16.4}$$

between the harvest rate h and the expenditure of 'fishing effort' E. Here q is a constant, called the 'catchability coefficient.' In a typical sea fishery, such as the yellowfin tuna (*Thunnus albacares*) fishery modelled by Schaefer (1954), units of effort E would be in terms of the number of vessels fishing at a given time. Note that eq. (16.4) implies that catch per unit effort (CPUE), h/E, is directly proportional to stock abundance, N. Indeed, CPUE is normally adopted as a standard index of stock abundance in fishery management, although it is recognized as being subject to a number of severe biases (Gulland, 1956; Clark, 1980).

16.2.1 *Maximum sustained yield*

Returning to eq. (16.2), let us for the moment treat the harvest rate h as a constant parameter. For certain values of $h > 0$ the equation has two equilibria, $N_1 < N_2$, with the first being unstable and the second stable (Fig. 16.1). As h increases, N_2 decreases and N_1 increases; when h

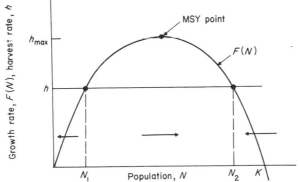

Fig. 16.1. The single-population harvesting model, eq. (16.2). For constant harvest rate $h < h_{max}$, there are two equilibria, N_1 (unstable) and N_2 (stable).

reaches a critical value h_{max} we see that N_1 and N_2 coincide ($F'(N_i) = 0$), and then disappear for larger h. The critical value of h, namely

$$h_{max} = \max F(N) \tag{16.5}$$

is referred to as the *maximum sustainable yield* (MSY) for the population in question. For the logistic population model this occurs at $N = \frac{1}{2}K$, but for other models it may occur elsewhere.

For many years the MSY concept dominated the practice of resource management in such fields as fisheries, forestry, and wildlife. For example, prior to the establishment of 200-mile 'economic zones' for marine fisheries, there were in existence some 39 international agencies for the management of marine fisheries, every one of which was required by its establishing convention to manage on the basis of the MSY objective

(Koers, 1973). As Larkin (1977) points out, MSY had taken on the role of a fundamental principle of resource management. Recently, however, MSY has come under criticism from a number of sources. Biologists (Larkin, 1977; Holt and Talbot, 1978; May *et al.*, 1979) have noted, among other things, that the concept overlooks all ecological aspects of populations and their environments. Also, MSY is hard to justify—or even to define—in the real world of fluctuating populations and uncertainty (May *et al.*, 1978). Finally, the MSY concept is clearly devoid of economic content; we will have more to say on this question later.

To bring some economics into our model, let us next postulate a constant price p per unit of resource harvested, and a constant cost c per unit of harvesting effort. (For the analysis of alternative economic models see Clark, 1976.) Net revenue is then given by

$$\Pi = ph - cE = (pqN - c)E. \qquad (16.6)$$

(It is a worthwhile mental exercise for the reader to specify realistic units for each of the symbols in these expressions, noting that Π is a revenue 'flow' with units of the form \$/unit time.)

Note that eq. (16.6) implies that net revenue resulting from a unit of effort is an increasing function of price p, and of the population abundance N (and also of the catchability of the population q), and a decreasing function of cost c. In particular, we can identify the 'zero-profit' population level $N = N_\infty$ given by

$$N_\infty = \frac{c}{pq}, \qquad (16.7)$$

which has the property that exploitation becomes unprofitable for $N < N_\infty$. This aspect of our model seems realistic for populations such as ocean fish, whose density decreases as the total population declines, but might not be realistic in other circumstances.

16.2.2   Competitive resource exploitation

Consider now the case of competitive exploitation. If the existing population level N is greater than N_∞, exploitation is profitable and will hence take place. Profits made by existing exploiters will attract other exploiters, with the result that the rate of exploitation will tend to increase. Unless this process is controlled in some manner, it can be predicted that exploitation (more precisely, effort E) will increase until the population has been reduced to the zero-profit level N_∞. At this level, biological production and the rate of exploitation will be in equilibrium, with $dN/dt = 0$ and

$$h = h_\infty = F(N_\infty) \tag{16.8}$$

$$E = E_\infty = F(N_\infty)/qN_\infty. \tag{16.9}$$

The zero-profit population level N_∞ has thus been referred to as the *bionomic equilibrium* of the unregulated, competitive biological resource industry (Gordon, 1954).

According to eq. (16.7), the bionomic equilibrium population level N_∞ decreases as the value p of the resource increases, or as the cost c of harvesting decreases, or as efficiency of harvesting as measured by q increases. If, as is usually the case in practice, these parameters shift over time, the equilibrium will shift accordingly. Thus as resource prices increase relative to costs, and as technology improves, we observe an increasing tendency towards depletion of renewable resource stocks, at least as long as exploitation remains competitive and un-restricted.

It may well occur that $N_\infty < N_{msy}$, in which case the competitive equilibrium results in *overexploitation*, or *biological depletion*, of the resource, in the sense that the population is reduced to a level below that which produces the maximum sustainable yield. Numerous in-stances of biological resource depletion attest to the realism of this pre-diction, a well known example being the Pacific sardine (*Sardinops caerula*) fishery: see Fig. 16.2. (Whether overfishing or natural environ-mental fluctuations were the primary cause of the collapse of this fishery is uncertain.) Murphy (1977) discusses the histories of this and six other clupeioid fisheries, including the famous Peruvian anchoveta (*Engraulis ringens*); see also Gulland (1977).

Depletion, or the threat of depletion, of renewable resources has led to numerous attempts at regulation. Inasmuch as competition is often a contributing cause of depletion, one approach is to establish individual *property rights* in resource ownership. It seems reasonable that the private owner of a renewable resource stock would not deliberately reduce it to a level of low productivity. We next investigate this question by formulating an optimization model. The problem of regulating resource exploitation in cases where private ownership is not feasible will be discussed later.

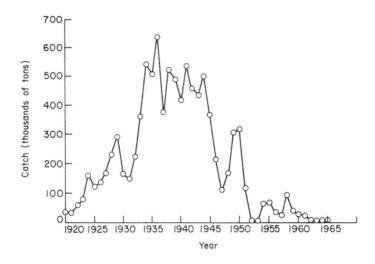

Fig. 16.2. The Pacific sardine fishery, 1920–1965. From Clark (1976).

16.2.3 *Optimization analysis*

It is a standard hypothesis of dynamic economic theory that firms behave in such a manner as to maximize the *present value* of their profits, discounting future revenues at a rate equal to the 'opportunity cost of capital.' This last term refers to the rate of interest that the firm's owners or stockholders would expect to earn on alternative investments. A typical rate quoted for the private opportunity cost of capital is 10 per cent per annum, but this figure may be questionable, in real (i.e. noninflationary) terms. It is noteworthy that interest rates that are, or should be, used in policy decisions are subject to such fuzziness, since it turns out—especially in natural resource economics—that policies selected are often highly sensitive to this particular parameter. From the social point of view, discounting implies a preferential treatment of present over future generations, and some commentators have therefore argued that discounting is socially indefensible (e.g. see Page, 1977).

For our resource model, future net revenues (or 'profits') are given by eq. (16.6) as $\Pi = \Pi(t)$; note that Π will vary over time if the population N, or effort E, changes. For simplicity, we assume here that p, c and q remain constant. We can express the discount factor conveniently in exponential form, $e^{-\delta t}$ (where δ is the 'instantaneous' discount rate), so that the present value of $\Pi(t)$ becomes $e^{-\delta t}\Pi(t)$. Thus the *total* present value is given by

$$PV = \int_0^\infty e^{-\delta t}\Pi(t)dt \qquad (16.10)$$

(here we are using an infinite time horizon; this simplifies the analysis without seriously affecting the main conclusions).

Our hypothesis, then, is that the private owner of the resource stock N would utilize a harvest effort policy $E = E(t)$ that would maximize eq. (16.10), under the condition that the population follows our equation

$$\frac{dN}{dt} = F(N) - qEN, \qquad N(0) = N_0 \qquad (16.11)$$

where N_0 represents the initial (i.e. current) population level. We assume that $E \geq 0$, and also (for simplicity) that there is a maximum effort level $E_{\max}$. Bringing together the ingredients of our private-ownership model, we thus obtain the following mathematical optimization problem:

$$\underset{[E(t)]}{\text{maximize}} \int_0^\infty e^{-\delta t}[pqN(t) - c]E(t)dt \qquad (16.12)$$

subject to

$$\frac{dN}{dt} = F(N) - qEN, \qquad N(0) = N_0 \qquad (16.13)$$

$$0 \leq E(t) \leq E_{\max}. \qquad (16.14)$$

This problem is one that the mathematician would recognize as a problem in 'optimal control theory.' Ordinarily such problems can be rather difficult, but in the present case a simple mathematical trick leads immediately to the solution.

Namely, we solve eq. (16.13) for $E = E(t)$, substitute into (16.12), and then perform an integration by parts:

$$\int_0^\infty e^{-\delta t}(pqN - c)Edt = \int_0^\infty e^{-\delta t}\left(p - \frac{c}{qN}\right)[F(N) - dN/dt]dt$$

$$= \int_0^\infty e^{-\delta t}\left[\left(p - \frac{c}{qN}\right)F(N) - \delta Z(N)\right]dt + Z(N_0) \qquad (16.15)$$

where

$$Z(N) = \int_{N_\infty}^{N} \left(p - \frac{c}{qx} \right) dx. \tag{16.16}$$

Ignoring the constant term $Z(N_0)$ in eq. (16.15), our problem now becomes

$$\text{maximize} \int_0^\infty e^{-\delta t} V[N(t)] dt \tag{16.17}$$

subject to the previous constraints, where

$$V(N) = \left(p - \frac{c}{qN} \right) F(N) - \delta Z(N). \tag{16.18}$$

Now, in order to maximize the integrated expression in eq. (16.17), we clearly must *maximize the function* $V(N)$. Thus if N^* denotes the maximizing value:

$$\max_N V(N) = V(N^*) \tag{16.19}$$

we see that N^* is the *optimal equilibrium population level*. Furthermore, if the initial population $N_0 \neq N^*$, then the optimal effort policy must be such as to force $N(t)$ towards N^* as rapidly as possible. In other words:

$$E^* = \begin{cases} E_{\max} & \text{if } N > N^* \\ F(N^*)/qN^* & \text{if } N = N^* \\ 0 & \text{if } N < N^* \end{cases} \tag{16.20}$$

(Note: strictly speaking, this conclusion is valid only under the assumption that $V(N)$ is a 'monodromic' function, i.e. that N^* is the only relative maximum of $V(N)$, $N \geq 0$. It is easily verified that this assumption holds in the logistic case.)

What is the significance of this result? First, we have shown that the renewable-resource owner who wishes to maximize profits (i.e. the present value thereof) will perceive an optimal stock level N^*, and a corresponding optimal sustained yield (OSY), $h^* = F(N^*)$. Initially he will wish to harvest any of the resource in excess of N^* as rapidly as possible; alternatively if the resource is in a depleted state ($N_0 < N^*$), the optimal policy will employ a moratorium on harvesting (i.e. an 'investment period') until the stock level has built up to N^*. The resource stock represents a *capital asset* to its owner, and N^* is the ultimate optimal size of this asset. (The particularly simple harvest-effort policy of eq. (16.20) is a consequence of the simplicity of our bioeconomic optimization model. If more complex models are employed, various complications may also enter the solution—see Clark, 1976, and Clark *et al.*, 1979).

For later reference, we also write down the necessary condition (16.19) for N^* in the differentiated form $dV/dN = 0$; after simplification this becomes:

$$F'(N) + \frac{\partial \Pi / \partial N}{\partial \Pi / \partial h} = \delta \qquad (16.21)$$

where $\Pi(N,h) = (p - c/qN)h$ as in eq. (16.6). Equation (16.21) is in a form recognizable by students of capital theory.

As an illustration, let us consider the following simplistic model of a highly K-selected species, the Antarctic blue whale (*Balaenoptera musculus*) population (see Clark, 1975):

$$F(N) = rN(1 - N/K)$$
$$r = \cdot 05 \text{ per year}$$
$$K = 150,000 \text{ whales}$$
$$N_\infty = 20,000 \text{ whales.}$$

The figures for r and K are taken from International Whaling Commission studies of the population dynamics of Antarctic whales. The zero-profit level $N_\infty = 20,000$ is a guess (economic data not being available), but is related to the fact that, during the years 1925–1965, Antarctic blue whales were depleted to a level somewhat less than 10,000. The current population, which is completely protected under IWC rules, is still thought to be less than 10,000. Nevertheless, Japan continues to press for a quota on this population (for 'scientific purposes'). The current wholesale price for whale meat in Japan renders the value of a single 80-ton blue whale carcass something in excess of $100,000.

The optimal population N^* for the case of the logistic growth model is easily calculated by solving the equation $dV/dN = 0$; the result is:

$$N^* = \frac{K}{4}\left[\frac{c}{pqK} + 1 - \frac{\delta}{r} + \sqrt{\left(\frac{c}{pqK} + 1 - \frac{\delta}{r}\right)^2 + \frac{8c\delta}{pqKr}}\right]. \qquad (16.22)$$

Calculated values of N^* and $h^* = F(N^*)$ for the blue whale model are shown in Table 16.1, as functions of the discount rate δ.

Table 16.1. Antarctic blue whale fishery: optimal population and optimal sustained yield.

Discount rate (per annum)	Optimal population (No. of whales)	Optimal sustained yield (No. of whales per annum)
0	85,000	1842
5%	44,050	1556
10%	31,200	1235
15%	26,954	1106
20%	25,000	1042
$+\infty$	20,000	867

Various features of the solution are worth noting. MSY occurs at $\frac{1}{2}K = 75,000$ whales, and the MSY itself is $F(\frac{1}{2}K) = 1875$ blue whales per annum. At zero discounting, the optimal population level is slightly greater than the MSY level, and OSY is correspondingly slightly below MSY. The introduction of a small positive discount rate, $\delta = 5$ per cent per annum, has a marked influence on the optimum, reducing the optimal population level by nearly 50 per cent. This also leads to a reduction in sustainable yield, but only by about 16 per cent. However, more 'realistic' discount rates, say 10 to 15 per cent per annum, have a much greater effect. Indeed, the 'optimal' equilibrium for $\delta = 15$ per cent is only marginally different from the bionomic equilibrium (which incidentally corresponds to $\delta = +\infty$).

Why does discounting have such a severe influence? And what are the implications for whale conservation? First, we recall that the blue whale population has a particularly low intrinsic growth rate: $r = 5$ per cent per annum. If $\delta > 5$ per cent blue whales become an 'inferior asset' at all population levels. Only the fact that harvesting costs become prohibitive at low population levels saves the whales from 'optimal extinction' by a profit-maximizing whaling industry. It can be announced as a general bioeconomic principle that: K-*selected species are especially vulnerable to depletion as a consequence of the discounting of future revenues* by profit-maximizing exploiters. Whales and other marine mammals provide classical examples of this phenomenon (Estes, 1979), but many other important biological resources are affected by the same principle. Trees, for example, are notoriously slow growing objects, and it is no surprise to find logging companies and conservationists in sharp disagreement over forest management practices. (Of course discounting effects are not the only source of conflict between exploiters and conservationists. It would be silly to assert that the whaling controversy is simply an argument over the appropriate 'social' discount rate, since much of the feeling behind the anti-whaling movement has a moralistic tone which seems to demand the permanent cessation of whaling. Nevertheless, the argument for 'preserving resource stocks for future generations' is fundamentally an argument for low discount rates—or at least for interfering with the depletion that may result from normal business practices.)

To continue a little longer with the whaling industry, we now have two 'explanations' for the depletion of the Antarctic whale populations: competitive exploitation on the one hand, and discounting at a rate greater than r on the other. Which is the correct explanation, and what difference does it make?

To answer the second question first, having a reasonably realistic underlying theory is a very important prerequisite to the development of *effective* management policies. If one believes (to quote an opinion formerly prevalent among some resource economists) that the private owner of a biological resource stock would never deliberately deplete the resource, then one will imagine that establishing private ownership rights will automatically resolve conservation problems. Perhaps the founding of the International Whaling Commission (the first official act of the newly established United Nations, in 1946) was based on some such hope. As will be noted later, multiple 'owners' of a common resource pool may be able, through a process of negotiation, to achieve agreement on management policies that will lead to conservative (and profitable) exploitation. For slow-growing (K-selected) species, however, our analysis shows that 'profitable' exploitation may not be conservative exploitation. For such species, the final result of profit maximizing management may be all but indistinguishable from the final result of an open competition. (The *history* of exploitation might be quite different under the two systems, however. For example, the annual quotas set by the IWC, and subsequently allocated to whaling states, probably served to slow down the depletion of whale stocks and to increase the profitability of whaling in the post-war decades.)

It seems clear from these arguments that discounting (at a sufficiently high rate) can be a leading cause of renewable resource depletion. Why does discounting take place? We cannot go into the details of capital theory here, and the reader is therefore referred to the professional literature (see, e.g. Herfindahl and Kneese, 1974, chapter 6). Let us just observe that real difficulties arise when the effect of discounting is combined with *irreversibility* of the depletion process, which occurs in many cases. An underdeveloped country, for example, may decide to cut down its tropical forests—or to deplete its wildlife populations—in order to obtain foreign currency with which to develop its economy. But various things may go wrong with the development scheme. Once the forests are gone, the people of the country may wind up with next to nothing until the trees have regrown—if indeed soil laterization does not completely prevent reforestation. A wiser policy would probably be to limit the extent of depletion of the renewable resource, and accept a slower rate of economic development, even if this contradicts calculations of 'optimal' economic growth.

16.3 Regulation

We have seen how important conflicts of interest are in the exploitation of common-property resource stocks. In the case of whales and other oceanic resources, these conflicts extend to the entire international community. In other cases, conflicts may be more localized.For example, a certain number of farmers may share a common water supply, and difficulties will arise when the supply of water is less than the total demand of the farmers—a common enough situation! In order to prevent wasteful competition—perhaps even violence—it is clear that some method must be found to regulate the use of the resource.

Two basic methods exist for the regulation of resource stocks:

(a) the government may *charge a fee* for the use of the resource, setting the fee sufficiently high to reduce demand to the available supply;

(b) the government may *ration* the resource, allocating specified amounts to each user.

If it is assumed that unrestricted transfers of ration quotas are allowed, then these two methods are easily seen to be equivalent in terms of economic efficiency; farmers who hold unused quotas will be prepared to sell them to farmers who want additional quotas. The price at which quotas are transferred will just suffice to clear the quota market—i.e. to equate supply and demand. Hence it will equal the price under (a). The fact that this price for water rights causes farmers to adopt optimal use patterns may be less obvious, but this will be demonstrated, in a slightly different setting, below.

Fee systems and rationing systems are of course not equivalent from the standpoint of users' incomes. Under a fee system the government collects the 'rent' from the resource, whereas a rationing system allocates rent to the initial recipients of quotas. Actually, by charging a fee for quotas which is below the demand price, the government can achieve any desired level of sharing of rents between itself and users. (It goes without saying that the users will be motivated to organize a lobby to minimize government fee charges!)

For the case of a biological resouce, unregulated competitive exploitation leads both to economic inefficiency, and frequently also to biological depletion. Since depletion is often more noticeable than inefficiency, resource management has often been directed primarily towards 'conservation' in the sense of preventing, or reversing, the

Table **16.2.** The Peruvian anchoveta fishery, 1959–1978.

Year	Number of boats	Number of fishing days	Catch (million tons)
1959	414	294	1·91
1960	667	279	2·93
1961	756	298	4·58
1962	1069	294	6·27
1963	1655	269	6·42
1964	1744	297	8·86
1965	1623	265	7·23
1966	1650	190	8·53
1967	1569	170	9·82
1968	1490	167	10·26
1969	1455	162	8·96
1970	1499	180	12·27
1971	1473	89	10·28
1972	1399	89	4·45
1973	1256	27	1·78
1974	n.a.	n.a.	4·0
1975	n.a.	n.a.	3·3
1976	n.a.	n.a.	4·3
1977	n.a.	n.a.	0·8
1978	n.a.	42	0·5

depletion of the resource. Not surprisingly, such management attempts are often quite ineffective from the economic viewpoint.

An unfortunately rather typical illustration is provided by the famous Peruvian anchoveta (*Engraulis ringens*) fishery, the catch history of which is shown in Table 16.2. During the years 1960–1972 this was the world's largest single fishery, and it also constituted a major segment of the Peruvian economy. Fishery experts advised the Peruvian authorities that MSY for this species was around 10 million tons annually, and catches were limited accordingly. No attempts were made, however, to limit the expansion of fishing or processing capacity. As a result, by the 1971 season capacity had become so great that the annual quota was caught in just three months' fishing, with fishing vessels and processing plants idle for the rest of the year. The catch could apparently have been handled with about 1/3 of the ultimate capacity.

The 1972–1973 crash of the anchoveta population has usually been attributed to an occurrence of 'El Niño'—an incursion of warm tropical water which interferes with upwelling and severely reduces the productivity of the Peruvian current. However, the main decline of the anchoveta actually preceded the 1973 El Niño, so that over-fishing seems to have been the main cause of the crash (Murphy, 1977). In spite of persistently low catches, the government has allowed fishing to continue every year since 1973.

One would think that the optimal management policy would be a moratorium of two or three seasons, in the hope that the anchoveta stocks would recover to their former levels. Unfortunately this policy seems not to be feasible, because of the political pressure exerted by some 20,000 people now dependent upon the anchoveta fishery for employment. Perhaps if the economic aspects of the fishery had been taken more seriously during the boom years, the present problems would not have arisen, or if so, would have been less difficult to resolve.

The method of total catch quotas has been used to manage many fisheries besides the Peruvian anchoveta. The method usually works as follows. The management agency formulates an estimate of the maximum sustainable yield, which is then adopted as the annual quota. On a specified day in the year the fishery is opened, and the accumulated catch is logged. When the total catch equals the quota the fishery is closed for the rest of the year. Obviously this method forces the fishermen to compete vigorously for their share of the catch. Annual quotas were used, for example, by the IWC from 1949–1960, and the ensuing scramble was commonly referred to as 'the whaling Olympics'. Eventually the whaling nations agreed to allocate quotas before the season opened, and some degree of sanity was restored.

16.3.1 Taxes versus allocated quotas

Let us now modify our bioeconomic model to include regulation. To be specific, we consider a fishery, in which m fishing vessels participate (m is not assumed fixed). We suppose that

$$h_i = qNE_i \qquad (16.23)$$

$$\Pi_i = ph_i - c_i(E_i) \qquad (i = 1, 2, \ldots, m) \qquad (16.24)$$

where $c_i(E_i)$ denotes the effort costs of the ith vessel (see Anderson, 1977 and Clark, 1979). For the competitive case, let us assume that each vessel utilizes an effort level that maximizes its own rate of profit; then:

$$c_i'(E_i) = pqN \quad (\text{if } E_i > 0).$$

More precisely, if we assume that the marginal cost curve $c_i'(E_i)$ has the traditional U-shape (Fig. 16.3), we see that

$$c_i'(E_i) = pqN \qquad \text{if } pqN > r_i \qquad (16.25)$$

$$E_i = 0 \qquad \text{if } pqN < r_i \qquad (16.26)$$

where r_i denotes the minimum average cost of vessel i:

$$r_i = \min \{c_i(E_i)/E_i\}.$$

Thus the greater the population level N, the greater the level of effort contributed by each vessel; if N falls below the zero-profit level $N_i^\infty = r_i/pq$, then vessel i leaves the fishery.

Let the vessels be ordered so that

$$r_1 \leq r_2 \leq \ldots \leq r_m. \qquad (16.27)$$

Then the 'bionomic equilibrium' of the unregulated, open-access fishery is determined by the conditions:

$$\frac{dN}{dt} = F(N) - qN \sum_{i=1}^{m} E_i = 0 \qquad (16.28)$$

$$c_i'(E_i) = pqN \qquad (i = 1, 2, \ldots, m) \qquad (16.29)$$

$$E_m = E_m^r \qquad (16.30)$$

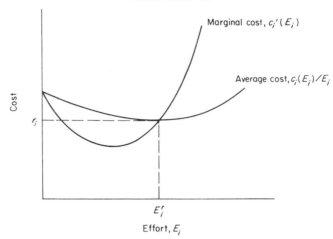

Fig. 16.3. Marginal and average cost-of-effort curves for a typical vessel. When marginal return to effort, pqx, is less than r_i, the ith vessel cannot fish profitably; $\pi_i \leq 0$. Otherwise the vessel employs a level of effort given by $c_i'(E_i) = pqx$. From Anderson, 1977, and Clark, 1980.

where E_m^r is the zero-profit effort level for vessel m (Fig. 16.3). The marginal (mth) vessel barely breaks even, but inframarginal vessels, because of their greater efficiency ($r_i < r_m$), earn a positive profit Π_i at equilibrium. The total profit $\sum_{i=1}^{m}\Pi_i$ is referred to as 'producers' surplus'.

The (social) optimum, on the other hand, would be determined by the conditions

$$\text{maximize} \int_0^\infty e^{-\delta t} \sum_{i=1}^{m} \Pi_i dt \qquad (16.31)$$

$$\text{subject to } \frac{dN}{dt} = F(N) - qN \sum_{i=1}^{m} E_i. \qquad (16.32)$$

Solution of this optimal control problem requires an application of the Pontrjagin maximum principle (c.f. Clark, 1976, ch. 4); we omit details. The optimal *equilibrium* solution is characterized by

$$\frac{dN}{dt} = F(N) - qN \sum_{i=1}^{m} E_i = 0 \qquad (16.33)$$

$$c_i'(E_i) = (p - \lambda)qN \qquad (i = 1, 2, \ldots, m) \qquad (16.34)$$

$$E_m = E_m^r \qquad (16.35)$$

$$F'(N) + \frac{\partial \Pi_m / \partial x}{\partial \Pi_m / \partial h_m} = \delta. \qquad (16.36)$$

Here λ is a positive 'multiplier' which arises in solving the optimization problem; in economic terms, λ is a 'shadow price' for the resource stock N. (Note that the above $m + 3$ equations just suffice to determine $m + 3$ unknowns $N, E_1, \ldots, E_m, m,$ and λ.)

It follows readily from this analysis that, by imposing an appropriate *catch tax*, the management authority can force the competitive fishery to operate at the social optimum in terms of population N, number of vessels m, and effort per vessel E_i. For suppose a tax τ is charged per unit of fish caught. The ith vessel's net profit is then $\Pi_i^0 = (p - \tau)h_i - c_i(E_i)$. Equation (16.29) is therefore replaced by

$$c_i'(E_i) = (p - \tau)qN \qquad (i = 1, 2, \ldots, m). \tag{16.37}$$

If $\tau = \lambda$, the shadow price determined by eq. (16.33)–(16.36), then eqs. (16.28)–(16.30) become identical to eqs. (16.33)–(16.35), and hence have the same solutions for m, N, E_i. To repeat: by imposing a catch tax equal to the 'shadow price' of the resource, the management authority can force the competitive fishery to operate in the socially optimal mode. This is the bioeconomic version of a standard theorem in welfare economics; similar results apply to situations such as pollution control (see Mills, 1978).

What about 'rationing' as an alternative to taxation? Suppose vessel i is granted a quota Q_i which it is not allowed to exceed:

$$h_i \le Q_i \qquad (i = 1, 2, \ldots, m). \tag{16.38}$$

Some vessel owners may want larger quotas than they have received. If transfers of quotas are allowed, it seems likely that a quota market would become established. Let μ denote the price on this market. A moment's consideration shows that the ith vessel owner will then wish to purchase additional quota units if and only if $\partial \Pi_i / \partial h_i \, (N, Q_i) > \mu$ where $\Pi_i = \Pi_i(h_i, N)$ is given by eq. (16.24). The equation

$$\frac{\partial \Pi_i}{\partial h_i} = \mu \tag{16.39}$$

therefore determines the ith vessel's demand function for quotas h_i at the quota price μ. The total demand for quotas is the sum of these demand functions. The total supply of quotas, on the other hand, is fixed at $\bar{Q} = \sum_{i=1}^{m} Q_i$. Equality of supply and demand then implies a specific quota price μ. It is easy to see, for example, that μ is an increasing function of the population level N, as one would expect.

Under the transferable vessel quota system, we therefore obtain the following equilibrium conditions:

$$\frac{dN}{dt} = F(N) - qN\sum_{i=1}^{m}E_i = 0 \qquad (16.40)$$

$$c_i'(E_i) = (p-\mu)qN \qquad (i = 1,2,\ldots,m) \qquad (16.41)$$

$$E_m = E_m^r \qquad (16.42)$$

$$\sum_{i=1}^{m}h_i = \bar{Q} = F(N) \qquad (16.43)$$

where (16.41) is an immediate consequence of eq. (16.39). Assuming that the authorities fix $\bar{Q}$ at the optimal level, we see that this new set of equilibrium conditions is equivalent to the previous system (16.33)–(16.36), and also that the equilibrium quota price μ is equal to the shadow price λ. Thus we have formally established the economic equivalence of taxes and transferable vessel quotas.

In fact, we can go a little further. Suppose a tax $\tau < \lambda$ is charged, in conjunction with the quota system. We then obviously get, in equilibrium,

$$\mu = \lambda - \tau, \qquad (16.44)$$

i.e. the quota price automatically makes up the deficit between the shadow price and the tax rate. By combining taxes and vessel quotas, the government can ensure efficiency of the fishery and at the same time achieve any desired level of sharing of the 'economic rent' between fishermen and the public coffers.

16.3.2   Resource jurisdiction

The above arguments have assumed the existence of a 'government' or other 'authority' with the power to impose taxes, or to set quotas on harvest. For many resource stocks, including pelagic fish and marine mammals, migratory birds, etc., no such central authority exists. If the stock is geographically restricted to areas under the jurisdiction of two or three states, it may be possible to negotiate effective management agreements. The simplest and most often used procedure consists of an agreed allocation of the total catch between states. This method can be successful provided that each state can feel confident that other states will adhere to the agreement. Taxes or quota systems can then be used to manage the national allocations.

Multinational resource stocks provide a much greater degree of difficulty. Individual states have historically shown a uniform disinclination to assign even limited jurisdiction to international agencies. The International Whaling Commission, for example, can only recommend total catches for the various stocks of whales, and each member state has a veto. (The IWC clearly has no control over non-members. Most present-day harvesting of large baleen whales is undertaken by 'pirate' whalers, Japan being the reported destination for most of the illicit catches.) At present it can only be admitted that most of the problems of managing the international commons remain wholly unresolved (Hardin and Baden, 1977).

16.4 Some complications

We conclude this chapter by discussing very briefly some of the numerous complications that can arise in practice in renewable resource management. For the most part it will be necessary to refer the reader to the literature for details. Published studies have by no means exhausted all of the interesting possibilities.

16.4.1 *Economic complications*

Time variations

Our basic bioeconomic model (see 16.2) assumed that all economic (and other) parameters are fixed, unchanging constants. Even though future trends in prices and costs may be hard to predict, no one really expects them to remain fixed at present levels. A question of interest is then, to what extent are current management decisions, based on current data, likely to prove to have been mistakes from some future vantage point?

Suppose, for example, that the economic parameters p, c, and δ of our basic model are in fact functions of time. Assume first that these functions are known. Thus

$$\Pi = \Pi(t) = [p(t)qN(t) - c(t)]E(t) \tag{16.45}$$

and

$$PV = \int_0^\infty \alpha(t)\Pi(t)dt \tag{16.46}$$

where the discount factor $\alpha(t) = \exp\left(-\int_0^t \delta(s)ds\right)$. It can be shown (Clark, 1976, section 3.3) that, instead of a fixed optimal equilibrium population level $N = N^*$, there now exists a 'moving' equilibrium $N = N^*(t)$, determined by the following extension of eq. (16.21):

$$F'(N) + \frac{\partial\Pi/\partial N}{\partial\Pi/\partial h} = \delta - \frac{1}{\Pi}\frac{\partial\Pi}{\partial t}. \tag{16.47}$$

The optimal adjustment policy from a nonoptimized initial population level $N(0) \neq N^*(0)$ is, with an important exception to be noted, the same 'bang-bang' policy, given by eq. (16.20).

It will be observed that eqs. (16.47) and (16.21) are identical except for the correction term $\dot\Pi/\Pi$ on the right (and the fact that the parameters in the equation are no longer constants). This term is clearly of the right form—if the profit $\Pi(t)$ is increasing at a given relative rate $\dot\Pi/\Pi$, there should be a corresponding correction to the discount rate δ. As a special case, if $\dot\Pi/\Pi$ = constant, as in an inflationary period, then the nominal rate δ is merely reduced by the inflationary rate to obtain the 'real' discount rate, as already noted.

Of greater interest is the fact that, to use the jargon of capital theory, the optimal varying stock level $N^*(t)$ is determined by a 'myopic' decision rule (16.47): the current optimum $N^*(t)$ depends only on current data $p(t)$, $c(t)$, $\delta(t)$, *and on their current rates of change.* This means that long-term movements in these economic parameters can be ignored—a reassuring result! Unfortunately there is an important exception: the myopic path $N^*(t)$ may not be 'feasible'.

Here is one obvious example (Clark and Munro, 1978). Under certain circumstances it can occur that the 'optimal' population level N^* as given in eq. (16.19) is zero—extinction is 'optimal'. (The reader should be sure he understands how this might arise.) This cannot be the case, however, if δ is sufficiently small. What if $\delta = \delta(t)$, with $\delta(t)$ large for t near zero, but eventually becoming much smaller? (Discount rates are supposed to decline, for example, as the economy of a country develops.) We could then find that

$$N^*(t) = 0 \text{ for } 0 < t < t_1$$

but

$$N^*(t) > 0 \text{ for } t > t_1.$$

This path is clearly not feasible—an extinct biological population cannot (normally) be resurrected! It would seem, under these conditions, that extinction should be avoided—we may want the population around later. Happily, this can be demonstrated rigorously: the optimal policy is to conserve a positive breeding stock N_b which will eventually allow harvesting at a sustained level (Clark and Munro, 1978).

Fixed costs

The bioeconomic model of section 16.2 treats costs in a highly stylized fashion, all costs being included in the single term cE. Actual costs of fishing (say) include wages, fuel, maintenance, insurance, and also the capital cost of the fishing vessel and its gear. Except for capital costs, it is perhaps not unreasonable to assume that cost is proportional to fishing effort, but capital costs are more realistically treated as fixed costs.

In natural resource industries, as elsewhere, fixed costs often have the characteristic of being irrecoverable. The cost of sinking a mine shaft is an example from exhaustible resource economics. Likewise, the cost of building a sophisticated trawler cannot be recovered if fish stocks later become depleted. Factory whaling ships made obsolete by the depletion of whale populations were then sold for scrap.

The following modification of our basic optimization model to incorporate fixed capital costs is achieved by supposing that maximum effort capacity E_{max} is itself proportional to total capital, G (see Clark, et al., 1979):

$$\text{maximize} \int_0^\infty e^{-\delta t}\{(pqN - c)E - c_1 I\}dt \qquad (16.48)$$

subject to
$$\frac{dN}{dt} = F(N) - qEN, \qquad N(0) = N_0 \qquad (16.49)$$

$$0 \le E(t) \le E_{max} = \alpha G(t) \qquad (16.50)$$

$$\frac{dG}{dt} = I - \gamma G, \qquad G(0) = G_0 \qquad (16.51)$$

$$I(t) \ge 0. \qquad (16.52)$$

Here $I(t)$ represents the rate of investment in vessel capital, c_1 is the cost of capital, and γ denotes the rate of depreciation of capital. For simplicity, it is assumed in eq. (16.52) that the investment is completely irreversible.

Solution of the above optimization problem is distinctly nontrivial, and we will not go into the details here. The optimal policy, it turns out, involves an initial phase of (apparent) overexpansion of capacity G and overexploitation of the resource stock N, but as capital depreciates there is an eventual recovery of the resource, until a long-term equilibrium is established $(dN/dt = dG/dt = 0)$. It is interesting that the general qualitative features of this 'optimal' solution have in fact frequently been observed, examples ranging from forestry to Antarctic whaling.

Variable demand

For most commodities there is an inverse relationship between the quantity Q supplied (per unit time) and the price p at which that quantity can be sold on the market. The relationship $p = p(Q)$ is called the 'demand schedule' for the given commodity. For our bio-economic model we treat the harvest rate h as the quantity produced, so that net revenue per unit time is now given by

$$\Pi = p(h)h - cE, \qquad h = qEN. \tag{16.53}$$

Previously we had simply assumed $p(h) = $ constant, i.e. that price was 'infinitely elastic'.

This change in our model leads to a technically more difficult optimization problem. In fact, there are now *two* optimization problems, the monopolist's optimum:

$$\text{maximize} \int_0^\infty e^{-\delta t} \Pi dt;$$

and, in contrast, the social welfare optimum:

$$\text{maximize} \int_0^\infty e^{-\delta t} [U(h) - cE] dt$$

where $U(h)$ denotes the 'social utility' function, defined as

$$U(h) = \int_0^h p(h) dh.$$

Without going into details (see Clark, 1976, section 5.3), let us just note that both problems have an optimal equilibrium $N = N^*$, $h = F(N^*)$. Assuming the same discount rate for both problems (this may not be realistic!), it turns out that the monopolist will always be a *more* conservative exploiter than the social welfare optimizer: the monopoly equilibrium N^* is larger than the social welfare equilibrium N^*. This is of course standard economic behaviour for the monopolist, who holds back on production in order to get a higher price. There is no question that such anti-social behaviour really does take place, in both exhaustible and renewable resource monopolies and cartels.

16.4.2   *Biological complications*

To an ecologist, our underlying model of population dynamics, $dN/dt = F(N)$, is almost pathologically oversimplified. While this model is useful as a general framework for the discussion of bioeconomic principles, many problems of practical interest demand a more realistic biological model. Also, it is important to stress that management decisions based on simplistic models are in danger of producing unexpected, and often undesirable results.

Fishing 'catastrophes'
Consider again our simple fishery model:

$$\frac{dN}{dt} = F(N) - h, \qquad h = qEN. \tag{16.54}$$

If $F(N)$ is logistic, or more generally any concave function ($F'' < 0$), eq. (16.54) has a unique stable equilibrium point $\bar{N}$ for any constant level of fishing effort E (Fig. 16.4). Moreover $\bar{N}$ is a continuous function of E.

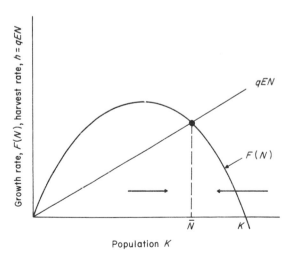

Fig. 16.4. The basic fishery model, eq. (16.54); the stable equilibrium at $\bar{N}$ varies continuously with changes in effort E.

The model thus predicts that the fish population will react in a continuous, reversible manner to changes in fishing intensity. Any errors made in formulating estimates of MSY can easily be taken care of by later reductions in effort levels. For many fisheries this prediction has proved to be realistic—reductions in fishing intensity have led to the recovery of depleted stocks. However, in the important class of clupeiod fisheries (herrings, sardines, anchovies) several spectacular population 'crashes' have occurred, following intensive fishing. In many cases, including the Pacific sardine (*Sardinops caerula*), Peruvian anchoveta, Hokkaido herring (*Clupea pellasi*), and North Sea herring (*C. harengus*), stocks have failed to recover even when fishing activity has been greatly reduced (Murphy, 1977).

It is usually not possible to specify with certainty the ultimate cause of a fishery collapse. People of a fatalistic philosophy will observe that the collapse was correlated with some alteration in the marine environment (the sea is always changing!), and will conclude that the collapse was probably inevitable. The failure of the population to recover after fishing has been reduced may be taken as corroboration of this viewpoint, although familiarity with nonlinear system theory would throw doubt on this argument.

A behavioural characteristic of many pelagic species, including especially the clupeoids, is the habit of forming large schools. The existence of an economically viable fishery often depends upon the concentrations of fish available in such schools. Furthermore, schools are readily detected by means of sonar devices and the like.

As the fish population is reduced by harvesting, either the number, or the average size of schools must decrease. For prey species, it can be shown that a decrease in the size of schools tends to increase the relative rate of mortality due to predation; this in turn implies a 'depensatory' relationship between stock and recruitment (Clark, 1976). In terms of our model, eq. (16.54), this means essentially that the growth curve $F(N)$ has a concave section, as shown in Fig. 16.5(a).

In some cases, school size has remained fairly constant, but the geographical area of the population has decreased as the fishery developed. In this case the catchability coefficient q in eq. (16.54) becomes a function of the population size, with $dq/dN < 0$—see Fig. 16.5(b).

Either of these changes in assumptions introduces the possibility of 'bifurcations,' or 'catastrophes' into our model. As indicated in Fig. 16.5(a and b), for small E there exist two positive equilibria $\bar{N}_1$ and

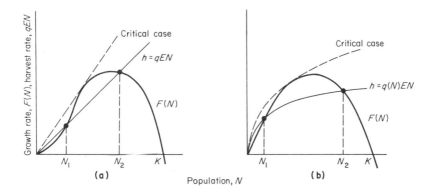

Fig. 16.5. Modifications of the basic fishery model: (a) depensation; (b) variable catchability. In both cases the model exhibits a bifurcation (or 'catastrophe') when effort E surpasses a citical level E_c.

$\bar{N}_2$ for eq. (16.54), with $\bar{N}_1$ being unstable and $\bar{N}_2$ stable. ($\bar{N} = 0$ is also a stable equilibrium.) As the effort level E increases, these equilibria coalesce, and then suddenly disappear at some critical level $E = E_c$. For $E \geq E_c$ the only equilibrium is at $N = 0$. The equilibrium of the model is thus no longer a continuous function of the control parameter E. Moreover, the effects of changes in E are not directly reversible: the population will not necessarily return to the stable equilibrium $\bar{N}_2$ when E is reduced to a subcritical level.

Such discontinuous behaviour is quite typical of nonlinear dynamic systems; examples are ubiquitous. In practice it is usually not possible to predict these catastrophic transitions, especially for the poorly understood natural systems involved in renewable resource industries. It would seem important, therefore, to design management procedures with built-in safety factors, allowing harvest rates to be quickly reduced at the first signs of an impending population crash. An example of this exists in the yellowfin tuna (*Thunnus albacares*) fishery of the Eastern Tropical Pacific Ocean, where the Director of Investigations of the Inter-American Tropical Tuna Commission is empowered to close the fishery as soon as catch per unit effort falls below a specified level.

Multispecies problems
Our simple fishery model, eq. (16.54), pertains to a single species or population. Many actual fisheries exploit several species simultaneously. The Antarctic whaling fleets, for example, exploited five species of

baleen whales (blue, fin, humpback, sei, Bryde's) and one toothed whale (sperm). The baleen whales all have a common main food source, Antarctic krill, while sperm whales feed at a higher trophic level. The population dynamics of the various species of whales are thus coupled in two ways in the fishery: ecologically and economically. A further bio-economic complication will arise if a commercial fishery on krill is established (May *et al.*, 1979).

Other well known examples of multispecies fisheries occur in the North Sea and the Northwest Atlantic. The ecological relationships between the various species in these fisheries are often amazingly complex—for example, fish of one species may feed on the young of a second species, which itself eats the eggs of the first species. When both species are commercially harvested, the system reaches an almost incomprehensible level of complexity. It is easy to criticize management policies that require MSY for each separate species in such a system (the standard approach in the past), but it is by no means obvious how a more reasonable policy ought to be formulated (May *et al.*, 1979).

When a multispecies system is exploited under competitive conditions, species at the higher trophic levels are often the first to be depleted. These species typically have lower r-values than those at lower trophic levels. In many cases, they are also more valuable and more easily caught. The ultimate (unregulated) equilibrium may involve the extinction, or near extinction, of such species. The need for management thus becomes transparent, even if the ideal methods are not.

From the economic viewpoint, the optimal management policy would be that which maximizes the present value of net economic returns from the resource. As noted in section 16.3, this can only be accomplished by some method of either taxing or rationing the resource. In principle there is no difficulty in applying taxes or quotas to the multispecies case—each species will carry an appropriate tax or quota. Operationally, however, the optimal taxes or quotas will be very difficult, if not impossible, to determine. The effects of a quota system would be more direct, and hence more predictable, than those of a tax system. Indeed, total (unallocated) catch quotas are currently in use in several multispecies fisheries. As we have seen, while such a system may be biologically effective, it generally increases the incentive for economic inefficiency. This tendency can be overcome by *allocating* quotas to fishermen. Especially in the multispecies situation, these allocated quotas should be transferable, in order to induce a maximum degree of flexibility.

Age of harvesting

The economic value of a tree, or of a fish, obviously depends on its size, and hence on its age. An optimal havest policy will therefore involve not only the rate of harvesting of a given resource stock, but also the appropriate age at which individuals are harvested. For many species the growth rate of individuals is quite low, and this means that discounting will have a significant influence both on the optimal age of harvest and on long-run sustained yield levels. For example, for Douglas fir (*Pseudotsuga menziesii*) on a typical stand in British Columbia, the optimal rotation age at zero discounting is 100 years, but this is reduced to 49 years at 10 per cent annual discounting; sustained yield by volume of timber would be reduced by 67 per cent (Clark, 1976, p. 261). It seems that such discrepancies are a continuing source of conflict between public forest services (which typically use zero discounting) and the forest industry (which presumably prefers a positive discount rate).

In fishery management, special precautions are often taken to prevent the harvesting of immature, small fish. The most common of these is a limitation on the size of mesh that can be used in nets. In lobster and crab fisheries minimum size limits are imposed, and sublegal captures must be returned to the sea. It is possible to work out the economic theory of size regulation, but we will not go into the details here. (The existing economic literature on this aspect is particularly unreliable, by the way. It is not true, as often claimed, that mesh-size regulation automatically results in increased economic benefits. Without rationing or taxes, the potential benefits will still be dissipated.)

This brief list by no means exhausts the possible biological complications; others include spatial distribution and aggregation of stocks, delay effects in population dynamics, and genetic properties of populations. All of these can have important economic consequences in particular cases; many details are yet to be worked out. It should be pointed out that the mathematical optimization problems corresponding to even slightly complex biological models can be exceedingly difficult, whether handled analytically or by computer—as anyone who has tussled with such problems will happily attest!

16.4.3 Uncertainty

Management decisions concerning biological resources are usually formulated under considerable uncertainty. The long-term biological effects of a newly introduced pesticide are seldom predictable. It is usually not known how long it will take a depleted population—or a clearcut forest—to recover. For example, Antarctic blue whales have been protected since 1965, but no one knows how much the population has increased, or even if it *has* increased in the interim. The Peruvian anchoveta stocks may be capable of recovering to former levels of abundance, but the fishing companies seem to hold a pessimistic view of the chances of a rapid recovery.

In most cases, too little is known about the structure of any particular ecosystem to be able to formulate precise quantitative predictions regarding the effects of harvesting certain species. For example, there is current speculation regarding the imminent development of a large-scale fishery for Antarctic krill (*Euphausia superba*), the main food source for whales, seals, fish, and sea birds in the southern ocean. Annual natural production of krill has been estimated at approximately 1500 million tons, or about 20 times the annual world catch of all marine species, including fish, mammals, crustaceans, mulluscs, and seaweeds. How much of this krill is 'surplus' to the Antarctic ecosystem? What effect would the harvest of, say, 100 million tons of krill annually have on the population dynamics of the important whale species? Clearly the current notion of achieving MSY for each species separately is biological nonsense, but reasonable alternatives are not easy to specify (May *et al.*, 1979).

Uncertainty has the effect of increasing the rate of discount applied to future revenues, and hence (as we have seen) adds to the incentive for depletion of resource stocks. Government policy in some areas, such as agriculture and forestry, has reflected this concern. Forests, for example, have often been retained in government ownership, possibly reflecting distrust of private owners' incentives to practice socially desirable levels of conservation.

The above examples involve uncertainty in its broadest sense, in which predictions simply cannot be made with any degree of confidence. A more limited form of uncertainty arises when the probability distribution of future events is assumed to be known, and the 'state of nature' that will exist at some future date is given in terms of this probability distribution. For example, annual recruitment to a certain fish population may fluctuate unpredictably about some mean value. Various possibilities arise: the mean and variance in recruitment may depend upon the size of the breeding stock, or on the other hand there may be no statistically significant relationship between stock and recruitment. A model pertaining to the former situation may be formulated as follows (Reed, 1974, 1978):

$$N_{t+1} = Z \cdot F(N_t) \qquad (16.55)$$

where Z is a random variable with mean 1, and where $F(\ )$ is the mean stock-recruitment relationship. If the population is exploited, this equation becomes

$$N_{t+1} = Z \cdot F(N_t - H_t) \qquad (16.56)$$

where H_t denotes the harvest in period t.

Three alternative types of harvest policy may be considered: (a) constant yield, $H_t =$ const., (b) constant effort, $E_t =$ const., and (c) constant escapement, $S_t = N_t - H_t =$ const. A constant yield policy, by definition, stabilizes the annual catch, but can be strongly destabilizing for the population (May et al., 1978). Indeed, if the constant yield requirement is taken literally, a sequence of bad years (below average recruitment) may result in the extinction of the population. While this may seem unrealistic, it has in practice sometimes proved difficult for the fishing industry to understand that a scientific estimate of 'maximum sustained yield' does not refer to a catch that can be taken each and every year regardless of natural fluctuations in recruitment.

A constant-effort policy is less destabilizing than a constant-yield policy, but probably has little practical value since 'effort' is a notoriously difficult variable to control. A constant escapement policy clearly provides maximum stability for the population, but will generally result in large fluctuations in the annual catch. (Whenever recruitment N_t is *below* the specified escapement level S, then of course H_t must be zero.)

What about the 'optimal' harvest policy for the stochastic model? If the price of fish, and cost of effort are both assumed constant, it can be demonstrated (Reed, 1974) that the harvest policy that maximizes the expected present value of future net revenes is in fact a constant escapement policy (see also Ludwig, 1979, who analyses the effect of noise on the optimal escapement level). In practice, constant escapement policies are currently used in the management of some fisheries, notably the Pacific salmon (*Oncorhynchus* spp.) fisheries. Substantial variation in annual catches is a characteristic feature of these fisheries, which are also especially subject to depletion unless closely regulated. Variation in the catches means that fishermen's incomes fluctuate correspondingly, but this seems to be accepted as a part of the fishing life.

16.5   Summary

Biological systems provide mankind with many essential products and services. In principle, the problem of managing these resource 'assets' so as to maximize social benefits differs from other asset management problems only in the additional complexity inherent in the biology. In practice, however, biological resources often involve characteristic difficulties centred largely around conflicts of interest. The exploitation of a biological stock by one user frequently inflicts externalities on other users or potential users. Whenever these externalites are significant, some form of regulation becomes necessary.

Regulation is itself costly, however. Furthermore, effective regulation requires enforcement, which is also costly. The simplest solution in principle is the institution of private property rights, but this solution is unfortunately not feasible for many biological resource stocks. And even where property rights are feasible, government intervention may remain necessary, for example to ensure socially beneficial conservation practices over the long run.

Where property rights are not feasible, alternative institutions are necessary, first to prevent depletion or over-use of the resource stock, and secondly to overcome economically wasteful exploitation practices. Two methods are available, taxation and rationing (allocated quotas). Both of these constitute proxies for full property rights: taxation retains rights to the taxing authority while quotas allocate rights to quota holders. By combining taxes and allocated quotas, the government can achieve a sharing of benefits between itself and the resource exploiters.

Both taxes and quotas have been quite commonly employed in certain resource management programmes, including public forests, grazing lands, and water resources. In other areas, particularly marine fisheries, only minor use has been made of these methods. Until recently, perhaps, the costs of setting up such systems of rights proxies have exceeded the potential benefits. But with 200-mile fishing zones now in existence, it seems certain that fishery management will evolve towards property rights or proxies thereof. As resource utilization intensifies with the growth of the human population, inefficient or shortsighted use of resources—renewable or exhaustible—will become more and more unwise.

References

Abrahamson, W.G., and Gadgil, M. 1973. Growth form and reproductive effort in Goldenrods (*Solidago*, Compositae). *Am. Nat.* **107**, 651–61.

Abrams, P. 1976a. Limiting similarity and the form of the competition coefficient. *Theoret. Pop. Biol.* **8**, 356–75.

Abrams, P. 1976b. Environmental variability and niche overlap. *Math. Biosci.* **28**, 357–72.

Acevedo, M.F. 1978. On a Markovian model of forest succession: its application to tropical forests. M.E. Thesis, Dept. Elect. Eng. and Comput. Sci., Univ. California.

Addicott, J.F. 1978. Competition for mutualists: aphids and ants. *Can. J. Zool.* **56**, 2093–6.

Aikman, D.P., and Watkinson, A.R. 1980. A model for growth and self-thinning in even-aged monocultures of plants. *Ann. Bot.* **45**, 419–27.

Akinlosotu, T.A. 1973. The Role of *Diaeretiella rapae* (McIntosh) in the Control of the Cabbage Aphid. Unpublished Ph.D. thesis, University of London.

Alexander, R.D. 1974. The evolution of social behaviour. *Ann. Rev. Ecol. Syst.* **5**, 325–83.

Alexander, R.D., and Sherman, P.W. 1977. Local mate competition and parental investment in social insects. *Science*, **196**, 494–500.

Alvarez, L.W., Alvarez, W., Asaro, F., and Michel, H.V. 1980. Extraterrestrial cause for the Cretaceous-Tertiary extinction. *Science*, **208**, 1095–108.

Anderson, E.S. 1957. The relations of bacteriophages to bacterial ecology. In *Microbial Ecology*. 7th Symposium of the Society for General Microbiology, pp. 189–217. London, Cambridge University Press.

Anderson, L.G. 1977. *The Economics of Fisheries Management*. Baltimore, Johns Hopkins University Press.

Anderson, M.C. 1966. Ecological groupings of plants. *Nature (Lond.)*, **212**, 54–6.

Anderson, R.C. 1971. Neurologic disease in reindeer (*Rangifer torandus torandus*) introduced into Ontario. *Can. J. Zool.* **49**, 159–66.

Anderson, R.C. 1972. The ecological relationship of meningeal worm and native cervids in North America. *J. Wildl. Dis.* **8**, 304–10.

Anderson, R.M. 1978. The regulation of host population growth by parasitic species. *Parasitology*, **76**, 119–58.

Anderson, R.M. 1979a. The influence of parasitic infection on the dynamics of host population growth. In R.M. Anderson, B.D. Turner and L.R. Taylor (eds.), *Population Dynamics*, pp. 245–81. Oxford, Blackwell Scientific Publications.

Anderson, R.M. 1979b. Parasite pathogenicity and the depression of host population equilibria. *Nature*, **279**, 150–2.

Anderson, R.M. 1980a. The control of infectious disease agents: strategic models. In G.R. Conway (ed.), *The Management of Pest and Disease Systems*. John Wiley. (In press.)

Anderson, R.M. 1980b. The dynamics and control of direct life cycle helminth parasites. *Lecture Notes in Biomathematics* **39**, pp. 278–322. Berlin, Springer-Verlag.

Anderson, R.M. 1980c. Depression of host population abundance by direct life-cycle macroparasites. *J. Theor. Biol.* **82**, 283–311.

Anderson, R.M., and May, R.M. 1978. Regulation and stability of host–parasite population interactions: I. Regulatory processes. *J. Anim. Ecol.* **47**, 219–49.

Anderson, R.M., Whitfield, P.J., and Dobson, A.P. 1978. Experimental studies on infection dynamics: infection of the definitive host by the cercariae of *Transversotrema patialense*. *Parasitology*, **77**, 189–200.

Anderson, R.M., and May, R.M. 1979. Population biology of infectious diseases. Part I. *Nature*, **280**, 361–7.

Anderson, R.M., and May, R.M. 1980. Infectious diseases and population cylces of forest insects. *Science*, **210**, 658–61.

Anderson, R.M., and May, R.M. 1981. The population dynamics of microparasites and their invertebrate hosts. *Phil. Trans. R. Soc.* (In press.)

Andrewartha, H.G. 1970. *Introduction to the Study of Animal Populations*. London, Chapman and Hall.

Andrewartha, H.G., and Birch, L.C. 1954. *The Distribution and Abundance of Animals*. Chicago, University of Chicago Press.

Annecke, D.P., and Moran, V.C. 1978. Critical reviews of biological pest control in South Africa. 2. The prickly pear, *Opuntia ficus-indica* (L.) Miller. *J. ent. Soc. sth. Afr.* **41**, 161–88.

Apple, J.L., and Smith, R.F. (eds.). 1976. *Integrated Pest Management*. New York, Plenum Press.

Armstong, N.E., Storrs, P.N., and Pearson, E.A. 1971. Development of a gross toxicity criterion in San Francisco Bay. In *Advances in Water Pollution Research: Proceedings of Vth International Conference*, pp.III 1–15. Oxford, Pergamon Press.

Aron, J.L., and May, R.M. 1981. Population of Malaria. In *Population Dynamics of Infectious Disease Agents*. R.M. Anderson Ed. London, Chapman and Hall. (In Press.)

Ashton, J.H., and Rowell, A.J. 1975. Environmental stability and species pro-lifⁿration in Late Cambrian Trilobite faunas: a test of the niche-variation hypothesis. *Paleobiology*, **1**, 161–74.

Auer, C. 1968. Erste Ergebnisse einfacher Modelluntersuchungen über die Ur-sachen der Populationsbewegung des grauen Larchenwicklers *Zieraphera diniana* Gn. (–*Z. griseana* Hb) im Oberengadin 1949/66. *Z. angew. Ent.* **62**, 202–35.

Augustine, D.L. 1923a. Investigations on the control of hookworm disease. XVI. Variation in the length of life of hookworm larvae from the stools of different individuals. *Am. J. Hyg.* **3**, 127–36.

Augustine, D.L. 1923b. Investigations on the control of hookworm disease. XXIII. Experiments on the factors determining the length of life of infective hookworm larvae. *Am. J. Hyg.* **3**, 420–43.

Auslander, D., Oster, G., and Huffaker, C. 1974. Dynamics of interacting popula-tions. *J. Franklin. Inst.* **297**, 345–76.

Austin, M.P., Aston, P.S., and Greig-Smith, P. 1972. The application of quanti-tative methods to vegetation survey. III. A re-examination of rain forest data from Brunei. *J. Ecol.* **60**, 305–24.

Axelrod, D.I., and Bailey, H.P. 1968. Cretaceous dinosaur extinction. *Evolution*, **22**, 595–611.

Ayala, F.J., Gilpin, M.E., and Ehrenfeld, J.G. 1973. Competition between species: theoretical models and experimental tests. *Theoret. Pop. Biol.* **4**, 331–56. [Referred to as Ayala *et al.*, 1973a.]

Ayala, F.J., Hedgecock, D., Zumwalt, G., and Valentine, J.W. 1973. Genetic variation in Tridacna maxima, an ecological analog of some unsuccessful evolutionary lineages. *Evolution*, **27**, 177–91. [Referred to as Ayala *et al.*, 1973b.]

Ayala, F.J., Valentine, J.W., and Zumwalt, G.S. 1975. Electrophoretic study of the Antarctic zooplankter *Euphausia superba*. *Limn. Oceanog.* **20**, 635–40.

Bailey, N.T.J. 1975. *The Mathematical Theory of Infectious Diseases* (2nd Edition). London, Griffin.

Bakker, K., Bagchee, S.N., Van Zwet, W.R., and Meelis, E. 1967. Host discrimination in *Pseudeucola bochei* (Hymenoptera: Cynipidae). *Entomologia exp. appl.* **10**, 295–311.

Bakker, R. 1974. Dinosaur bioenergetics: a reply to Bennett and Dalzell, and Feduccia. *Evolution*, **28**, 497–503.

Bakker, R. 1975a. Experimental and fossil evidence for the evolution of tetrapod bioenergetics. In D.M. Gates and R.B. Schmerl (eds.), *Perspectives in Biophysical Ecology*. New York, Springer-Verlag.

Bakker, R. 1975b. Dinosaur renaissance. *Sci. Amer.* **232(4)**, 58–78.

Bakker, R.T. 1977. Tetrapod mass extinctions—a model of the regulation of speciation rates and immigration by cycles of topographic diversity. In A. Hallam (ed.), *Patterns of Evolution*, pp. 439–68. Amsterdam. Elsevier.

Baltensweiler, W. 1964. *Zeiraphera griseana* Hubner (Lepidoptera: Tortricidae) in the European Alps. A contribution to the problem of cycles. *Can. Ent.* **96**, 792–800.

Baltensweiler, W. 1971. The relevance of changes in the composition of larch bud moth populations for the dynamics of its numbers. In P.J. den Boer and G.R. Gradwell (eds.), *Dynamics of Populations*, pp. 208–19. Wageningen, Centre for Agricultural publishing.

Bambach, R.K. 1977. Species richness in marine benthic habitats through the Phanerozoic. *Paleobiology*, **3**, 152–67.

Banerjee, B. 1979. A key-factor analysis of population fluctuations in *Andraca bipunctata* Walker (Lepidoptera: Bombycidae). *Bull. ent. Res.* **69**, 195–201.

Bang, F.B. 1973. Immune reactions among marine and other invertebrates. *Bio-Science*, **23**, 584–9.

Barbehenn, K.R. 1969. Host–parasite relationships and species diversity in mammals: an hypothesis. *Biotropica*, **1**, 29–35.

Barbour, A.D. 1978. Macdonald's model and the transmission of bilharzia. *Trans. R. Soc. Trop. Med. Hyg.* **72**, 6–15.

Barclay, H. 1975. Population strategies and random environments. *Can. J. Zool.* **53**, 160–5.

Barclay-Estrup, P. 1970. The description and interpretation of cyclical processes in a heath community. II. Changes in biomass and shoot production during the *Calluna* cycle. *J. Ecol.* **58**, 243–9.

Bateson, P. 1980. Optimal outbreeding and the development of sexual preferences in Japanese quail. *Z. Tierpsychol.* (In press.)

Bayne, C.J., and Kine, J.B. 1970. In vivo removal of bacteria from the haemolymph of the land snail *Helix pomatia* (Pulmonata, Sylommatophora). *Malacological Reviews*, **3**, 103–13.

Bazzaz, F.A. 1975. Plant species diversity in old-field successional ecosystems in southern Illinois. *Ecology*, **56**, 485–8.

Bazzaz, F.A., and Harper, J.L. 1976. Relationship between plant weight and numbers in mixed populations of *Sinapsis alba* (L.) Rabenh. and *Lepidium sativum* L. *J. Appl. Ecol.* **13**, 211–16.

Bazzaz, F.A., and Harper, J.L. 1977. Demographic analysis of the growth of *Linum usitatissimum*. *New Phytol.* **78**, 193–208.

Beauchamp, R.S.A., and Ullyett, P. 1932. Competitive relationships between certain species of fresh-water triclads. *J. Ecol.* **20**, 200–8.

Beddington, J.R. 1974. Age distribution and the stability of simple discrete time population models. *J. Theor. Biol.* **47**, 65–74.

Beddington, J.R. 1975. Mutual interference between parasites or predators and its effect on searching efficiency. *J. Anim. Ecol.* **44**, 331–40.

Beddington, J.R. 1978. On the risks associated with different harvesting strategies. *Int. Comm. Whaling Rep.* **28**, 165–7.

Beddington, J.R., and May, R.M. 1975. Time delays are not necessarily destabilizing. *Math. Biosci.* **27**, 109–17.

Beddington, J.R., Free, C.A., and Lawton, J.H. 1975. Dynamic complexity in predator–prey models framed in difference equations. *Nature, (Lond.),* **255**, 58–60.

Beddington, J.R., Free, C.A., and Lawton, J.H. 1976. Concepts of stability and resilience in predator–prey models. *J. Anim. Ecol.* **45**, 791–816.
[Referred to as Beddington *et al.*, 1976a]

Beddington, J.R., Hassell, M.P., and Lawton, J.H. 1976. The components of arthropod predation. II. The predator rate of increase. *J. Anim. Ecol.* **45**, 165–85.
[Referred to as Beddington *et al.*, 1976b.]

Beddington, J.R., and May, R.M. 1977. Harvesting natural populations in a randomly fluctuating environment. *Science,* **197**, 463–5.

Beddington, J.R., Free, C.A., and Lawton, J.H. 1978. Characteristics of successful natural enemies in models of biological control of insect pests. *Nature,* **273**, 513–19.

Beddington, J.R., and Lawton, J.H. 1978. On the structure and behaviour of ecosystems. *Journal de Physique,* **39c**, 5–39.

Beddington, J.R., and Grenfell, B. 1979. Risk and stability in whale harvesting. *Int. Comm. Whaling Rep.* **29**, 171–3.

Bell, A.D. 1974. Rhizome organisation in relation to vegetative spread in *Medeola virginiana*. *J. of the Arnold Arboretum,* **55**, 458–68.

Bell, A.D. 1976. Computerised vegetative mobility in rhizomatous plants. In A. Lindenmayer and G. Rozenberg (eds.), *Automata, Languages, Development*. North Holland Publishing Company.

Bell, A.D. 1979. The hexagonal branching pattern of rhizomes of *Alpina speciosa* L. (Zingiberaceae). *Ann. Bot.* **43**, 209–23.

Bengtsson, B.O. 1978. Avoiding inbreeding: at what cost? *J. Theor. Biol.* **73**, 439–44.

Benson, J.F. 1973. Laboratory studies of insect parasite behaviour in relation to population models. Unpublished D.Phil. thesis, Oxford University.

Benson, W.W., Brown, K.S., and Gilbert, L.E. 1976. Coevolution of plants and herbivores: passion flower butterflies. *Evolution,* **29**, 183–6.

Bentley, B.L. 1976. Plants bearing extrafloral nectaries and the associated ant community: interhabitat differences in the reduction of herbivore damage. *Ecology,* **54**, 815–20.

Bentley, B.L. 1977. The protective function of ants visiting the extrafloral nectaries of *Bixa orellana* L. *J. Ecol.* **65**, 27–38.

Berkner, L.V., and Marshall, L.C. 1964. The history of oxygenic concentration in the earth's atmosphere. *Discuss. Faraday Soc.* **37**, 122–41.

Bernoulli, D. 1760. Essai d'une nouvelle analyse de la mortalité causée par la petite vérole et des advantages de li'noculation pour la prevenir. *Mem. Math. Phys. Acad. Roy. Sci., Paris*, 1–45.

Berry, R.J. 1980. The great mouse festival. *Nature*, **283**, 15–16.

Bertram, B.C.R. 1975. Social factors influencing reproduction in wild lions. *J. Zool.* **177**, 463–82.

Bigger, M. 1976. Oscillations of tropical insect populations. *Nature*, **259**, 207–9.

Birch, L.C. 1957. The meanings of competition. *Am. Nat.* **91**, 5–18.

Birch, L.C. 1971. The role of environmental heterogeneity and genetical heterogeneity in determining distribution and abundance. In P.J. den Boer and G.R. Gradwell (eds.) *Dynamics of Populations*, pp. 109–28. Wageningen Centre for Agricultural Publishing.

Biswell, H.H. 1974. Effects of fire on chaparral. In T.T. Kozlowski and C.E. Ahlgren (eds.), *Fire and Ecosystems*, pp. 321–64. New York, Academic Press.

Biswell, H.H., Buchanan, H., and Gibbens, R.P. 1966. Ecology of the vegetation of a second-growth Sequoia forest. *Ecology*, **47**, 630–4.

Black, R. 1976. The effects of grazing by the limpet, *Acmaea insessa* on the kelp, *Egregia laevigata*, in the intestidal zone. *Ecology*, **57**, 265–77.

Blackith, R.E. 1974. Strategies and tactics in evolution. *Recherches Biologiques Contemporaines*, 427–35.

Blackwell, J., Freeman, J., and Bradley, D. 1980. Influence of H-2 complex on acquired resistance to *Leishmania donovani* infection in mice. *Nature*, **283**, 72–4.

Bliss, C.A., and Fisher, R.A. 1953. Fitting the negative binomial distribution to biological data and a note on the efficient fitting of the negative binomial. *Biometrics*, **9**, 176–200.

Blydenstein, J. 1967. Tropical savanna vegetation of the Llanos of Columbia. *Ecology*, **48**, 1–15.

Bonner, J.T. 1965. *Size and Cycle: an essay on the structure of biology*. Princeton, Princeton University Press.

Bonner, J.T. 1974. On Development. Princeton, Princeton University Press.

Bonner, J.T. 1980. *The Evolution of Culture in Animals*. Princeton, Princeton University Press.

Bonner, J.T., and May, R.M. 1981. *Introduction to C. Darwin, The Descent of Man*. Paperback reprint, Princeton University Press.

Borg, K. 1962. Predation on roe deer in Sweden. *J. Wildl. Mgmt.* **26**, 133–6.

Boscher, J. 1979. Modified reproduction strategy of leek *Allium porrum* in response to a phytophagous insect, *Acrolepiopsis assectella*. *Oikos*, **33**, 451–6.

Botkin, D.B., Janak, J.F., and Wallis, J.R. 1972. Some ecological consequences of a computer model of forest growth. *J. Ecol.* **60**, 849–72.

Bradley, D.J. 1977. Human pest and disease problems: contrasts between developing and developed countries. In J.M. Cherritt, and G.R. Sagar (eds.), *Origins of Pest, Parasite, Disease and Weed Problems*, pp. 239–45. Oxford, Blackwell Scientific Publications.

Bradley, D.J., and May, R.M. 1978. Consequences of helminth aggregation for the dynamics of schistosomiasis. *Trans. R. Soc. Trop. Med. Hyg.* **72**, 262–73.

Brenchley, W.E. 1958. *The Park Grass Plots at Rothamsted*. Rothamsted Exptl. Station, Harpenden.

Brénière, J., and Walker, P.T. 1970. Host: *Oryza sativa* (rice); Organism:

Maliorpha separatella (African white rice borer). In *Crop Loss Assessment Methods*. Rome, Food and Agriculture Organization.

Bretsky, P.W. and Lorenz, D.M. 1970. Adaptive response to environmental stability: a unifying concept in paleoecology. *Proc. N. Amer. Paleont. Convention, Chicago* 1969 (*part E*), 522–50.

Bretsky, P.W., and Bretsky, S.S. 1975. Succession and repetition of Late Ordovician fossil assemblages from the Nicolet River Valley, Quebec. *Paleobiology*, **1**, 225–37.

Brian, M.V. 1952. Structure of a dense natural ant population. *J. Anim. Ecol.* **21**, 12–24.

Briand, F., and Yodzis, P. 1980. The phylogenetic distribution of obligate mutualism. *Nature* (in press).

Bristow, C.M. 1981. The role of mutualists in structuring the honeydew-producing guild on *Veronia noveboracensis*. Ph.D. thesis, Princeton University.

Brougham, R.W. 1955. A study in rate of pasture growth. *Aust. J. Agric. Res.* **6**, 804–12.

Brown, E.S. 1962. *The African Army Worm*, Spodoptera exempta [*Walker*]. [*Lep., Noctuidae*]: *A Review of the literature.* Commonwealth Institute of Entomology.

Brown, J.H. 1971. Mammals on mountaintops: Non-equilibrium insular biogeography, *Am. Nat.* **105**, 467–78.

Brown, J.H., and Lieberman, G.A. 1973. Resource utilization and coexistence of seed-eating desert rodents in sand dune habitats. *Ecology*, **54**, 788–97.

Brown, W.L., and Wilson, E.O. 1956. Character displacement. *Syst. Zool.* **5**, 49–64.

Browning, J.A., Simons, M.D., and Torres, E. 1977. Managing host genes: epidemiologic and genetic concepts. In J.G. Horsfall and E.B. Cowling (eds.), *Plant Disease: An Advanced Treatise, Vol. 1: How Disease is Managed*, pp. 191–212. New York, Academic Press.

Bryant, J.P. 1980. The regulation of snowshoe hare feeding behavior during winter by plant antiherbivore chemistry. In K. Meyers (ed.), *Proceedings of the First International Lagomorph Conference*, pp. 69–98. Guelph, Guelph Univ. Press.

Buell, M.F., Buell, H.F., and Small, J.A. 1954. Fire in the history of Mettler's Woods. *Bull. Torrey Bot. Club*, **81**, 1–3.

Burnett, T. 1956. Effects of natural temperatures on oviposition of various numbers of an insect parasite (Hymenoptera, Chalcididae, Tenthredinidae). *Ann. ent. Soc. Am.* **49**, 55–9.

Burnett, T. 1958. Dispersal of an insect parasite over a small plot. *Can. Ent.* **90**, 279–83.

Buss, L.W. 1979. Habitat selection, directional growth and spatial refuges. Why colonial animals have more hiding places. In J. Larwood and B.R. Rosen (eds.), *Biology and Systematics of Colonial Organisms*, pp. 459–96. London, Academic Press.

Buss, L.W., and Jackson, J.B.C. 1979. Competitive networks: nontransitive competitive relationships in cryptic coral reef environments. *Am. Nat.* **113**, 223–34.

Butlin, N.G. 1962. Distribution of sheep population: preliminary statistical picture, 1860–1957. In A. Barnard (ed.), *The Simple Fleece*, pp. 281–307. Melbourne, Melbourne University Press.

Campbell, C.A., and Valentine, J.W. 1977. Comparability of modern and ancient marine faunal provinces. *Paleobiology*, **3**, 49–57.

Canter, H.M., and Lund, J.W.G. 1968. The importance of Protozoa in controlling the abundance of planktonic algae in lakes. *Proc. Linn. Soc. Lond.* **179**, 203–19.

Cantlon, J.E. 1969. The stability of natural populations and their sensitivity to technology. *Brookhaven Symp. Biol.* **22**, 197–207.

Caplan, A.L. (ed.) 1978. *The Sociobiology Debate.* New York, Harper & Row.

Caraco, T. 1979a. Time budgeting and group size: a theory. *Ecology*, **60**, 611–17.

Caraco, T. 1979b. Time budgeting and group size: a test of theory. *Ecology*, **60**, 618–27.

Carleton, T.J., and Maycock, P.F. 1978. Dynamics of boreal forest south of James Bay. *Canadian J. of Bot.* **56**, 1157–73.

Carlile, M.J. 1980. From prokaryote to eukaryote: gains and losses. In G.W. Gooday, D. Lloyd, and A.P.J. Trinci (eds.), *The Eukaryotic Microbial Cell.* Soc. Gen. Microbiol. Symp. **30**, 1–40.

Carpenter, F.L. 1979. Competition between hummingbirds and insects for nectar. *Amer. Zool.* **19**, 1105–14.

Caswell, H. 1976. Community structure: a neutral model analysis. *Ecol. Monog.* **46**, 327–54.

Cates, R.G., and Orians, G.H. 1975. Successional status and the palatability of plants to generalized herbivores. *Ecology*, **56**, 410–18.

Caughley, G. 1970. Eruption of ungulate populations, with emphasis on Himalayan thar in New Zealand. *Ecology*, **51**, 53–72.

Caughley, G. 1976. Wildlife management and the dynamics of ungulate populations. In T.H. Coaker (ed.), *Applied Biology*, **I**, 183–246.

Cavalier-Smith, T. 1978. Nuclear volume control by nucleoskeletal DNA, selection for cell volume and cell growth rate, and the solution of the DNA C-value paradox. *J. Cell. Sci.* **34**, 247–78.

Cavalli-Sforza, L.L., and Bodmer, W.F. 1971. *The Genetics of Human Populations.* San Francisco, Freeman.

Cavalli-Sforza, L.L., and Feldman, M. 1978. Darwinian selection and 'altruism'. *Theor. Pop. Biol.* **14**, 268–80.

Cavalli-Sforza, L.L., and Feldman, M. 1981. *Cultural Transmission and Evolution.* Princeton, Princeton University Press.

Charlesworth, B. and Charlesworth, D. 1978. A model for the evolution of dioecy and gynodioecy. *Am. Nat.* **112**, 975–97.

Charnov, E.L. 1973. Optimal foraging: some theoretical explorations. Ph.D. Thesis, Seattle, University of Washington.

Charnov, E.L. 1975. Optimal foraging: attack strategy of a mantid. *Am. Nat.* **110**, 141–51.

Charnov, E.L. 1978. Eusocial behavior: offspring choice or parental parasitism? *J. Theor. Biol.* **75**, 451–65.

Charnov, E.L. 1979. Natural selection and sex change in pandalid shrimp: test of a life-history theory. *Am. Nat.* **113**, 715–33.

Charnov, E.L., and Schaffer, W.M. 1973. Life history consequences of natural selection: Cole's result revisited. *Am. Nat.* **107**, 791–3.

Charnov, E.L., Maynard Smith, J., and Bull, J.J. 1976. Why be an hermaphrodite? *Nature*, **263**, 125–6.

Chitty, H. 1950. Canadian arctic wildlife enquiry, 1945–49, with a summary of results since 1933. *J. Anim. Ecol.* **19**, 180–93.

Chua, T.C. 1975. Population studies on the cabbage aphid. *Brevicoryne brassicae*

426   REFERENCES

(L.), and its parasites, with special reference to synchronization. Unpublished Ph.D. thesis, London University.

Clapham, W.B. 1973. *Natural Ecosystems*. New York, Collier-Macmillan.

Clark, C.W. 1975. The economics of whaling: a two-species model. In G.S. Innis (ed.), *New Directions in the Analysis of Ecological Systems*. Simulation Councils Proc. Ser. 5 (No. 1), pp. 111–19.

Clark, C.W. 1976. *Mathematical Bioeconomics: The Optimal Management of Renewable Resources*. New York, Wiley-Interscience.

Clark, C.W. 1979. Towards a predictive model for the economic regulation of commercial fisheries. *Can. J. Fish. Aqua. Sciences*, 37 (7), 1111–29.

Clark, C.W. 1980. Concentration profiles and the production and management of marine fisheries. *Univ of B.C. Inst. App. Math. Stat. Tech. Report* No. 80–8.

Clark, C.W., and Munro, G.R. 1978 Renewable resource management and extinction. *J. Envir. Econ. Manag.* 5, 198–205.

Clark, C.W., Clarke, F.H., and Munro, G.R. 1979. The optimal exploitation of renewable resource stocks: problems of irreversible investment. *Econometrica*, 47, 25–49.

Clark, C.W., and Mangel, M. 1979. Aggregation and fishery dynamics: a theoretical study of schooling and the purse seine tuna fisheries. *Fish. Bull.* 77, 317–37.

Clatworthy, J.N., and Harper, J.L. 1961. The comparative biology of closely related species living in the same area: V. Inter- and intraspecific interference within cultures of Lemna spp. and *Salvinia natans*. *J. Exp. Bot.* 13(38), 307–24.

Clausen, C.P. (ed.) 1978. *Introduced Parasites and Predators of Arthropod Pests and Weeds: A World Review*. Washington D.C., U.S.D.A.

Clegg, L.M. 1978. The morphology of clonal growth and its relevance to the population dynamics of perennial plants. Ph.D. thesis, U.C.N.W., Bangor.

Clements, F.E. 1916. Plant succession: an analysis of the development of vegetation. *Carnegie Inst. Wash. Publ.* 242.

Clements, R.O., Gibson, R.W., Henderson, I.F., and Plumb, R.T. 1978. Ryegrass: pest and virus problems. *ARC Res. Rev.* 1978, *Natnl. Grassland Issue*, 7–10.

Clutton-Brock, T.H., and Harvey, P.H. 1977. Primate ecology and social organisation. *J. Zool.* 183, 1–39.

Clutton-Brock, T.H., Harvey, P.H., and Rudder, B. 1977. Sexual dimorphism, socionomic sex ratio and body weight in primates. *Nature*, 269, 797–800.

Clutton-Brock, T.H., and Harvey, P.H. (eds.) 1978. *Readings in Sociobiology*. San Francisco, Freeman.

Cody, M.L. 1966. A general theory of clutch size. *Evolution*, 20, 174–84.

Cody, M.L. 1968. On the methods of resource division in grassland bird communities. *Am. Nat.* 102, 107–48.

Cody, M.L. 1969. Convergent characteristics in sympatric populations: A possible relation to interspecific territoriality. *Condor*, 71, 222–39.

Cody, M.L. 1974. *Competition and the Structure of Bird Communities*. Princeton, Princeton University Press.

Cody, M.L. 1975. Towards a theory of continental species diversities. In M.L. Cody and J.M. Diamond (eds.), *Ecology and Evolution of Communities*, pp. 214–57. Cambridge, Mass., Harvard University Press.

Cohen, J.E. 1973. Selective host mortality in a catalytic model applied to schistosomiasis. *Am. Nat.* 107, 199–212.

Cohen, J.E. 1976. Schistosomiasis, a human-parasite system. In R.M. May (ed.),

Theoretical Ecology: Principles and Applications, 1st edition. Oxford, Blackwell Scientific Publications.

Cohen, J.E. 1978. *Food Webs and Niche Space*. Princeton, Princeton University Press.

Cohen, J.E., and Singer, B. 1979. Malaria in Nigeria: constrained continuous-time Markov models for discrete-time longitudinal data on human mixed-species infections. In S.A. Levin (ed.), *Some Mathematical Questions in Biology*. Vol. 12. Providence, U.S.A., American Mathematical Society.

Cole, L.C. 1960. Competitive exclusion. *Science*, **312**, 348–9.

Colinvaux, P.A. 1973. *Introduction to Ecology*. New York, Wiley.

Colwell, R.K. 1973. Competition and coexistence in a simple tropical community. *Am. Nat.* **107**, 737–60.

Colwell, R.K., and Futuyma, D.J. 1971. On the measurement of niche breadth and overlap. *Ecology*, **52**, 567–76.

Colwell, R.K., and Fuentes, E.R. 1975. Experimental studies of the niche. *A. Rev. Ecol. Syst.* **6**, 281–310.

Comins, H.N. 1977a. The development of insecticide resistance in the presence of migration. *J. Theor. Biol.* **64**, 177–97.

Comins, H.N. 1977b. The management of pesticide resistance. *J. Theor. Biol.* **65**, 399–420.

Comins, H.N. 1980. The management of pesticide resistance. *Proc. Conf. Pesticide Resistance*. Rockefeller Foundation. (In press.)

Comins, H.N., and Hassell, M.P. 1976. Predation in multi-prey communities. *J. Theor. Biol.* **62**, 93–114.

Comins, H.N., and Hassell, M.P. 1979. The dynamics of optimally foraging predators and parasitoids. *J. Anim. Ecol.* **48**, 335–51.

Comins, H.N., Hamilton, W.D., and May, R.M. 1980. Evolutionarily stable dispersal strategies. *J. Theor. Biol.* **82**, 205–30.

Connell, J.H. 1961. The influence of interspecific competition and other factors on the distribution of the barnacle *Chthamalus stellatus*. *Ecology*, **42**, 710–23.

Connell, J.H. 1972. Community interactions on marine rocky intertidal shores. *A. Rev. Ecol. Syst.* **3**, 169–92.

Connell, J.H. 1974. Field experiments in marine ecology. In R. Mariscal (ed.), *Experimental Marine Biology*. New York, Academic Press.

Connell, J.H. 1975. Some mechanisms producing structure in natural communities: a model and evidence from field experiments. In M.L. Cody and J.M. Diamond (eds.), *Ecology and Evolution of Communities*, pp. 460–90, Cambridge, Mass., Harvard University Press.

Connell, J.H. 1978. Diversity in tropical rain forests and coral reefs. *Science*, **199**, 1302–10.

Connell, J.H., and Slayter, R.O. 1977. Mechanisms of succession in natural communities and their role in community stability and organization. *Am. Nat.* **111**, 1119–44.

Connor, E.F., and Simberloff, D. 1980. The assembly of species communities: chance or competition? *Ecology*. (In press.)

Conway, G.R. 1971. *Pests of Cocoa in Sabah and their Control*. Ministry of Agriculture and Fisheries, Sabah, Malaysia.

Conway, G.R. 1972. Ecological aspects of pest control in Malaysia. In M.T. Farvar, and J.P. Milton (eds.), *The Careless Technology*. New York, The Natural History Press.

Conway, G.R. 1973. Aftermath of the Green Revolution. In N. Calder (ed.), *Nature in the Round*, pp. 226–35. London, Weidenfeld and Nicolson.

Conway, G.R., Norton, G.A., Small, N.J., and King, A.B.S. 1975. A systems approach to the control of the sugar cane froghopper. In G.E. Dalton (ed.), *Study of Agricultural Systems*, pp. 193–229. London, Applied Science Publishers.

Cook, R.M., and Hubbard, S.F. 1977. Adaptive searching strategies in insect parasites. *J. Anim. Ecol.* 46, 115–75.

Cooper, C.F. 1960. Changes in vegetation, structure and growth of southwestern pine forests since white settlement. *Ecol. Monogr.* **30**, 129–64.

Cooper-Driver, G.A., and Swain, T. 1976. Cyanogenic polymorphism in bracken in relation to herbivore predation. *Nature*, **260**, 604.

Corbett, J.R. 1978. The future of pesticides and other methods of pest control. In T.H. Coaker (ed.), *Applied Biology* Vol. 3, pp. 229–330. Academic Press.

Cornell, H. 1974. Parasitism and distributional gaps between allopatric species. *Am. Nat.* **108**, 880–3.

Crisler, L. 1956. Observations of wolves hunting caribou. *J. Mammal.* **37**, 337–46.

Crocker, R.L. and Major, J. 1955. Soil development in relation to vegetation and surface age at Glacier Bay, Alaska. *J. Ecol.* **42**, 427–48.

Crofton, H.D. 1971. A quantitative approach to parasitism. *Parasitology*, **62**, 179–94.

Crombie, A.C. 1947. Interspecific competition. *J. Anim. Ecol.* **16**, 44–73.

Crowell, K.L. 1962. Reduced interspecific competition among the birds of Bermuda. *Ecology*, **43**, 75–88.

Cullen, J.M. 1978. Evaluating the success of the programme for the biological control of *Chondrilla juncea* L. In T.E. Freeman (ed.), *Proc. IV Int. Symp. Biol. Control. Weeds*, pp. 117–21. Gainesville, University of Florida.

Culver, D.C. 1970. Analysis of simple cave communities: niche separation and species packing. *Ecology*, **51**, 949–58.

Cuppy, W. 1941. *How to Become Extinct*. New York, Farrar & Rinehart.

Dahlsten, D.L. 1967. Preliminary life tables for pine sawflies in the *Neodiprion fulviceps* complex (Hymenoptera: Diprionidae). *Ecology*, **48**, 275–89.

Dajoz, R. 1974. *Dynamique des Populations*. Paris, Masson et cie.

Dale, M. 1977. Graph theoretical analysis of the phytosociological structure of plant communities: an application to mixed forest. *Vegetatio*, **35**, 35–46.

Daly, M., and Wilson, M. 1978. *Sex, Evolution & Behaviour*. North Scituate, Mass., Duxbury.

Dammerman, K.W. 1948. The fauna of Krakatau 1883–1933. *Verhandel. Koninkl. Ned. Akad. Wetenschap. Afdel. Natuurk.* **44(2)**, 1–594.

Darwin, C. 1859. *The Origin of Species*. London, John Murray.

Darwin, C. 1871. *The Descent of Man, and Selection in Relation to Sex*. London, John Murray.

Dauphiné, T.C. 1975. The disappearance of Caribou reintroduced to Cape Breton-Highlands national park. *Can. Field Naturalist*, **89**, 299–310.

Dausset, J., and Colombani, J. (eds.) 1972. *Histocompatibility Testing*. Copenhagen, Munksgraad.

Davidson, J.L., and Donald, C.M. 1958. The growth of swards of subterranean clover with particular reference to leaf area. *Aust. J. Agric. Res.* **9**, 53–72.

Davis, J. 1973. Habitat preferences and competition of wintering juncos and golden-crowned sparrows. *Ecology*, **54**, 174–80.

Dawkins, R. 1976. *The Selfish Gene*. Oxford, Oxford University Press.

Dawkins, R. 1979. Twelve misunderstandings of kin selection. *Z. Tierpsychol.* **51**, 184–200.

Day, P.R. 1972. Crop resistance to pests and pathogens. *Pest Control Strategies for the Future*, pp. 257–71. Washington D.C., Natl. Acad. Sci.

Dayton, P.K. 1971. Competition, disturbance, and community organization: the provision and subsequent utilization of space in a rocky intertidal community. *Ecol. Monogr.* **41**, 351–89.

Dayton, P.K. 1975. Experimental evaluation of ecological dominance in a rocky intertidal algal community. *Ecol. Monogr.* **45**, 137–59.

De Angelis, D.L. 1975. Stability and connectance in food web models. *Ecology*, **56**, 238–43.

De Angelis, D.L., Gardner, R.H., Mankin, J.B., Post, W.M., and Carney, J.H. 1978. Energy flow and the number of trophic levels in ecological communities. *Nature*, **273**, 406–7.

De Angelis, D.L., Travis, C.C., and Post, W.M. 1979. Persistence and stability of seed-dispersed species in a patchy environment. *Theor. Pop. Biol.* **16**, 107–25.

DeBach, P. 1966. The competitive displacement and coexistence principles. *A. Rev. Ent.* **11**, 183–212.

DeBach, P. 1974. *Biological Control by Natural Enemies*. Cambridge, Cambridge University Press.

DeBach, P., and Smith, H.S. 1941. The effect of host density on the rate of reproduction of entomophagous parasites. *J. econ. Ent.*, **34**, 741–5.

DeBach, P., and Schlinger, E.I. (eds.) 1964. *Biological Control of Insect Pests and Weeds*. London, Chapman and Hall.

Debach, P., Rosen, D., and Kennett, C.E. 1971. Biological control of coccids by introduced natural enemies. In C.B. Huffaker (ed.), *Biological Control*, pp. 165–94. New York, Plenum Press.

De Beer, G. 1958. *Embryos and Ancestors*. Oxford, Clarendon Press.

Deevey, E.S. 1965. Environments of the geological past. *Science*, **147**, 592–4.

Delaney, M.J., and Happold, D.C.D. 1979. *Ecology of African Mammals*. London and New York, Longman.

DeLong, K.T. 1966. Population ecology of feral house mice: interference by *Microtus*. *Ecology*, **47**, 481–4.

Dempster, J.P. 1975. *Animal Population Ecology*. London, New York, Academic Press.

Den Boer, P.J. 1968. Spreading of risk and stabilization of animal numbers. *Acta. Biother.* **18**, 165–94.

Diamond, J.M. 1969. Avifaunal equilibria and species turnover rates on the Channel Islands of California. *Proc. Natl. Acad. Sci. U.S.* **64**, 57–63.

Diamond, J.M. 1971. Comparison of faunal equilibrium turnover rates on a tropical island and a temperate island. *Proc. Natl. Acad. Sci. U.S.* **68**, 2742–5.

Diamond, J.M. 1972. Biogeographic kinetics: Estimation of relaxation times for avifaunas of Southwest Pacific Islands. *Proc. Natl. Acad. Sci. U.S.* **69**, 3199–203.

Diamond, J.M. 1973. Distributional ecology of New Guinea birds. *Science*, **179**, 759–69.

Diamond, J.M. 1974. Colonization of exploded volcanic islands by birds: the Supertramp strategy. *Science*, **184**, 803–6.

Diamond, J.M. 1975a. The island dilemma: Lessons of modern biogeographic studies for the design of natural reserves. *Biol. Conservation*, **7**, 129–46.

Diamond, J.M. 1975b. Assembly of species communities. In M.L. Cody and J.M.

Diamond (eds.), *Ecology and Evolution of Communities*, pp. 342–444. Cambridge, Mass., Harvard University Press.

Diamond, J.M. 1976a. Relaxation and differential extinction on land-bridge islands: applications to natural reserves. *Proc. 16th International Ornith. Congress, Canberra, August,* 1974. pp. 616–28.

Diamond, J.M. 1976b. Island biogeography and conservation: strategy and limitations. *Science,* **193,** 1027–9.

Diamond, J.M., and Mayr, E. 1976. The species-area relation for birds of the Solomon Archipelago. *Proc. Natl. Acad. Sci. U.S.* **73,** 262–6.

Diamond, J.M., and Marshall, A.G. 1977. Distributional ecology of New Hebridean birds: a species kaleidoscope. *J. Anim. Ecol.* **46,** 703–27.

Diamond, J.M., and May, R.M. 1977. Species turnover rates on islands: dependence on census interval. *Science,* **197,** 266–70.

Dietz, K. 1974. Transmission and control of arbovirus diseases. In *Epidemiology,* pp. 104–21. Philadelphia, SIAMS.

Dietz, K. 1976. The incidence of infectious diseases under the influence of seasonal fluctuations. In J. Berger, W. Bühler, R. Repges, and P. Tautu (eds.), *Mathematical Models in Medicine, Lecture Notes in Biomathematics,* **11,** 1–15.

Dietz, K., Molineaux, L., and Thomas, A. 1974. A malaria model tested in the African Savannah. *Bull. Wld. Hlth. Org.* **50,** 347–57.

Dingle, H. 1974. The experimental analysis of migration and life history strategies in insects. In L. Barton-Browne (ed.), *Experimental Analysis of Insect Behaviour,* pp. 329–42. Berlin, Springer-Verlag.

Dirzo, R. 1979. Studies on plant–animal interactions: terrestrial molluscs and their food plants. Unpublished Ph.D. thesis, University of Wales.

Dixon, A.F.G. 1959. An experimental study of the searching behaviour of the predatory coccinellid beetle *Adalia decempunctata* (L.). *J. Anim. Ecol.* **28,** 259–81.

Dixon, A.F.G. 1970. Quality and availability of food for a sycamore aphid population. In A. Watson (ed.), *Symp. Brit. Ecol. Soc.* **10,** 271–87.

Dixon, A.F.G. 1971a. The role of aphids in wood formation. I. The effect of the sycamore aphid, *Drepanosiphum platanoides* (Schr.) (Aphididae), on the growth of sycamore, *Acer pseudoplatanus* (L.). *J. Appl. Ecol.* **8,** 165–79.

Dixon, A.F.G. 1971b. The role of aphids in wood formation. II. The effect of the lime aphid, *Eucallipterus tiliae* L. (Aphididae), on the growth of lime, *Tilia x vulgaris* Hayne. *J. Appl. Ecol.* **8,** 393–9.

Dixon, A.F.G., and Logan, M. 1973. Leaf size and availability of space to the sycamore aphid *Drepanosiphum platanoides. Oikos,* **24,** 58–63.

Dobben, W.H. van, and Lowe-McConnell, R.H. (eds.), 1975. *Unifying Concepts in Ecology.* The Hague, W. Junk.

Docters van Leeuwen, W.M. 1936. Krakatau, 1883–1933. *Ann. Jard. Botan. Buitenzorg,* **56–7,** 1–506.

Dodd, A.P. 1940. *The Biological Campaign against Prickly-Pear.* Brisbane, Government Printer.

Dogiel, V.A. 1964. *General Parasitology.* Edinburgh, Oliver and Boyd.

Doubleday, W.G. 1976. Environmental fluctuations and fisheries management. *Int. Comm. Northw. Atlant. Fish. Sel. Pap.* **1,** 141–50.

Dowell, R.V., and Horn, D.J. 1977. Adaptive strategies of larval parasitoids of the alfalfa weevil (Coleoptera: Curcculionidae). *Can. Ent.* **109,** 641–8.

Dransfield, R. 1975. The Ecology of Grassland and Cereal Aphids. London, unpublished Ph.D. thesis.

Drury, W.H., and Nisbet, I.C.T. 1971. Inter-relations between developmental models in geomorphology, plant ecology, and animal ecology. *Gen. Syst.* **16**, 57–68.

Drury, W.H., and Nisbet, I.C.T. 1973. Succession. *J. Arnold Arboretum Harvard University*, **54**, 331–68.

Dueser, R.D., and Shugart, H.H., Jr. 1979. Niche pattern in a forest floor small-mammal fauna. *Ecology*, **60**, 108–18.

Dunford, C. 1977. Kin selection for ground squirrel alarm calls. *Am. Nat.* **III**, 782–5.

Dunham, A.E. 1980. An experimental study of interspecific competition between the iguanid lizards *Sceloporus merriamii* and *Urosaurus ornatus*. *Ecol. Monogr.* **50**, 309–30.

Dyer, M.I. 1980. Mammalian epidermal growth factor promotes plant growth. *Proc. Natl. Acad. Sci. U.S.* **77**, 4836–7.

Edmunds, G.F. Jr., and Alstad, D.N. 1978. Coevolution in insect herbivores and conifers. *Science*, **199**, 941–5.

Egler, F.E. 1954. Vegetation science concepts. 1. Initial floristic composition—a factor in old field vegetation development. *Vegetatio*, **4**, 412–17.

Ehler, L.E. 1977. Natural enemies of cabbage looper on cotton in Sanjoaquin Valley. *Hilgardia*, **45**, 73–106.

Ehler, L.E., and van den Bosch, R. 1974. An analysis of the natural biological control of *Trichoplusia ni* (Lepidoptera: Noctuidae) on cotton in California. *Can. Ent.* **106**, 1067–73.

Ehrlich, P., and Gilbert, L.E. 1973. Population structure and dynamics of the tropical butterfly, *Heliconius ethilla*. *Biotropica*, **5**, 69–82.

Ehrlich, P.R., White, R.R., Singer, M.C., McKechnie, S.W., and Gilbert, L.E. 1975. Checkerspot butterflies: A historical perspective. *Science*, **188**, 221–8.

Eisenberg, R.M. 1970. The role of food in the regulation of the pond snail, *Lymnaea elodes*. *Ecology*, **51**, 680–4.

Eldredge, N., and Gould, S.J. 1972. Punctuated equilibria: an alternative to phyletic gradualism. In T.J.M. Schopf (ed.), *Models in Paleobiology*, pp. 82–115. San Francisco, Freeman, Cooper and Company.

Elliott, H. 1971. Wandering albatross. In J. Gooders (ed.), *Birds of the World*, **1**, pp. 29–31. London, I.P.C.

Elliott, P.F. 1975. Longevity and the evolution of polygamy. *Am. Nat.* **109**, 281–7.

Elton, C.S. 1927. *Animal Ecology*, London, Sidgwick and Jackson.

Elton, C.S. 1942. *Voles, Mice and Lemmings: problems in population dynamics.* Oxford, Oxford University Press.

Elton, C.S. 1958. *The Ecology of Invasion by Animals and Plants.* London, Methuen.

Embree, D.G. 1965. The population dynamics of the winter moth in Nova Scotia 1954–1962. *Mem. Ent. Soc. Can.* **46**, 1–57.

Emden, H.F., van, and Way, M.J. 1973. Host plants in the population dynamics of insects. *Symp. Roy. Ent. Soc. London*, **6**, 181–99.

Emlen, J.M. 1966. The role of time and energy in food preference. *Am. Nat.* **100**, 611–17.

Emlen, J.M. 1968. Optimal choice in animals. *Am. Nat.* **102**, 385–90.

Emlen, J.M. 1973. *Ecology: an evolutionary approach.* Addison-Wesley.

Emlen, S.T., and Oring, L.W. 1977. Ecology, sexual selection, and the evolution of mating systems. *Science*, **197**, 215–23.

Estes, J.A. 1979. Exploitation of marine mammals: *r*-selection or *K*-strategists? *J. Fish. Res. Board Canada*, **36(8)**, 1009–17.

Estes, J.A., and Palmisano, J.F. 1974. Sea otters: their role in structuring near-shore communities. *Science*, **185**, 1058–60.

Evans, H.F. 1973. A study of the predatory habits of *Anthocoris* species (Hemiptera-Heteroptera). Unpublished D.Phil. Thesis, Oxford University.

Faaborg, J. 1976. Patterns in the structure of West Indian bird communities. *Am. Nat.* **111**, 903–16.

Fager, E.W. 1968. The community of invertebrates in decaying oak wood. *J. Anim. Ecol.* **37**, 121–42.

Farnworth, E.G., and Golley, F.B. (eds.) 1974. *Fragile Ecosystems: evaluation of research and applications in the neotropics*. New York, Springer-Verlag.

Feeney, P. 1976. Plant apparency and chemical defence. *Rec. Adv. Phytochem.* **10**, 1–40.

Feigenbaum, M.J. 1978. Quantitative universality for a class of non-linear transformations. *J. Stat. Phys.* **19**, 25–52.

Feigenbaum, M.J. 1979. The onset spectrum of turbulence. *Physics Letters*, **74A**, 375–8.

Feinsinger, P. 1976. Organization of a tropical guild of nectarivorous birds. *Ecol. Monogr.* **46**, 257–91.

Feldman, M., and Roughgarden, J. 1975. A population's stationary distribution and chance of extinction in a stochastic environment with remarks on the theory of species packing. *Theoret. Pop. Biol.* **7**, 197–207.

Fellows, D.P., and Head, W.B. 1972. Factors affecting host plant selection in desert adapted cactiphilic *Drosophila*. *Ecology*, **53**, 850–8.

Fenchel, T. 1974. Intrinsic rate of natural increase: the relationship with body size. *Oecologia*, **14**, 317–26.

Fenchel, T. 1975. Character displacement and coexistence in mud snails (Hydrobiidae). *Oecologia*, **20**, 19–32.

Fenchel, T., and Christiansen, F.B. 1976. *Theories of Biological Communities*. New York, Springer-Verlag.

Fenner, F. 1948. The epizootic behaviour of mouse pox (infectious ectromelia of mice). II. The course of events in long-continued epidemics. *J. Hyg.* **46**, 383–96.

Fenner, F. 1949. Mouse pox (infectious ectromelia of mice): A review. *J. Immunol.* **63**, 341–73.

Fenner, F., and Ratcliffe, F.N. 1965. *Myxomatosis*. London, Cambridge University Press.

Fisher, R.A. 1930. *The Genetical Theory of Natural Selection*. Oxford, Clarendon Press.

Force, D.C. 1972. *r*- *K*- strategists in endemic post-parasitoid communities. *Bull. Ent. Soc. Am.* **18**, 135–7.

Forcier, L.K. 1975. Reproductive strategies and the co-occurrence of climax tree species. *Science*, **189**, 808–10.

Ford, E.D. 1975. Competition and stand structure in some even-aged plant monocultures. *J. Ecol.* **63**, 311–33.

Forsyth, A.B., and Robertson, R.J. 1975. *K*-reproductive strategy and larval behaviour of the pitcher plant sarcophagid fly, *Blaseoxipha fletcheri*. *Can. J. Zool.* **53**, 174–9.

Fortey, R.A. 1980. Generic longevity in Lower Ordovician trilobites: relation to environment. *Paleobiology*, **6**, 24–31.

Fox, J.F. 1977. Alternation and coexistence of tree species. *Am. Nat.* **111**, 69–89.

Fox, J.F. 1979. Intermediate-disturbance hypothesis. *Science*, **204**, 1344–5.

Frauenthal, J.C., and Swick, K.E. 1980. Limit cycle oscillations in human population dynamics, *Theor. Pop. Biol.* (in press.)

Free, C.A., Beddington, J.R., and Lawton, J.H. 1977. On the inadequacy of simple models of mutual interference for parsitism and predation. *J. Anim. Ecol.* **46**, 543–54.

Freeland, W.J., and Janzen, D.H. 1974. Strategies in herbivory by mammals: the role of plant secondary compounds. *Am. Nat.* **108**, 269–89.

Fretwell, S.D. 1977. The regulation of plant communities by the food chains exploiting them. *Perspectives Biol. Med.* **20**, 169–85.

Fujii, K. 1968. Studies on interspecies competition between the azuki bean weevil and the southern cowpea weevil: III, some characteristics of strains of two species. *Res. Popul. Ecol.* **10**, 87–98.

Fuller, W.A. 1962. The biology and management of the bison of Wood Buffalo National Park. *Wld. Manag. Bull. Ottowa* Series 1 (16).

Gadgil, M., and Solbrig, O.T. 1972. The concept of r- & K-selection: evidence from wild flowers and some theoretical considerations. *Am. Nat.* **106**, 14–31.

Gamboa, G.J. 1978. Intraspecific defense: advantage of social cooperation among paper wasp foundresses. *Science*, **199**, 1463–5.

Gardner, M.R., and Ashby, W.R. 1970. Connectance of large dynamical (cybernetic) systems: critical values for stability. *Nature (Lond.)*, **288**, 784.

Gates, D.M., and Schmerl, R.B. (eds.), 1975. *Perspectives in Biophysical Ecology*. New York, Springer-Verlag.

Gatz, A.J. 1979. Community organisation in fishes as indicated by morphological features. *Ecology*, **60**, 711–18.

Gause, G.F. 1934. *The Struggle for Existence*. Baltimore, Williams and Wilkins. Reprinted, 1964, by Hafner, New York.

Geisel, T.S. 1955. *On Beyond Zebra*. New York, Random House.

George, D.G., and Edwards, R.W. 1974. Population dynamics and production of *Daphnia hyalina* in an eutrophic reservoir. *Freshwat. Biol.* **4**, 445–65.

Ghiselin, M.T. 1974. *The Economy of Nature and the Evolution of Sex*. Berkeley, University of California Press.

Gibo, D.L. 1978. The selective advantage of foundress associations in *Polistes fuscatus* (Hymenoptera: Vespidae): a field study of the effects of predation on productivity. *Can. Entomol.* **110**, 519–40.

Gilbert, L.E. 1975. Ecological consequences of a coevolved mutualism between butterflies and plants. In L.E. Gilbert and P.H. Raven (eds.), *Coevolution of Animals and Plants*, pp. 210–40. Austin, Texas, University of Texas Press.

Gilbert, L.E. 1977. The role of insect–plant coevolution in the organization of ecosystems. *Coll. Int. C.N.R.S.* **265**, 399–413.

Gill, F.B., and Wolf, L.L. 1975a. Economics of feeding and territoriality in the golden-winged sunbird. *Ecology*. **56**, 333–45.

Gill, F.B., and Wolf, L.L. 1975b. Foraging strategies and energetics of east African sunbirds at mistletoe flowers. *Am. Nat.* **109**, 491–510.

Gilpin, M.E. 1972. Enriched prey–predator systems: theoretical stability. *Science*, **177**, 902–4.

Gilpin, M.E. 1974. A Liapunov function for competition communities. *J. Theoret. Biol.* **44**, 35–48.

Gilpin, M.E. 1975a. Limit cycles in competition communities. *Am. Nat.* **109**, 51–60.

Gilpin, M.E. 1975b. *Group Selection in Predator–prey Communities.* Princeton, Princeton University Press.

Gilpin, M.E. 1975c. Stability of feasible predator–prey systems. *Nature,* **254**, 137–8.

Gilpin, M.E., and Justice, K.E. 1972. Reinterpretation of the invalidation of the principle of competitive exclusion. *Nature (Lond.),* **236**, 273–4, and 229–301.

Gilpin, M.E., and Ayala, F.J. 1973. Global models of growth and competition. *Proc. Natl. Acad. Sci., U.S.* **70**, 3590–3.

Gilpin, M.E., and Case, T. 1976. Multiple domains of attraction in competition communities. *Nature,* **261**, 40–2.

Gilpin, M.E., and Diamond, J.M. 1976. Calculation of immigration and extinction curves from the species-area-distance relation. *Proc. Natl. Acad. Sci. U.S.* **73**, 4130–4.

Gilpin, M.E., and Diamond, J.M. 1981. Immigration and extinction probabilities for individual species: relation to incidence functions and species colonization curves. *Proc. Natl. Acad. Sci. U.S.* (In press).

Glass, L., and Mackay, M.C. 1979. Pathological conditions resulting from instabilities in physiological control systems. *Ann. N.Y. Acad. Sci.* **316**, 214–35.

Gleason, H.A. 1926. The individualistic concept of the plant association. *Bull. Torrey Bot. Club,* **53**, 331–68.

Gleason, H.A. 1927. Further views of the succession concept. *Ecology,* **8**, 299–326.

Goeden, R.D., and Louda, S.M. 1976. Biotic interference with insects imported for weed control. *Ann. Rev. Ent.* **21**, 325–42.

Goh, B.S. 1979a. Robust stability concepts for ecosystems models. In E. Halfon (ed.), *Theoretical Systems Ecology,* pp. 467–87. New York, Academic Press.

Goh, B.S. 1979b. Stability in models of mutualism. *Am. Nat.* **113**, 261–75.

Goh, B.S., and Jennings, L.S. 1977. Feasibility and stability in randomly assembled Lotka–Volterra models. *Ecol. Model.* **3**, 63–71.

Golley, F.B. 1968. Secondary productivity in terrestrial ecosystems. *Am. Zool.* **8**, 53–62.

Goodman, D. 1974. Natural selection and a cost ceiling on reproductive effort. *Am. Nat.* **108**, 247–68.

Gordon, H.S. 1954. The economic theory of a common-property resource: the fishery. *J. Political Economy,* **62**, 124–42.

Gordon, M.H. 1963. Nutrition and helminthosis in sheep. *Proc. Aust. Soc. Anim. Prod.* **3**, 93–104.

Gorham, E. 1979. Shoot height, weight and standing crop in relation to density of monospecific plant stands. *Nature,* **279**, 148–50.

Gough, H.C. 1946. Studies on wheat bulb fly *Leptohylemia coarctata* Fall. II Numbers in relation to crop damage. *Bull. ent. Res,* **37**, 439–54.

Gould, S.J. 1974. The great dying. *Nat. Hist.* **83 (8)**, 22–7.

Gould, S.J. 1975. Catastrophes and steady state earth. *Nat. Hist.* **84 (2)**, 14–18.

Gould, S.J. 1977. *Ontogeny and Phylogeny.* Cambridge, Mass. Harvard University Press.

Gould, S.J. 1980. Is a new and general theory of evolution emerging? *Paleobiology,* **6**, 119–30.

Gould, S.J., and Eldredge, N. 1977. Punctuated equilibria: the tempo and mode of evolution reconsidered. *Paleobiology,* **3**, 115–51.

Gould, S.J., Raup, D.M., Sepkoski, J.J. Jr., Schopf, T.J.M., and Simberloff, D.S.

1977. The shape of evolution: a comparison of real and random clades. *Paleobiology*, **3**, 23–40.

Goulden, C.E., and Hornig, L.L. 1980. Population oscillations and energy reserves in planktonic cladocera and their consequences to competition. *Proc. Natl. Acad. Sci. U.S.* **77**, 1716–20.

Gradwell, G.R. 1974. The effect of defoliators on tree growth. In M.G. Morris and F.H. Perring (eds.), *The British Oak*, pp. 182–93. Faringdon, E.W. Classey.

Grant, P.R. 1968. Bill size, body size and the ecological adaptations of bird species to the competitive situations on islands. *Syst. Zool.* **17**, 319–33.

Grant, P.R. 1972a. Convergent and divergent character displacement. *Biol. J. Linn. Soc.* **4**, 39–68.

Grant, P.R. 1972b. Interspecific competition between rodents. *A. Rev. Ecol. Syst.* **3**, 79–106.

Greenwood, M., Bradford-Hill, A., Topley, W.W.C., and Wilson, J. 1936. *Experimental Epidemiology*. Medical Research Council Special Report, No. 209, London., H.M.S.O.

Greenwood, P.J., Harvey, P.H., and Perrins, C.M. 1978. Inbreeding and dispersal in the great tit. *Nature*, **271**, 52–7.

Griffiths, D.A. 1974. A catalytic model of infection for measles. *Appl. Statist.* **23**, 330–9.

Griffiths, K.J. 1959. Development and diapause in *Pleophus basizonus* (Hymenoptera: Ichneumonidae). *Can. Ent.* **101**, 907–14.

Grime, J.P. 1974. Vegetation classification by reference to strategies. *Nature*, **250**, 26–31.

Grime, J.P. 1977. Evidence for the existence of three primary strategies in plants and its relevance to ecological and evolutionary theory. *Am Nat.* **111**, 1169–94.

Grinnell, J. 1917. The niche relationships of the California thrasher. *Auk*, **21**, 364–82.

Grinnell, J. 1924. Geography and evolution. *Ecology*, **5**, 225–9.

Grinnell, J. 1928. The presence and absence of animals. *Univ. Calif. Chronicle*, **30**, 429–50.

Guckenheimer, J. 1979. The bifurcation of quadratic functions. *Ann. N.Y. Acad. Sci.* **316**, 78–85.

Guckenheimer, J., Oster, G.F., and Ipaktchi, A. 1976. The dynamics of density dependent population models. *J. Math. Biol.* **4**, 101–47.

Gulland, J.A. 1956. The study of fish populations by the analysis of commercial catches. *Cons. Perm. Int. Explor. Mer Rapp. Proc-Verb.* **140**, 21–7.

Gulland, J.A. 1975. The stability of fish stocks. *J. du Conseil Inter. Explor. Mer*, **37**, 199–204.

Gurney, W.S.C., and Nisbet, R.M. 1979. Ecological stability and social hierarchy. *Theor. Pop. Biol.* **16**, 48–80.

Gurney, W.S.C., Blythe, S.P., and Nisbet, R.M. 1980. Nicholson's blowflies revisited. *Nature*, **287**, 17–21.

Gutierrez, A.P., Wang, Y., and Jones, R.E. 1979. Systems analysis applied to crop protection. *EPPO Bull.* **9**, 133–48.

Gutierrez, A.P., Wang, Y., and Regev, U. 1979. An optimization model for *Lygus hesperus* (Heteroptera: Miridae) damage in cotton: the economic threshold revisited. *Can. Ent.* **111**, 41–54.

Hairston, N.G. 1951. Interspecies competition and its probable influence upon the vertical distribution of Appalachian salamanders of the genus *Plethodon*. *Ecology*, **32**, 266–74.

Hairston, N.G. 1965. An analysis of age-prevalence data by catalytic models. *Bull. Wld. Hlth. Org.* **33**, 163–75.

Hairston, N.G., Smith, F.E., and Slobodkin, L.B. 1960. Community structure, population control, and competition. *Am. Nat.* **94**, 421–5.

Hairston, N.G. *et al.* 1968. The relationship between species diversity and stability: an experimental approach with protozoa and bacteria. *Ecology,* **49**, 1091–101.

Halbach, U. 1979. Introductory remarks: strategies in population research exemplified by rotifer population dynamics. *Fortschritte der Zoologie,* **25**, 1–27.

Haldane, J.B.S. 1932. *The Causes of Evolution.* London, Longmans.

Haldane, J.B.S. 1949. Disease and evolution. *Riorca Sci. (Suppl.),* **19**, 3–10.

Halfon, E. (ed.) 1979. *Theoretical Systems Ecology: Advances and Case Studies.* New York, Academic Press.

Hallam, A. 1969. Faunal realms and facies in the Jurassic. *Palaeontology,* **12**, 1–18.

Hallam, A. 1975. *Jurassic Environments.* Cambridge, Cambridge University Press.

Hallett, J.G., and Pimm, S.L. 1979. Direct estimation of competition. *Am. Nat.* **113**, 593–600.

Halter, A.N., and Dean, G.W. 1972. *Decisions Under Uncertainty with Research Application.* Cincinnati, South-Western Pub. Co.

Hamilton, W.D. 1963. The evolution of altruistic behaviour. *Am. Nat.* **97**, 354–6.

Hamilton, W.D. 1964. The genetical theory of social behaviour. *J. Theor. Biol.* **7**, 1–52.

Hamilton, W.D. 1967. Extraordinary sex ratios. *Science,* **156**, 477–88.

Hamilton, W.D. 1972. Altruism and related phenomena, mainly in social insects. *A. Rev. Ecol. Syst.* **3**, 193–232.

Hamilton, W.D., and May, R.M. 1977. Dispersal in stable habitats. *Nature,* **269**, 578–81.

Hanes, T.L. 1971. Succession after fire in the chaparral of southern California. *Ecol. Monogr.* **41**, 27–52.

Hansen, T. 1978. Larval dispersal and species longevity in Lower Tertiary gastropods. *Science.* **199**, 885–7.

Hardin, G. 1960. The competitive exclusion principle. *Science,* **131**, 1292–7.

Hardin, G., and Baden, J. 1977. *Managing the Commons.* San Francisco, Freeman.

Harper, J.L. 1969. The role of predation in vegetational diversity. *Brookhaven Symp. Biol.* **22**, 48–62.

Harper, J.L. 1977. *Population Biology of Plants.* London, Academic Press.

Harper, J.L., and Ogden, J. 1970. The reproductive strategy of higher plants. [i] the concept of strategy with special reference to *Senecio vulgaris* L. *J. Ecol.* **58**, 681–98.

Harper, J.L., and White, J. 1974. The demography of plants. *A. Rev. Ecol. Syst.* **5**, 419–63.

Harper, J.L., and Bell, A.D. 1979. 2. The population dynamics of growth forms in organisms with modular construction. In R.M. Anderson, B.D. Turner, and L.R. Taplor (eds.), *Symp. Brit. Ecol. Soc.* 20, *Population Dynamics.* Oxford, Blackwell Scientific Publications.

Harris, P. 1973. Insects in the population dynamics of plants. *Symp. Roy. Ent. Soc. London.* **6**, 201–9.

Harris, P., Wilkinson, A.T.S., Thompson, L.S., and Neary, M. 1978. Interaction between the cinnabar moth, *Tyria jacobaeae* L. (Lep: Arctiidae) and ragwort. *Senecio jacobaea* L. (Compositae) in Canada. In T.E. Freeman (ed.), *Proc. IV*

Int. Symp. Biol. Control. Weeds, pp. 174–80. Gainesville, University of Florida.

Harrison, G. 1978. *Mosquitoes, Malaria and Man: A History of the Hostilities Since 1880.* New York, E.P. Dutton.

Harte, J., and Levy, D. 1975. On the vulnerability of ecosystems disturbed by man. In W.H. van Dobben and R.H. Lowe-McConnell (eds.), *Unifying Concepts in Ecology*, pp. 208–23. The Hague, W. Junk.

Hartnett, D.C., and Abrahamson, W.G. 1979. The effects of stem gall insects on life history patterns in *Solidago canadensis. Ecology,* **60,** 910–17.

Hassell, M.P. 1971. Mutual interference between searching insect parasites. *J. Anim. Ecol.* **40,** 473–86.

Hassell, M.P. 1977. Some practical implications of recent theoretical studies of host–parasitoid interactions. *Proc. XV Int. Congr. Ent.* 608–16.

Hassell, M.P. 1978. *The Dynamics of Arthropod Predator–Prey Systems.* Princeton, Princeton University Press.

Hassell, M.P. 1979. Non-random search in predator–prey models. *Fortschr. Zool.* **25,** 311–30.

Hassell, M.P. 1980. Foraging strategies, population models and biological control: a case study. *J. Anim. Ecol.* **49,** 603–28.

Hassell, M.P. 1981. Host–parasitoid models and biological control. In G.R. Conway (ed.), *The Management of Pest and Disease Problems. Int. Inst. Appl. Syst. Anal.,* Vienna. (In press.)

Hassell, M.P., and Huffaker, C.B. 1969. Regulatory processes and population cyclicity in laboratory populations of *Anagasta kuehniella* (Zeller) (Lepidoptera: Phycitidae). III. The development of population models. *Researches Popul. Ecol. Kyoto University,* **11,** 186–210.

Hassell, M.P., and Varley, G.C. 1969. New inductive population model for insect parasites and its bearing on biological control. *Nature (Lond.),* **223,** 1133–6.

Hassell, M.P., and Rogers, D.J. 1972. Insect parasite responses in the development of population models. *J. Anim. Ecol.* **41,** 661–76.

Hassell, M.P., and May, R.M. 1973. Stability in insect host–parasite models. *J. Anim. Ecol.* **42,** 693–726.

Hassell, M.P., and May, R.M. 1974. Aggregation in predators and insect parasites and its effect on stability. *J. Anim. Ecol.* **43,** 567–94.

Hassell, M.P., Lawton, J.H., and Beddington, J.R. 1976. The components of arthropod predation. I. The prey death-rate. *J. Anim. Ecol.* **45,** 135–64. [Referred to as Hassell *et al.*, 1976b]

Hassell, M.P., Lawton, J.H., and May, R.M. 1976. Patterns of dynamical behaviour in single species populations. *J. Anim. Ecol.* **45,** 471–86. [Referred to as Hassell *et al.*, 1976a]

Hassell, M.P., and Moran, V.C. 1976. Equilibrium levels and Biological Control. *J. Ent. Soc. Sth. Afr.* **39,** 357–66.

Hassell, M.P., Lawton, J.H., and Beddington, J.R. 1977. Sigmoid functional responses in invertebrate predators and parasitoids. *J. Anim. Ecol.* **46,** 249–62.

Hassell, M.P., and Comins, H.N. 1978. Sigmoid functional responses and population stability. *Theor. Pop. Biol.* **9,** 202–21.

Haukioja, E., and Hakala, T. 1975. Herbivore cycles and periodic outbreaks: formulation of a general hypothesis. *Rep. Kevo Subarctic Res. Stat.* **12,** 1–9.

Haukioja, E., and Niemela, P. 1979. Birch leaves as a resource for herbivores: seasonal occurrence of increased resistance in foliage after mechanical damage of adjacent leaves. *Oecologia,* **39,** 151–9.

Haven, S.B. 1973. Competition for food between the intertidal gastropods *Acmaea scabra* and *Acmaea digitalis*. *Ecology*, **54**, 143–51.

Heal, O.W., and MacLean, S.F. 1975. Comparative productivity in ecosystems: secondary productivity. In W.H. van Dobben and R.H. Lowe-McConnell (eds.), *Unifying Concepts in Ecology*, pp. 89–108. The Hague, W. Junk.

Heatwole, H., and Levins, R. 1972. Trophic structure stability and faunal change during recolonization. *Ecology*, **53**, 531–4.

Hebert, P.D.N. 1978. The population biology of *Daphnia* (Crustacea, Daphnidae). *Biol. Rev.* **53**, 387–426.

Hedgpeth, J.W. (ed.), 1957. *Treatise on Marine Ecology and Paleoecology, Vol. 1 Ecology*, Mem. Geol. Soc. Am. No. 67.

Heinrich, B. 1975. Bee flowers: a hypothesis on flower variety and blooming times. *Evolution*, **29**, 325–34.

Heinrich, B., and Raven, P.H. 1972. Energetics and pollination ecology. *Science*, **176**, 597–602.

Heinselman, M.L. 1963. Forest sites, bog processes and peatland types in the glacial Lake Agassiz region, Minnesota. *Ecol. Monogr.* **33**, 327–74.

Heinselman, M.L. 1973. Fire in the virgin forests of the Boundary Waters Canoe Area, Minnesota. *Quarternary Res.* **3**, 329–82.

Henry, J.D., and Swan, J.M.A. 1974. Reconstructing forest history from live and dead plant material—an approach to the study of forest succession in southwest New Hampshire. *Ecology*, **55**, 772–83.

Herfindahl, O.C., and Kneese, A.V. 1974. *Economic Theory of Natural Resources*. Columbus, Ohio, Bobbs Merrill.

Heron, A.C. 1972. Population ecology of a colonizing species: the pelagic tunicate *Thalia democratica*. *Oecologia*, **10**, 269–93 and 294–312.

Hespenheide, H.A. 1973. Ecological inferences from morphological data. *A. Rev. Ecol. Syst.* **4**, 213–29.

Hickman, J.C. 1975. Environmental unpredictability and plastic energy allocation strategies in the annual *Polygonum cascadense* (Polygonaceae). *J. Ecol.* **63**, 689–701.

Hill, M.G., and Blackmore, P.J.M. 1980. Interactions between ants and the coccid, *Icerya seychellarum*, on Aldabra Atoll. *Oecologia*, **45**, 360–5.

Hill, R.B. 1926. The estimation of the number of hookworms harboured by the use of the dilution egg count method. *Am. J. Hyg.* **6** (*suppl.*), 19–41.

Hirsch, M.W., and Smale, S. 1974. *Differential Equations, Dynamical Systems, and Linear Algebra*, New York, Academic Press.

Hirst, S.M. 1965. Ecological aspects of big game predation. *Fauna Flora Pretoria*, **16**, 3–15.

Hoagland, K.E., and Schad, G.A. 1978. *Necator americanus* and *Ancylostoma duodenale*: Life history parameters and epidemiological implications of two sympatric hookworms of humans. *Exp. Parasitol.* **44**, 36–49.

Hoffman, A. 1978. System concepts and the evolution of benthic communities. *Lethaia*, **11**, 179–83.

Holling, C.S. 1959a. The components of predation as revealed by a study of small mammal predation of the European pine sawfly. *Can. Ent.* **91**, 293–320.

Holling, C.S. 1959b. Some characteristics of simple types of predation and parasitism. *Can. Ent.* **91**, 385–98.

Holling, C.S. 1973. Resilience and stability of ecological systems. *A. Rev. Ecol. Syst.* **4**, 1–24.

Holling, C.S., Jones, D.D., and Clark, W.C. 1976. Ecological policy design: a case study of forest and pest management. *Int. Inst. Appl. Systems Anal. Conf.* **1**, 139–58.

Holmes, J.C. 1979. Parasite population and host community structure. In B.B. Nickol (ed). *Host–Paeasite Interfaces*, pp. 27–48. New York. Academic Press,

Holmes, R.J., Schultz., J.C., and Nothnagle, P. 1979. Bird predation on forest insects: an exclosure experiment. *Science*, **206**, 462–3.

Holmes, R.T., Bonney, R.E. Jr., and Pacala, S.W. 1979. Guild structure of the Hubbard Brook bird community: a multivariate approach. *Ecology*, **60**, 512–20.

Holt, S.J., and Talbot, L.M. 1978. New principles for the conservation of wild living resources. *Wild. Monog.* No. 59.

Hoogland, J.L. 1980. Nepotism and cooperative breeding in the Black-tailed Prairie Dog (Sciuridae: *Cynomys lucovicianus*). In R.D. Alexander and D.W. Tinkle (eds.), *Natural Selection and Social Behavior: Recent Research and Theory*. New York, Chiron. (In press.)

Hooykaas, R. 1963. *The Principle of Uniformity*. Leiden, E.J. Brill.

Hopkinson, J.M. 1964. Studies on the expansion of the leaf surface. IV. The carbon and phosphorus economy of a leaf. *J. Exp. Bot.* **15**, 125–37.

Horn, H.S. 1971. *The Adaptive Geometry of Trees*. Princeton, Princeton University Press.

Horn, H.S. 1974. The ecology of secondary succession. *A. Rev. Ecol. Syst.* **5**, 25–37.

Horn, H.S. 1975a. Forest succession. *Scientific American*, **232 (5)**, 90–8.

Horn, H.S. 1975b. Markovian properties of forest succession. In M.L. Cody and J.M. Diamond (eds.), *Ecology and Evolution of Communities*, pp. 196–211. Cambridge, Mass., Harvard University Press.

Horn, H.S. 1978. Optimal tactics of reproduction and life-history. In J.R. Krebs and N.B. Davies (eds.), *Behavioural Ecology: an Evolutionary Approach*, pp. 411–29. Oxford, Blackwell Scientific Publications.

Horn, H.S. 1979. Adaptation from the perspective of optimality. In O.T. Sobrig *et al.* (eds.), *Topics in Plant Population Biology*, pp. 48–61. New York, Columbia University Press.

Horn, H.S. 1981. Some theories about dispersal. In I.R. Swingland and P.J. Greenwood (eds.), *The Ecology of Animal Movement*. Oxford, Oxford University Press.

Horn, H.S., and MacArthur, R.H. 1972. Competition among fugitive species in a harlequin environment. *Ecology*, **53**, 749–52.

Horn, H.S., and May, R.M. 1977. Limits to similarity among coexisting competitors. *Nature*, **270**, 660–1.

Hornocker, M.G. 1970. An analysis of mountain lion predation upon mule deer and elk in the Idaho primitive area. *Wildl. Monog.* **21**, 1–38.

Horowitz, D.L. 1975. Ethnic identity. In N. Glazer and D.P. Moynihan (eds.), *Ethnicity: theory and experience*. Cambridge, Mass., Harvard University Press.

Horwood, J.W., Knight, P.J., and Overy, R.W. 1979. Harvesting of whale populations subject to stochastic variability. *Int. Comm. Whaling Rep.* **29**, 219–29.

Hotton, N. 1955. A survey of adaptive relationships of dentition to diet in the North American Iguanidae. *Am. Midl. Natur.* **53**, 88–114.

Houston, D.B. 1973. Wildfires in northern Yellowstone National Park. *Ecology*, **54**, 1111–17.

Howell, D.J. 1979. Flock foraging in nectar-feeding bats: advantages to the bats and to the host plants. *Am. Nat.* **114**, 23–49.

440   REFERENCES

Howell, D.J., and Hartl, D.L. 1980. Optimal foraging in Glossophagine bats:
    when to give up. *Am. Nat.* **115**, 696–704.
Hrdy, S.B. 1977. *The Langurs of Abu: Female and Male Strategies of Reproduc-
    tion.* Cambridge, Mass., Harvard University Press.
Hsü, H.J. 1980. Terrestrial catastrophe caused by cometary impact at the end of
    Cretaceous. *Nature*, **285**, 201–3.
Hubbell, S.P. 1979. Tree dispersion, abundance, and diversity in a tropical dry
    forest. *Science*, **203**, 1299–1309.
Huey, R.B., and Pianka, E.R. 1974. Ecological character displacement in a lizard.
    Am. Zool. **14**, 1127–36.
Huey, R.B., Pianka, E.R., Egan, M.E., and Coons, L.W. 1974. Ecological shifts in
    sympatry: Kalahari fossorial lizards (*Typhlosaurus*). *Ecology*, **55**, 304–16.
Huey, R.B., and Pianka, E.R. 1977a. Patterns of niche overlap among broadly
    sympatric versus narrowly sympatric Kalahari lizards (Scincidae: Mabuya).
    Ecology, **58**, 119–28.
Huey, R.B., and Pianka, E.R. 1977b. Seasonal variation in thermoregulatory
    behavior and body temperature of diurnal Kalahari lizards. *Ecology*, **58**,
    1066–75.
Huey, R.B., and Pianka, E.R. 1980. Ecological consequences of foraging mode.
    Ecology. (In press.)
Huffaker, C.B. 1958. Experimental studies on predation: dispersion factors and
    predator–prey oscillations. *Hilgardia*, **27**, 343–83.
Huffaker, C.B. (ed.) 1971. *Biological Control.* New York, Plenum Press.
Huffaker, C.B., and Kennett, C.E. 1966. Studies of two parasites of olive scale,
    Parlatoria oleae (Colvee). IV. Biological control of *Parlatoria oleae* (Colvee)
    through the compensatory action of two introduced parasites. *Hilgardia*, **37**,
    283–335.
Huffaker, C.B., and Messenger, P.S. (eds.), 1976. *Theory and Practice of Biological
    Control.* New York, Academic Press.
Humphreys, W.F. 1979. Production and respiration in animal populations. *J.
    Anim. Ecol.* **48**, 427–53.
Hunt, G.J., Jr., and Hunt, M.W. 1974. Trophic levels and turnover rates: the
    avidauna of Santa Barbara Island, California. *Condor*, **76**, 363–9.
Hunt, J.H. (ed.) 1980. *Selected Readings in Sociobiology.* New York, McGraw-Hill.
Hunt, R. 1978. *Plant Growth Analysis.* London, Arnold.
Hutchings, M.J. 1979. Weight-density relationships in ramet populations of clonal
    perennial herbs, with special reference to the –3/2 thinning law. *J. Ecol.* **67**,
    21–34.
Hutchinson, G.E. 1948. Circular causal systems in ecology. *Ann. N.Y. Acad. Sci.*
    50, 221–46.
Hutchinson, G.E. 1951. Copepodology for the ornithologist. *Ecology*, **32**, 571–7.
Hutchinson, G.E. 1953. The concept of pattern in ecology. *Proc. Acad. Nat. Sci.*,
    Philadelphia, **105**, 1–12.
Hutchinson, G.E. 1957. Concluding remarks. *Cold Spring Harbor Symp. Quant.
    Biol.* **22**, 415–27.
Hutchinson, G.E. 1959. Homage to Santa Rosalia, or why are there so many
    kinds of animals? *Am. Nat.* **93**, 145–59.
Hutchinson, G.E. 1961. The paradox of the plankton. *Am. Nat.* **95**, 137–45.
Hutchinson, G.E. 1975. Variations on a theme by Robert MacArthur. In M.L.
    Cody and J.M. Diamond (eds.), *Ecology and Evolution of Communities*, pp.
    492–521. Cambridge, Mass., Harvard University Press.

Hutchinson, G.E. 1978. *An Introduction to Population Ecology*. New Haven, Yale University Press.

Hutchinson, G.E., and MacArthur, R.H. 1959. A theoretical ecological model of size distributions among species of animals. *Am. Nat.* **93**, 117–25.

Ikusima, I., Shinozaki, K., and Kira, T. 1955. Intraspecific competition among higher plants. III. Growth of duckweed, with a theoretical consideration on the C-D effect. *J. Inst. Polytech. Osaka City Univ.* **6**, 107–19.

Inger, R., and Colwell, R.K. 1977. Organization of contiguous communities of amphibians and reptiles in Thailand. *Ecol. Monogr.* **47**, 229–53.

Ivlev, V.S. 1961. *Experimental Ecology of the Feeding of Fishes*. New Haven, Yale University Press.

Jackson, J.B.C. 1977. Competition on marine land substrata: the adaptive significance of solitary and colonial strategies. *Am. Nat.* **111**, 743–67.

Jackson, J.B.C. 1979. Overgrowth competition between encrusting cheilostome ectoprocts in a Jamaican cryptic reef environment. *J. Anim. Ecol.* **48**, 805–23.

Jacques, R.P. 1969. Leaching of the Nuclear-Polyhedrosis Virus of *Trichoplusiani* from soil. *J. Invert. Pathol.* **13**, 256–63.

Jacques, R.P. 1977. Stability of Entomopathogenic viruses. *Miscellaneous Publications of the Entomological Society of America*, **10**, 99–116.

Jaeger, R.G. 1970. Potential extinction through competition between two species of terrestrial salamanders. *Evolution*, **24**, 632–42.

Jaeger, R.G. 1971. Competitive exclusion as a factor influencing the distributions of two species of terrestrial salamanders. *Ecology*, **52**, 632–7.

Janetos, A.C. 1980. Strategies of female mate choice: a theoretical analysis. *Behav. Ecol. Sociobiol.* **7**, 107–12.

Janzen, D.H. 1970. Herbivores and the number of tree species in tropical forests. *Am. Nat.* **104**, 501–28.

Janzen, D.H. 1971. Euglossine bees as long-distance pollinators of tropical plants. *Science*, **171**, 203–5.

Janzen, D.H. 1977. Why are there so many species of insects? In *Proceedings of the XV Congress of Entomology*, pp. 84–94. Entomological Society of America, College Park, Md.

Järvinen, O. 1979. Geographical gradients of stability in European land bird communities. *Oecologia*, **38**, 51–69.

Jarvis, M.J.F. 1974. The ecological significance of clutch size in the South African Gannet (*Sula capensis* (Lichtenstein)). *J. Anim. Ecol.* **43**, 1–17.

Jenkins, D., Watson, A., and Miller, G.R. 1963. Population studies on red grouse, *Lagopus lagopus scoticus* (Lath.) in north-east Scotland. *J. Anim. Ecol.* **32**, 317–76.

Jenkins, D., Watson, A., and Miller, G.R. 1964. Predation and red grouse populations. *J. Appl. Ecol.* **1**, 183–95.

Joern, A., and Lawlor, L.R. 1980. Food and microhabitat utilization by grasshoppers: comparision with neutral models. *Ecology*, **61**, 591–9.

Johnson, C.G. 1969. *Migration and Dispersal of Insects by Flight*, London, Methuen.

Johnson, E.A. 1979. Fire recurrence in the sub-arctic and its implications for vegetation composition. *Canadian J. of Bot.* **57**, 1374–9.

Jolly, A. 1972. *The Evolution of Primate Behavior*. New York, Macmillan.

Jones, E.W. 1945. Structure and reproduction of the virgin forest of the north temperate zone. *New Phytol.* **44**, 130–48.

Jones, H.L., and Diamond, J.M. 1976. Short-time-base studies of turnover in breeding bird population on the California Channel Islands. *Condor*, **78**, 526–49.

Joule, J., and Jameson, D.L. 1972. Experimental manipulation of population density in three sympatric rodents. *Ecology*, **53**, 653–60.

Joule, J., and Cameron, G.N. 1975. Species removal studies. I. Dispersal strategies of sympatric *Sigmodon hispidus* and *Reithrodontomys fulvescens* populations. *J. Mammal.* **56**, 378–96.

Karns, P.D. 1967. *Pneumostronglyus tenuis* in deer in Minnesota and implications for moose. *J. Wild. Mgmt.* **31**, 299–303.

Kauffman, E.G. 1972. Evolutionary rates and patterns of North American Cretaceous Mollusca. *24th Int. Geol. Cong. Section*, **7**, 174–89.

Kay, M. 1947. Analysis of stratigraphy. *Bull. Am. Ass. Petroleum Geologists*, **31**, 162–8.

Kayll, A.J. 1974. Use of fire in land management. In T.T. Kozlowski and C.E. Ahlgren (eds.), *Fire and Ecosystems*, pp. 483–511. New York, Academic Press.

Kays, S., and Harper, J.L. 1974. The regulation of plant and tiller density in a grass sward. *J. Ecol.* **62**, 97–105.

Keith, L.B. 1963. *Wildlife's Ten-Year Cycle*. Madison, University of Wisconsin Press.

Kelsall, J.P., Prescott, J.R. 1971. Moose and deer behaviour in snow in Fundy National Park, New Brunswick. *Ca. Wildl. Serv. Rep.* Ser. No. 15, Information, Ottawa, Canada.

Kempton, R.A. 1979. The structure of species abundance and measurement of diversity. *Biometrics*, **35**, 307–21.

Kennedy, J.S. 1975. Insect dispersal. In D. Pimentel (ed.), *Insects, Science & Society*, pp. 103–19. New York, Academic Press.

Kermack, W.O., and McKendrick, A.G. 1927. Contributions to the mathematical theory of epidemics. *Proc. Roy. Soc. A*, **115**, 700–21.

Kiester, A.R., and Barakat, R. 1974. Exact solutions to certain stochastic differential equation models of population growth. *Theoret. Pop. Biol.* **6**, 199–216.

Kilham, P. 1971. A hypothesis concerning silica and the freshwater planktonic diatoms. *Limnol. Oceanogr.* **16**, 10–18.

Kincade, R.T., Laster, M.L., and Brazzel, J.R. 1970. Effect on cotton yield of various levels of simulated *Heliothis* damage to squares and bolls. *J. econ. Ent.* **63**, 613–15.

King, C.E., Gallaher, E.E., and Levin, D.A. 1975. Equilibrium diversity in plant-pollinator systems. *J. Theoret. Biol.* **53**, 263–75.

Kira, T., Ogawa, H., and Shinozaki, K. 1953. Intraspecific competition among higher plants. 1. Competition-density-yield inter-relationships in regularly dispersed populations. *J. Inst. Polytech. Osaka Cy. Univ. D* **4**, 1–16.

Klahn, J.E. 1979. Philopatric and nonphilopatric foundress associations in the social wasp *Polistes fuscatus*. *Behav. Ecol. Sociobiol.* **5**, 417–24.

Klein, D.R. 1968. The introduction, increase, and crash of reindeer on St Matthew Island. *J. Wildl. Mgmt.* **32**, 350–67.

Knight, D.H. 1975. A phytosociological analysis of species-rich tropical forest on Barro Colorado Island, Panama. *Ecol. Monogr.* **45**, 259–84.

Kodric-Brown, A., and Brown, J.H. 1979. Competition between distantly related taxa in the coevolution of plants and pollinators. *Amer. Zool.* **19**, 1115–27.

Koers, A.W. 1973. *International Regulation of Marine Fisheries*. London, Fishing News.

Kok, L.T., and Surles, W.W. 1975. Successful biocontrol of musk thistle by an introduced weevil. *Rhinocyllus conicus*. *Env. Ent.* **4**, 1025–7.

Kolmogorov, A.N. 1936. Sulla teoria di Volterra della lotta per l'esistenza. *Giorn. Instituto Ital. Attuari*, **7**, 74–80.

Koplin, J.R., and Hoffman, R.S. 1968. Habitat overlap and competitive exclusion in voles (*Microtus*). *Am. Midl. Natur.* **80**, 494–507.

Korns, P.D. 1967. *Pneumostrongylus tenuis* in deer in Minnesota and implications for moose. *J. Wildl. Mgmt.* **31**, 299–303.

Kostitzin, V.A. 1939. *Mathematical Biology*. London, Harrap.

Kozlovsky, D.G. 1968. A critical evaluation of the trophic level concept: I, ecological efficiencies. *Ecology*, **49**, 48–60.

Krebs, C.J. 1978. *Ecology, the Experimental Analysis of Distribution and Abundance* (2nd. Ed.). New York, Harper and Row.

Krebs, J.R., and Davies, N.B. (eds.) 1978. *Behavioural Ecology: an Evolutionary Approach*. Oxford, Blackwell Scientific Publications.

Krupp, I.M. 1961. Effects of crowding and of superinfection on habitat selection and egg production in *Ancylostoma caninum*. *J. Parasitol.* **47**, 957–61.

Kuhn, T.S. 1962. *The Structure of Scientific Revolutions*. Chicago, University of Chicago Press.

Kulman, H.M. 1971. Effects of insect defoliation on growth and mortality of trees. *Ann. Rev. Ent.* **16**, 289–324.

Kuno, E., and Hokyo, N. 1970. Comparative analysis of the population dynamics of rice leafhoppers, *Nephotettix cincticeps* Uhler and *Nilaparvata lugens* Stål, with special reference to natural regulation of their numbers. *Res. Pop. Ecol.* **12**, 154–84.

Lack, D. 1954. *The Natural Regulation of Animal Numbers*. London, Oxford University Press.

Lack, D. 1966. *Population Studies of Birds*. Oxford, Clarendon Press.

Lack, D. 1968. *Ecological Adaptations for Breeding in Birds*. London, Methuen.

Lack, D. 1973. The numbers and species of hummingbirds in the West Indies. *Evolution*, **27**, 326–37.

Lack, D. 1976. *Island Birds*. Oxford, Blackwell Scientific Publications.

Ladd, H.S. 1957. *Treatise on Marine Ecology and Paleoecology, Vol. 2 Paleoecology*. Mem. Geol. Soc. Am. No. 67.

Lance, A.N. 1978. Territories and the food plant of individual red grouse. II. Territory size compared with an index of nutrient supply in heather. *J. Anim. Ecol.* **47**, 307–13.

Larkin, P.A. 1977. An epitaph for the concept of maximum sustained yield. *Trans. Amer. Fish. Soc.* **196**, 1–11.

Law, R. 1979. The cost of reproduction in annual meadow grass. *Am. Nat.* **113**, 3–16.

Lawlor, L.R. 1978. A comment on randomly constructed model ecosystems. *Am. Nat.* **112**, 445–7.

Lawlor, L.R. 1979. Direct and indirect effects of *n*-species competition. *Oecologia*. **43**, 355–64

Lawlor, L.R. 1980. Structure and stability in natural and randomly-constructed competitive communities. *Am. Nat.* **116**, 394–408.

Lawlor, L.R., and Maynard Smith, J. 1976. Coevolution and stability of competing species. *Am. Nat.* **110**, 79–99.

Laws, R. 1980. The dynamics of a colonising population of *Poa annua. J. Ecol.* (In press.)

Lawton, J.H. 1974. The structure of the arthropod community on bracken (*Pteridium aquilinium* (L.) Kuhn). In F.H. Perring (ed.), *The Biology of Bracken*. New York, Academic Press.

Lawton, J.H., and Pimm, S.L. 1978. Population dynamics and the length of food chains. *Nature*, **272**, 190.

Lawton, J.H., and McNeill, S. 1979. Between the devil and the deep blue sea: on the problem of being a herbivore. In R.M. Anderson, B.D. Turner, and L.R. Taylor (eds.), *Symp. Brit. Ecol. Soc. 20, Population Dynamics*, pp. 223–44. Oxford, Blackwell Scientific Publications

Lawton, J.R., and Ralliston, S.P. 1979. Stability and diversity in grassland communities. *Nature*, **279**, 351.

Leak, W.B. 1970. Successional change in northern hardwoods predicted by birth and death simulation. *Ecology*, **51**, 794–801.

Leck, C.F. 1979. Avian extinctions in an isolated tropical wet-forest preserve, Ecuador. *Auk*, **96**, 343–52.

Leigh, E. 1975. Population fluctuations and community structure. In M.L. Cody and J.M. Diamond (eds.), *Ecology of species and communities*, pp. 74–80. Cambridge, Mass., Harvard University Press.

Leith, H. 1975. Primary productivity in ecosystems: comparative analysis of global patterns. In W.H. van Dobben and R.H. Lowe-McConnell (eds.), *Unifying Concepts in Ecology*, pp. 67–88. The Hague, W. Junk.

Leon, J.A. 1974. Selection in contexts of interspecific competition. *Am. Nat.* **108**, 739–57.

Leopold, A., Sowls, L.K., and Spencer, D.L. 1947. A survey of over-populated deer ranges in the United States. *J. Wildl. Mgmt.* **11**, 162–77.

Leslie, P.H. 1945. On the use of matrices in certain population mathematics. *Biometrika*, **33**, 183–212.

Leslie, P.H. 1948. Some further notes on the use of matrices in population mathematics. *Biometrika*, **35**, 213–45.

Leuthold, W. 1977. *African Ungulates, a Comparative Review of their Ethology and Behavioural Ecology*. Berlin, Springer-Verlag.

Levin, S.A. 1970. Community equilibria and stability, and an extension of the competitive exclusion principle. *Am. Nat.* **104**, 413–23.

Levin, S.A. 1974. Dispersion and population interactions. *Am. Nat.* **108**, 207–28.

Levine, S.H. 1976. Competitive interactions in ecosystems. *Am. Nat.* **110**, 903–10.

Levins, R. 1968. *Evolution in Changing Environments*. Princeton, Princeton University Press.

Levins, R. 1969. The effects of random variations of different types on population growth. *Proc. Natl. Acad. Sci. U.S.* **62**, 1061–5.

Levins, R. 1975. Evolution in communities near equilibrium. In M.L. Cody and J.M. Diamond (eds.), *Ecology and Evolution of Communities*, pp. 16–50. Cambridge, Mass., Harvard University Press.

Levins, R., and Culver, D. 1971. Regional coexistence of species and competition between rare species. *Proc. Natl. Sci. U.S.* **68**, 1246–8.

Levinton, J.S. 1970. The paleoecological significance of opportunistic species. *Lethaia*, **3**, 69–78.

Lewontin, R.C. 1978. Adaptation. *Scientific American*, **239 (3)**, 212–30.

Lewontin, R.C., and Cohen, D. 1969. On population growth in a randomly varying environment. *Proc. Natl. Acad. Sci. U.S.* **62**, 1056–60.

Li, T.-Y., and Yorke, J.A. 1975. Period three implies chaos. *Am. Math. Monthly*, **82**, 985–92.

Libchaper, A., and Maurer, J. 1980. *Une expérience de Rayleihg-Bénard de geometric réduite. J. de Phys.* **41,** C3–51.

Liem, K.F. 1973. Evolutionary strategies and morphological innovations: cichlid pharyngeal jaws. In S.J. Gould (ed.), ICSEB. Symposium on evolutionary development of form and symmetry. *Syst. Zool.* **22,** 425–41.

Lloyd, M. 1967. Mean Crowding. *J. Anim. Ecol.* **36,** 1–30.

Lomnicki, A. 1977. Evolution of plant resistance and herbivore population cycles. *Am. Nat.* **111,** 198–200.

London, W.P., and Yorke, J.A. 1973. Recurrent outbreaks of measles, chicken pox and mumps. I: Seasonal variation in contact rates. *Amer. J. Epidemiol.* **98,** 453–68.

Lotka, A.J. 1925. *Elements of Physical Biology.* Baltimore, Williams and Wilkins.

Loucks, O.L. 1970. Evolution of diversity, efficiency, and community stability. *Am. Zool.* **10,** 17–25.

Lovat, Lord. 1911. Moor Management. In Lovat (ed.), *The Grouse in Health and in Disease,* pp. 372–91. London.

Loya, Y. 1976. *Stylophora pistillata*: an *r*-strategist among the Red Sea Corals. *Nature (Lond.)* **259,** 478–80.

Lubchenco, J. 1978. Plant species diversity in a marine intertidal community: importance of herbivore food preference and algal competitive abilities. *Am. Nat.* **112,** 23–39.

Luckinbill, L.S. 1973. Coexistence in laboratory populations of *Paramecium aurelia* and its predator *Didinium nasutum. Ecology,* **54,** 1320–7.

Luckinbill, L.S. 1979. Selection and the *r/K* continuum in experimental populations of Protozoa. *Am. Nat.* **113,** 427–37.

Ludwig, D. 1975. Persistence of dynamical systems under random perturbations. *SIAM Review,* **17,** 605–40.

Ludwig, D.A. 1979. Harvesting strategies for a randomly fluctuating population. *SIAM J. Appl. Math.* **37,** 166–84.

Ludwig, D., Jones, D.D., and Holling, C.S. 1978. Qualitative analysis of insect outbreak systems: the spruce budworm and forest. *J. Anim. Ecol.* **47,** 315–32.

Lumsden, C.J., and Wilson, E.O. 1980. Translation of epigenetic rules of individual behavior into ethnographic patterns. *Proc. Natl. Acad. Sci. U.S.* **77,** 4382–6.

Lumsden, C.J., and Wilson, E.O. 1981. *Genes, Mind, and Culture: The Coevolutionary Process.* Cambridge, Mass., Harvard University Press.

MacArthur, R.H. 1961. Community. In P. Gray (ed.), *The Encylcopedia of the Biological Sciences,* pp. 262–4. New York, Reinhold.

MacArthur, R.H. 1965. Ecological consequences of natural selection. In T.H. Waterman and H.J. Morowitz (eds.), *Theoretical and Mathematical Biology,* pp. 388–97. New York, Blaisdell.

MacArthur, R.H. 1968. The theory of the niche. In R.C. Lewontin (ed.), *Population Biology and Evolution,* pp. 159–76. Syracuse, Syracuse University Press.

MacArthur, R.H. 1969. Species packing and what competition minimizes. *Proc. Natl. Acad. Sci. U.S.* **64,** 1369–71.

MacArthur, R.H. 1970. Species packing and competitive equilibrium for many species. *Theoret. Pop. Biol.* **1,** 1–11.

MacArthur, R.H. 1972. *Geographical Ecology.* New York, Harper and Row.

MacArthur, R.H., and MacArthur, J.W. 1961. On bird species diversity. *Ecology,* **42,** 594–8.

446

MacArthur, R.H., and Wilson, E.O. 1963. An equilibrium theory of insular zoo-geography. *Evolution*, **17**, 373–87.

MacArthur, R.H., and Pianka, E.R. 1966. On optimal use of a patchy environ-ment. *Am. Nat.* **100**, 603–9.

MacArthur, R.H., and Levins, R. 1967. The limiting similarity, convergence and divergence of coexisting species. *Am. Nat.* **101**, 377–85.

MacArthur, R.H., and Wilson, E.O. 1967. *The Theory of Island Biogeography.* Princeton, Princeton University Press.

McCauley, E., and Briand, F. 1979. Zooplankton grazing and phytoplankton species richness: field tests of the predation hypothesis. *Limnol. Oceanogr.* **24**, 243–52.

McCormick, J., and Buell, M.F. 1968. The plains: pigmy forests of the New Jersey pine barrens, a review and annotated bibliography. *Bull. N.J. Acad. Sci.* **13**, 20–34.

Macdonald, G. 1952. The analysis of equilibrium in malaria. *Trop. Dis. Bull.* **49**, 813–29.

Macdonald, G. 1957. *The Epidemiology and Control of Malaria.* London, Oxford University Press.

Macdonald, G. 1961. Epidemiological models in studies of vector borne diseases. *Publ. Hlth. Rep.* **76**, 753–64.

Macdonald, G. 1965. The dynamics of helminth infections, with special reference to schistosomes. *Trans. R. Soc. Trop. Med. Hyg.* **59(5)**, 489–506.

Macdonald, G., Cuellar, C.B., and Foll, C.V. 1968. The dynamics of malaria. *Bull. Wld. Hlth. Org.* **38**, 743–55.

MacDonald, N. 1978. The prevalence of chaos. *Nature*, **271**, 305–6.

McIntosh, R.P. 1980. The relationship between succession and the recovery pro-cess in ecosystems. In J. Cairns, Jr. (ed.), *The Recovery Process in Damaged Ecosystems*, pp. 11–62. Ann Arbor, Michigan, Ann Arbor Science Publishers Inc.

Mack, R.N., and Harper, J.L. 1977. Interference in dune annuals: spatial pattern and neighbourhood effects. *J. Ecol.* **65**, 345–63.

Mackintosh, N.J. 1979. Book review: On Human Nature, by E.O. Wilson. *Science*, **204**, 735–7.

McMurtrie, R.E. 1975. Determinants of stability of large, randomly connected systems. *J. Theoret. Biol.* **50**, 1–11.

McMurtrie, R.E. 1978. Persistence and stability of single species and prey–predator systems in spatially heterogeneous environments. *Math. Biosci.* **39**, 11–51.

McNaughton, S.J. 1975. r- & K-selection in *Typha. Am. Nat.* **109**, 251–61.

McNaughton, S.J. 1978. Stability and diversity of ecological communities. *Nature*, **274**, 251–3.

McNaughton, S.J. 1979a. Grazing as an optimization process: grass ungulate relationships in the Serengeti. *Am. Nat.* **113**, 691–703.

McNaughton, S.J. 1979b. Stability and diversity in grassland communities—reply. *Nature*, **279**, 351–2.

McNaughton, S.J., and Wolf, L.L. 1973. *General Ecology.* New York, Holt, Rinehart and Winston.

McNeill, S. 1973. The dynamics of a population of *Leptoterna dolabrata* (Hepterop-tera: Miridae) in relation to its food resources. *J. Anim. Ecol.* **42**, 495–507.

McNeill, S., and Southwood, T.R.E. 1978. The role of nitrogen in the development of insect/plant relationships. In J.B. Harborne (ed.), *Biochemical Aspects of Plant and Animal Coevolution*, pp. 77–98. London, Academic Press.

Maillett, L. 1979. The Structural Dynamics of Tree Growth. Phd thesis University of Wales.

Maiorana, V.C. 1978. An explanation of ecological and developmental constants. *Nature*, **273**, 375–7.

Maissurow, D.K. 1941. The role of fire in the perpetuation of virgin forests of Northern Wisconsin. *J. Forestry*, **39**, 201–7.

Maramorosch, K., and Shope, R.E. 1975. *Invertebrate Immunity*. New York, Academic Press.

Margalef, R. 1968. *Perspective in Ecological Theory*. Chicago, University of Chicago Press.

Markl, H. (ed.) 1980. *Evolution of Social Behaviour*. Berlin, Dahlem Konferenzen.

Matthes, F.E., posthumously prep. by F. Fryxell. 1965. Glacial reconnaissance of Sequoia National Park California. *U.S. Geol. Surv. Prof. Paper* **504**-A, 1–58.

May, R.M. 1972a. Limit cycles in predator–prey communities. *Science*, **177**, 900–2.

May, R.M. 1972b. Will a large complex system be stable? *Nature* (*Lond*), **238**, 413–14.

May, R.M. 1973a. Time-delay versus stability in population models with two and three trophic levels. *Ecology*, **54**, 315–25.

May, R.M. 1973b. Stability in randomly fluctuating versus deterministic environments. *Am. Nat.* **107**, 621–50.

May, R.M. 1973c. Qualitative stability in model ecosystems. *Ecology*, **54**, 638–41.

May, R.M. 1973. *Stability and Complexity in Model Ecosystems*. Princeton, Princeton University Press.

May, R.M. 1974a. Biological populations with nonoverlapping generations: stable points, stable cycles and chaos. *Science*, **186**, 645–7.

May, R.M. 1974b. Ecosystem patterns in randomly fluctuating environments. In R. Rosen and F. Snell (eds.), *Progress in Theoretical Biology*, pp. 1–50. New York, Academic Press.

May, R.M. 1974c. How many species: some mathematical aspects of the dynamics of populations. In J.D. Cowan (ed.), *Some Mathematical Problems in Biology. Vol. 4*, pp. 64–98. Providence, R.I., The American Mathematical Society.

May, R.M. 1974d. On the theory of niche overlap. *Theoret. Pop. Biol.* **5**, 297–332.

May, R.M. 1975a. *Stability and Complexity in Model Ecosystems*. (Second edition). Princeton, Princeton University Press.

May, R.M. 1975b. Biological populations obeying difference equations: stable points, stable cycles and chaos. *J. Theoret. Biol.*, **49**, 511–24.

May, R.M. 1975c. Stability in ecosystems: some comments. In W.H. van Dobben and R.H. Lowe-McConnell (eds.), *Unifying Concepts in Ecology*, pp. 161–8. The Hague, W. Junk.

May, R.M. 1975d. Some notes on estimating the competition matrix, α. *Ecology*, **56**, 737–41.

May, R.M. 1975e. Group selection. *Nature* (*Lond.*), **254**, 485.

May, R.M. 1975f. Patterns of species abundance and diversity. In M.L. Cody and J.M. Diamond (eds.), *Ecology and Evolution of Communities*, pp. 81–120. Cambridge, Mass., Harvard University Press.

May, R.M. 1975g. Island biogeography and the design of wildlife preserves. *Nature* (*Lond.*), **254**, 177–8.

May, R.M. 1976a. Mathematical aspects of the dynamics of animal populations. In S.A. Levin (ed.), *Studies in Mathematical Biology*. Providence, R.I., American Mathematical Society.

May, R.M. 1976b. Estimating r: a pedagogical note. *Am. Nat.* **110**, 469–99

May, R.M. 1976c. Simple mathematical models with very complicated dynamics. *Nature*, **261**, 459–67.

May, R.M. 1977a. Thresholds and breakpoints in ecosystems with a multiplicity of stable states. *Nature*, **269**, 471–7.

May, R.M. 1977b. Togetherness among schistosomes: its effect on the dynamics of the infection. *Math. Biosc.* **35**, 301–43.

May, R.M. 1977c. Predators that switch. *Nature*, **269**, 103–4.

May, R.M. 1978a. Host–parasitoid systems in patchy environments: a pheno-menological model. *J. Anim. Ecol.* **47**, 833–44.

May, R.M. 1978b. The dynamics and diversity of insect faunas. In L.A. Mound and N. Waloff (eds.), *Diversity of Insect Faunas*, (Royal Ent. Soc. Symposium. London, Sept., 1977), pp. 188–204. Oxford, Blackwell Scientific Publications.

May, R.M. 1979. The structure and dynamics of ecological communities. In R.M. Anderson, B.D. Turner and L.R. Taylor (eds.), *Symp. Brit. Ecol. Soc.* 20, *Population Dynamics*, pp. 385–407. Oxford, Blackwell Scientific Publications.

May, R.M. 1980a. Mathematical models in whaling and fisheries management. In G.F. Oster (ed.), *Some Mathematical Questions in Biology, Vol.*13, pp. 1–64. Providence, R.I., The American Mathematical Society.

May, R.M. 1980b. Nonlinear phenomena in ecology and epiderminology. *Ann. N.Y. Acad. Sci.* (In press.)

May, R.M., and MacArthur, R.H. 1972. Niche overlap as a function of environmental variability. *Proc. Natl. Acad. Sci. U.S.* **69**, 1109–13.

May, R.M., Conway, G.R., Hassell, M.P., and Southwood, T.R.E. 1974. Time delays, density dependence, and single species oscillations. *J. Anim. Ecol.* **43**, 747–70.

May, R.M., and Leonard, W.J. 1975. Nonlinear aspects of competition between three species. *SIAM J. Appl. Math.* **29**, 243–53.

May, R.M., and Oster, G.F. 1976. Bifurcations and dynamic complexity in simple ecological models. *Am. Nat.* **110**, 573–99.

May, R.M., and Anderson, R.M. 1978. Regulation and stability of host–parasite population interactions: II. *J. Anim. Ecol.* **47**, 249–67.

May, R.M., Beddington, J.R., Horwood, J.W., and Shepherd, J.G. 1978. Exploiting natural populations in an uncertain world. *Math. Biosciences*, **42**, 219–52.

May, R.M., and Anderson, R.M. 1979. Population biology of infectious diseases: Part II. *Nature*, **280**, 455–61.

May, R.M., Beddington, J.R., Clark, C.W., Holt, S.J., and Laws, R.M. 1979. Management of multispecies fisheries. *Science*, **205**, 267–77.

May, R.M., and Hassell, M.P. 1980. The dynamics of multi-parasitoid–host interactions. *Am. Nat.* (In press.)

May, R.M., and Oster, G.F. 1980. Period doubling and the onset of turbulence: an analytic estimate of the Feigenbaum ratio. *Physics Letters*, **78A**, 1–3.

May, R.M., and Robertson, M. 1980. Just so stories and cautionary tales. *Nature*, **286**, 327–8.

Maynard Smith, J. 1976. Evolution and the theory of games *Am. Scient.* **64**, 41–5.

Maynard Smith, J. 1977. Parental investment: a prospective analysis *Anim. Behav.* **25**, 1–9.

Maynard Smith, J. 1978a. *The Evolution of Sex*. Cambridge, Cambridge University Press.

Maynard Smith, J. 1978b. The ecology of sex. In J.R. Krebs and N.B. Davies (eds.), *Behavioural Ecology: an Evolutionary Approach*, pp. 159–79. Oxford, Blackwell Scientific Publications.

Maynard Smith, J., and Price, G.R. 1973. The logic of animal conflict. *Nature*, **246**, 15–18.

Mayr, E. 1969. *Principles of Systematic Zoology*. New York, McGraw-Hill.

Mazanov, A. 1978. Acceptor control in model ecosystems. *J. Theor. Biol.* **71**, 21–38.

Mech, L.D. 1966. *The Wolves of Isle Royale*. Fauna of National Parks of U.S. Fauna Series 7. Washington, U.S. Government Printing Office.

Meijden, E. van der 1979. Herbivore exploitation of a fugitive plant species: local survival and extinction of the cinnabar moth and ragwort in a heterogeneous environment. *Oecologia*, **42**, 307–23.

Menge, B.A. 1972. Competition for food between two intertidal starfish species and its effect on body size and feeding. *Ecology*, **53**, 635–44.

Menge, J.L., and Menge, B.A. 1974. Role of resource allocation, aggression and spatial heterogeneity in coexistence of two competing intertidal starfish. *Ecol. Monogr.* **44**, 189–209.

Metcalf, R.A., and Whitt, G.S. 1977a. Intra-nest relatedness in a social wasp *Polistes metricus*. A genetic analysis. *Behav. Ecol. Sociobiol.* **2**, 339–51.

Metcalf, R.A., and Whitt, G.S. 1977b. Relative inclusive fitness in the social wasp *Polistes metricus*. *Behav. Ecol. Sociobiol.* **2**, 353–60.

Michelakis, S. 1973. A study of the laboratory interaction between *Coccinella septempunctata* larvae and its prey *Myzus persicae*. Unpublished M.Sc. thesis. University of London.

Miles, J. 1979. *Vegetation Dynamics*. London, Chapman and Hall.

Miller, C.A. 1959. The interaction of the spruce budworm, *Choristoneura fumiferana* (Clem.), and the parasite *Apanteles fumiferanae* Vier. *Can. Ent.* **91**, 457–77.

Miller, G.R., and Watson, A. 1978. Territories and the food plant of individual red grouse. I. Territory size, number of mates and brood size compared with the abundance, production and diversity of heather. *J. Anim. Ecol.* **47**, 293–305.

Miller, R.S. 1967. Pattern and process in competition. *Adv. Ecol. Res.* **4**, 1–74.

Mills, E.S. 1978. *The Economics of Environmental Quality*. New York, Norton.

Milne, A. 1961. Definition of competition among animals. In F.L. Milthorpe (ed.), *Mechanisms in Biological Competition, Symp. Soc. Exp. Biol.* **15**, 40–61.

Milthorpe, F.L. (ed.) 1961. *Mechanisms in Biological Competition, Symp. Soc. Exp. Biol.* 15. Cambridge, Cambridge University Press.

Mitchell, W.C., and Mau, R.F.L. 1971. Response of the female Southern green stink bug and its parasite, *Trichopoda pennipes*, to male stink bug pheromones. *J. Econ. Ent.* **64**, 856–9.

Mogi, M. 1969. Predation response of the larvae of *Harmonia axyridis* Pallas (Coccinellidae) to the different prey density. *Jap. J. appl. Ent. Zool.* **13**, 9–16.

Monro, J. 1967. The exploitation and conservation of resources by populations of insects. *J. Anim. Ecol.* **36**, 531–47.

Monro, J. 1975. Environmental variation and efficiency of biological control— *Cactoblastis* in the southern hemisphere. *Proc. Ecol. Soc. Australia*, **9**, 204–12.

Morowitz, H.J. 1968. *Energy Flow in Biology*, New York, Academic Press.

Morrow, P.A., and La Marche, V.C. Jr. 1978. Tree ring evidence for chronic insect suppression of productivity in subalpine *Eucalyptus*. *Science*, **201**, 1244–6.

450 REFERENCES

Moser, J.W. 1972. Dynamics of an uneven-aged forest stand. *Forest Sci.* **18**,184–91.

Mountford, M.D. 1973. The significance of clutch size. In M.S. Bartlett and R.W. Hiorns (eds.), *The Mathematical Theory of the Dynamics of Biological Populations*, pp. 315–23. London, Academic Press.

Muench, H. 1959. *Catalytic Models in Epidemiology*. Cambridge, Mass., Harvard University Press.

Muggleton, J., and Benham, B.R. 1975. Isolation and the decline of the Large Blue Butterfly (*Maculinea arion*) in Great Britain. *Biol. Conserv.* **7**, 119–28.

Muller, R. 1975. *Worms and Diseases*. London, Heinemann.

Murdie, G., and Hassell, M.P. 1973. Food distribution, searching success and predator–prey models. In M.S. Bartlett and R.W. Hiorns (eds.), *The Mathematical Theory of the Dynamics of Biological Populations*, pp. 87–101. London, Academic Press.

Murdoch, W.W., and Oaten, A. 1975. Predation and population stability. *Adv. Ecol. Res.* **9**, 2–131.

Murie, O.J. 1930. An epizootic disease of elk. *J. Mammal.* **11**, 214–22.

Murphy, G.J. 1977. Clupeoids. In J.A. Gulland (ed.), *Fish Population Dynamics*, pp. 283–308. New York, Wiley-Interscience.

Murton, R.K., Isaacson, A.J., and Westwood, N.J. 1966. The relationships between wood pigeons and their clover food supply and the mechanism of population control. *J. Appl. Ecol.* **3**, 55–93.

Mutch, R.W. 1970. Wildland fires and ecosystems – a hypothesis. *Ecology*, **51**, 1046–51.

Nakasuji, F., Hokyo, N., and Kiritani, K. 1966. Assessment of the potential efficiency of parasitism in two competitive scelionid parasites of *Nezara viridula* L. (Hemiptera: Pentatomidae). *Appl. Ent. Zool.* **1**, 113–19.

Namkoong, G., and Roberds, J.H. 1974. Extinction probabilities and the changing age structure of redwood forests. *Am. Nat.* **108**, 355–68.

Nassell. I., and Hirsch, W.M. 1973. The transmission dynamics of schistosomiasis. *Communications on Pure and Applied Mathematics*, **26**, 395–453.

Nawalinski, T., Schad, G.A., and Chowdhury, A.B. 1978a. Population biology of hookworms in children in rural West Bengal. I. General parasitological observations. *Am. J. Trop. Med. Hyg.* **27**, 1152–61.

Nawalinski, T., Schad, G.A., and Chowdhury, A.B. 1978b. Population biology of hookworms in children in rural West Bengal. II. Acquisition and loss of hookworms. *Am. J. Trop. Med. Hyg.* **27**, 1162–73.

Neill, W.E. 1972. Effects of size-selective predation on community structure in laboratory aquatic microcosms. Ph.D. Thesis, University of Texas, Austin.

Neill, W.E. 1974. The community matrix and interdependence of the competition coefficients. *Am. Nat.* **108**, 399–408.

Neill, W.E. 1975. Experimental studies of microcrustacean competition, community composition and efficiency of resource utilization. *Ecology*, **56**, 809–26.

Nevo, E., Gorman, G., Soule, M., Yang, S.Y., Clover, R., and Jovanociv, V. 1972. Competitive exclusion between insular *Lacerta* species (Sauria, Lacertidae). Notes on experimental introductions. *Oecologia*, **10**, 183–90.

Newbury, D. McC. 1980a. Interactions between the coccid, *Icerya seychellarum* (Westw.) and its host tree species on Aldabra Atoll. I. *Euphorbia pyrifolia* Lam. *Phil. Trans. R. Soc. London. B.* (In press.)

Newbury, D. McC. 1980b. Interaction between the coccid, *Icerya seychellarum* (Westw.), and its host tree species on Aldabra Atol. II. *Scaevola taccada* (Gaertn.) Roxb. *Phil. Trans. R. Soc. London. B.* (In press.)

Newell, N.D. 1973. The very last moment of the Paleozoic Era. *Mem. Can. Soc. Petroleum Geologists*, **2**, 1–10.

Newsome, A.E. 1969. A population study of house-mice temporarily inhabiting a South Australian wheatfield. *J. Anim. Ecol.* **38**, 341–59.

Neyman, J., Park, T., and Scott, E.L. 1956. Struggle for existence. The *Tribolium* model: biological and statistical aspects. In *Proc. 3rd Berkeley Symp. on Mathematical Statistics and Probability*, Vol. IV, pp. 41–79. Berkeley, University of California Press.

Nicholson, A.J. 1933. The balance of animal populations. *J. Anim. Ecol.* **2**, 131–78.

Nicholson, A.J. 1947. Fluctuations of animal populations. *Rep. 26th Meeting Aust. N.Z. Assn. Advnt. Sci., Perth.*

Nicholson, A.J., and Bailey, V.A. 1935. The balance of animal populations. Part I. *Proc. zool. Soc. Lond.*, 551–98.

Nobel, J.C., Bell, A.D., and Harper, J.L. 1979. The population dynamics of plant growth I. the morphology and structural demography of *Carex arenaria*. *J. Ecol.* **67**, 983–1008.

Noonan, K.M. 1978. Sex ratio of parental investment in colonies of the social wasp *Polistes fuscatus*. *Science*, **199**, 1354–6.

Norton, G.A. 1975. Multiple cropping and pest control, an economic perspective. *Meded. Fac. Landbouww Rijks. Univ. Gent*, **40**, 219–28.

Norton, G.A. 1976a. Pest control decision making—an overview. *Ann. Appl. Biol.* **84**, 444–7.

Norton, G.A. 1976b. The analysis of decision making in crop protection. *Agro-ecosystems*, **3**, 27–44.

Noyes, J.S. 1974. The Biology of the Leek Moth, *Acrolepia assectella* (Zeller). Unpublished Ph.D. thesis, University of London.

Noy-Meir, I. 1975. Stability of grazing systems: an application of predator–prey graphs. *J. Ecol.* **63**, 459–81.

Nunney, L. 1980. The stability of complex model ecosystems. *Am. Nat.* **115**, 639–49.

Nunney, L. 1981. Density compensation, isocline shape and single-level competition models. *J. Theoret. Biol.* (In press.)

Odum, E.P. 1968. Energy flow in ecosystems: a historical review. *Am. Zool.* **8**, 11–18.

Odum, E.P. 1969. The strategy of ecosystem development. *Science*, **164**, 262–70.

Oliver, C.D. 1975. The development of northern red oak (*Quercus rubra* L.) in mixed species, even-aged stands in central New England. Doctoral dissertation. New Haven, Yale University.

Oliver, R.T.D. 1978. HLA–D locus and susceptibility to disease. *Nature*, **274**, 14–16.

Olson, J.S. 1958. Rates of succession and soil changes on southern Lake Michigan sand dunes. *Bot. Gazette*, **119**, 125–70.

Orians, G.H. 1969. On the evolution of mating systems in birds and mammals. *Am. Nat.* **103**, 589–603.

Orians, G.H. 1975. Diversity, stability and maturity in natural ecosystems. In W.H. van Dobben and R.H. Lowe-McConnell (eds.), *Unifying Concepts in Ecology*, pp. 139–50. The Hague, W. Junk.

Orians, G.H., and Horn, H.S. 1969. Overlap in foods and foraging of four species of blackbirds in the Potholes of central Washington. *Ecology*, **50**, 930–8.

Orians, G.H., and Wilson, M.F. 1964. Interspecific territories of birds. *Ecology*, **45**, 736–45.

Osman, R.W. 1975. The influence of seasonality and stability on the species equilibrium. Ph.D. Dissertation, Dept. Geology, University of Chicago.

Oster, G., and Ipaktchi, A. 1978. In H. Eyring and D. Henderson (eds.), *Theoretical Chemistry: Periodicities in Chemistry and Biology*, pp. 111–32. New York, Academic Press.

Oster, G.F., and Wilson, E.O. 1978. *Caste and Ecology in the Social Insects.* Princeton, Princeton University Press.

Owen, D.F., and Chanter, D.O. 1972. Species diversity and seasonal abundance in *Charaxes* butterflies (Nymphalidae). *J. Ent. (A)*, **46**, 135–43.

Owen, D.F., and Wiegert, R.G. 1976. Do consumers maximise plant fitness? *Oikos*, **27**, 488–92.

Packer, C. 1979. Inter-troop transfer and inbreeding avoidance in *Papio anubis*. *Anim. Behav.* **27**, 1–36.

Page, T. 1977. *Conservation and Economic Efficiency.* Baltimore, Johns Hopkins University Press.

Paine, R.T. 1966. Food web complexity and species diversity. *Am. Nat.* **100**, 65–75.

Paine, R.T. 1974. Intertidal community structure. *Oecologia*, **15**, 93–120.

Paine, R.T. 1980. Food webs: interaction strength, linkage and community infrastructure. *J. Anim. Ecol.* **49**, 667–85.

Park, T. 1948. Experimental studies of interspecies competition. I. Competition between populations of the flour beetles, *Tribolium confusum* (Duval) and *Tribolium castaneum* (Herst.). *Ecological Monographs*, **18**, 267–307.

Park, T. 1954. Experimental studies of interspecies competition. II. Temperature, humidity, and competition in two species of *Tribolium*. *Physiol. Zool.* **27**, 177–238.

Park, T. 1962. Beetles, competition, and populations. *Science*, **138**, 1369–75.

Park, T., Leslie, P.H., and Mertz., D.B. 1964. Genetic strains and competition in populations of *Tribolium*. *Physiol. Zool.* **37**, 97–162.

Parker, G.A., and Macnair, M.R. 1979. Models of parent–offspring conflict. IV. Suppression: evolutionary retaliation by the parent. *Anim. Beh.* **27**, 1210–35.

Parker, G.A., and Rubenstein, D.I. 1981. Role assessment, reserve strategy, and acquisition of information in asymmetric animal conflicts. *Anim. Behav.* **29**, (In press.)

Parker, R.A. 1968. Simulation of an aquatic ecosystem. *Biometrics*, **24**, 803–21.

Parker, W.S., and Pianka, E.R. 1973. Notes on the ecology of the iguanid lizard, *Sceloporus magister*. *Herpetologica*, **29**, 143–52.

Parker, W.S., and Pianka, E.R. 1974. Further ecological observations on the western banded gecko, *Coleonyx variegatus*. *Copeia*, 1974, 528–31.

Parker, W.S., and Pianka, E.R. 1975. Comparative ecology of populations of the; lizard, *Uta stansburiana*. *Copeia*, 1975, 615–32.

Parker, W.S., and Pianka, E.R. 1976. Ecological observations on the leopard lizard *Crotaphytus wislizeni* in different parts of its range. *Herpetologica*, **32**, 95–114.

Pathak, M.D. 1975. Utilization of insect–plant interactions in pest control. In D. Pimentel (ed.), *Insects, Science and Society*, pp. 121–48. New York and London, Academic Press.

Patrick, R. 1973. Use of algae, especially diatoms, in the assessment of water quality. *American Society for Testing and Materials*, Special Tech. Publ. **528**, 76–95.

Patrick, R. 1975. Stream communities. In M.L. Cody and J.M. Diamond (eds.),

Ecology and Evolution of Communities, pp. 445–59. Cambridge, Mass., Harvard University Press.

Patten, B.C. 1961. Competitive exclusion. *Science*, **134**, 1599–601.

Pearson, O.P. 1966. The prey of carnivores during one cycle of mouse abundance. *J. Anim. Ecol.* **35**, 217–33.

Pease, J.L., Vowles, R.H., and Keith, L.B. 1979. Interaction of snowshoe hares and woody vegetation. *J. Wildl. Mgmt.* **43**, 43–60.

Peden, L.M., Williams, J.S., and Frayer, W.E. 1973. A Markov model for stand projection. *Forest Sci.* **19**, 303–14.

Pella, J.J., and Tomlinson, P.K. 1969. A generalized stock production model. *Bull. Inter-Amer. Trop. Tuna Comm.* **13**, 421–96.

Pentecost, A. 1980. Aspects of competition in saxicolous lichen communities. *Lichenologist*, **12**, 135–44.

Perkins, B.D. 1978. Enhancement of effect of *Neochetina eichhorniae* for biological control of waterhyacinth. In T.E. Freeman (ed.), *Proc. IV Int. Symp. Biol. Control Weeds*, pp. 87–93. Gainesville, University of Florida.

Peterkin, G.F., and Tubbs, C.R. 1965. Woodland regeneration in the New Forest, Hampshire, since 1650. *J. Appl. Ecol.* **2**, 159–70.

Peterman, R.M., Clark, W.C., and Holling, C.S. 1979. The dynamics of resilience: shifting stability domains in fish and insect systems. In R.M. Anderson, B.D. Turner, and L.R. Taylor (eds.), *Symp. Brit. Ecol. Soc.* 20. *Population Dynamics*, pp. 321–41. Oxford, Blackwell Scientific Publications.

Peters, B. 1980. The demography of leaves in a permanent pasture. Ph.D. thesis, University of Wales.

Peters, W. 1978. Medical Aspects: Comments and discussion II. In A.E. Taylor and R. Muller (eds.), *The Relevance of Parasitology to Human Welfare Today. Symp. Brit. Soc. Parasitol.*, 16. Oxford, Blackwell Scientific Publications.

Petrides, G.A., and Bryant, C.R. 1951. An analysis of the 1949–50 Fowl Cholera epizootic in Texan panhandle waterfowl. *Trans. N. Amer. Wildlife Conf.* **16**, 193–216.

Petrusewicz, K. 1967. *Secondary Productivity in Terrestrial Ecosystems.* Warsaw, Polish Acad. Sciences.

Phillips, O.M. 1978. The equilibrium and stability of simple marine biological systems. III. Fluctuations and survival. *Am. Nat.* **112**, 745–57.

Pianka, E.R. 1969. Sympatry of desert lizards (*Ctenotus*) in Western Australia. *Ecology*, **50**, 1012–30.

Pianka, E.R. 1970. On *r*- and *K*-selection. *Am. Nat.* **104**, 592–7.

Pianka, E.R. 1972. *r*- and *K*-selection or *b* and *d* selection? *Am. Nat.* **106**, 581–8.

Pianka, E.R. 1973. The structure of lizard communities. *A. Rev. Ecol. Syst.* **4**, 53–74.

Pianka, E.R. 1974. Niche overlap and diffuse competition. *Proc. Natl. Acad. Sci. U.S.* **71**, 2141–5.

Pianka, E.R. 1975. Niche relations of desert lizards. In M. Cody and J. Diamond (eds.), *Ecology and Evolution of Communities*, pp. 291–314. Cambridge, Mass., Harvard University Press.

Pianka, E.R. 1976. Natural selection of optimal reproductive tactics. *Am. Zool.* **16**, 775–84.

Pianka, E.R. 1978. *Evolutionary Ecology*, 2nd edition. New York, Harper and Row.

Pianka, E.R. 1980. Guild structure in desert lizards. *Oikos*, **35**, 194–201.

Pianka, E.R., and Pianka, H.D. 1976. Comparative ecology of twelve species of nocturnal lizards (Gekkonidae) in the Western Australian desert. *Copeia*, **1976,** 125–42.

Pianka, E.R., Huey, R.B., and Lawlor, L.R. 1979. Niche segregation in desert lizards. In D.J. Horn, R. Mitchell, and G.R. Stairs (eds.), *Analysis of Ecological Systems, Chapter 4,* pp. 67–115, Ohio State Univ. Press.

Pickett, S.T.A. 1976. Succession: an evolutionary interpretation. *Am. Nat.* **110,** 107–19.

Pico, M.M., Maldonado, D., and Levins, R. 1965. Ecology and genetics of Puerto Rican *Drosophila:* I. Food preferences of sympatric species. *Carib. J. Sci.* **5,** 29–37.

Pielou, E.C. 1972. Niche width and niche overlap: a method of measuring them. *Ecology,* **53,** 687–92.

Pielou, E.C. 1975. *Ecological Diversity.* New York, Wiley-Interscience.

Pillemer, E.A., and Tingey, W.M. 1976. Hooked trichomes: a physical plant barrier to a major agricultural pest. *Science,* **193,** 482–4.

Pimentel, D. 1973. Extent of pesticide use, food supply and pollution. *J.N.Y. Entomol. Soc.* **81,** 13–37.

Pimentel, D., Levin, S.A., and Soans, A.B. 1975. On the evolution of energy balance in some exploiter-victim systems. *Ecology,* **56,** 381–90.

Pimm, S.L. 1980. Properties of food webs. *Ecology,* **61,** 219–25.

Pimm, S.L., and Lawton, J.M. 1977. Number of trophic levels in ecological communities. *Nature,* **268,** 329–31.

Pimm, S.L., and Lawton, J.H. 1978. On feeding on more than one trophic level. *Nature,* **275,** 542–3.

Pitelka, F.A. 1967. Some characteristics of microtine cycle in the arctic. In H.P. Hanson (ed.), *Arctic Biology* (2nd. edn.), pp. 153–84. Corvallis, Oregon State University Press.

Pleszczynska, W.K. 1978. Microgeographic prediction of polygyny in the lark bunting. *Science,* **201,** 935–7.

Podoler, H., and Rogers, D. 1975. A new method for the identification of key factors from life-table data. *J. Anim. Ecol.* **44,** 85–114.

Poluektov, R.A. (ed.) 1974. *Dynamical Theory of Biological Populations.* Moscow, Science Pubs. (In Russian).

Pontin, A.J. 1969. Experimental transplantation of nest-moulds of the ant *Lasius flavus* (F.) in a habitat containing also *L. niger* (L.) and *Myrmica scabrinodis* Nyl. *J. Anim. Ecol.* **38,** 747–54.

Porter, K.G. 1973. Selective grazing and differential digestion of algae by zooplankton. *Nature,* **244,** 179–80.

Prager, E.M., Wilson, A.C., Lowenstein, J.M., and Sarich, V.M. 1980. Mammoth albumin. *Science,* **209,** 287–9.

Pratt, D.M. 1943. Analysis of population development in *Daphnia* at different temperatures. *Biol. Bull.* **85,** 116–40.

Preston, F.W. 1948. The commonness, and rarity, of species. *Ecology,* **29,** 254–83.

Preston, F.W. 1962. The canonical distribution of commonness and rarity. *Ecology,* **43,** 185–215 and 410–32.

Price, P.W. 1972. Parasitoids utilizing the same host: adaptive nature of differences in size and form. *Ecology,* **53,** 190–5.

Price, P.W. 1973. Reproductive strategies in parasitoid wasps. *Am. Nat.* **107,** 684–93.

Pruszynski, S. 1973. The influence of prey density on prey consumption and oviposition of *Phytoseiulus persimilis* Athias-Henriot (Acarina: Phytoseiidae). SROP/WPRS Bulletin: Integrated Control in Glasshouses. 1973–74, pp. 41–6.

Pulliam, H.R. 1975. Coexistence of sparrows: a test of community theory. *Science*, **189**, 474–6.

Pyke, G.H. 1979. The economics of territory size and time budget in the golden-winged sunbird. *Am. Nat.* **114**, 131–45.

Rabinovich, J.E. 1974. Demographic strategies in animal populations: a regression analysis. In F.B. Golley and E. Medina (eds.) *Tropical Ecological Systems*, pp. 19–40. Ecological Studies 11. New York, Springer-Verlag.

Rabinowitz, A. 1979. Bimodal distributions of seedling weight in relation to density of *Festuca paradoxa* Desv. *Nature*, **277**, 297–8.

Ralls, K., Brugger, K., and Ballou, J. 1979. Inbreeding and juvenile mortality in small populations of Ungulates. *Science*, **206**, 1101–3.

Rappoldt, C., and Hogeweg, P. 1980. Niche packing and number of species. *Am. Nat.* **116**, 480–92.

Rathcke, B.J., and Poole, R.W. 1975. Coevolutionary race continues: butterfly larval adaptation to plant trichomes. *Science*, **187**, 175–6.

Raup, D.M. 1972. Taxonomic diversity during the Phanerozoic. *Science*, **177**, 1065–71.

Raup, D.M. 1975. Taxonomic diversity estimation using rarefaction. *Paleobiology*, **1**, 333–42.

Raup, D.M. 1976. Species diversity in the Phanerozoic: an interpretation. *Paleobiology*, **2**, 289–97.

Raup, D.M., Gould, S.J., Schopf, T.J.M., and Simberloff, D. 1973. Stochastic models of phylogeny and the evolution of diversity. *J. Geol.* **81**, 525–42.

Raup, D.M., and Gould, S.J. 1974. Stochastic simulation and evolution of morphology: towards a nomothetic paleontology. *Syst. Zool.* **23**, 305–22.

Raup, H.M. 1964. Some problems in ecological theory and their relation to conservation. *J. Ecol.* **52**, (*Suppl.*), 19–28.

Raup, H.M. 1975. Species versatility in shore habitats. *J. Arnold Arboretum Harvard University*, **56**, 126–63.

Rausher, M.D. 1978. Search image for leaf shape in a butterfly. *Science*, **200**, 1071–3.

Recher, H.F. 1969. Bird species diversity and habitat diversity in Australia and North America. *Am. Nat.* **103**, 75–80.

Redfern, M., and Cameron, R.A.D. 1978. Population dynamics of the yew gall midge *Taxomyia taxi* (Inchbald) (Diptera: Cecidomyiidae). *Ecol. Ent.* **3**, 251–63.

Reed, W.J. 1974. A stochastic model for the economic management of a renewable animal resource. *Math. Biosciences*, **22**, 313–37.

Reed, W.J. 1978. The steady state of a stochastic harvesting model. *Math. Biosciences*, **41**, 273–307.

Rejmanek, M., and Stary, P. 1979. Connectance in real biotic communities and critical values for stability of model ecosystems. *Nature*, **280**, 311–13.

Reynolds, H.T., Adkisson, P.L., and Smith, R.F. 1975. Cotton insect pest management. In R.L. Metcalf and W.H. Luckmann (eds.), *Introduction to Insect Pest Management*, pp. 379–443. New York, Wiley-Interscience.

Reynoldson, T.B. 1964. Evidence for intraspecific competition in field populations of triclads. *J. Anim. Ecol.* **33**, 187–207.

Reynoldson, T.B. 1975. Food overlap of lake-dwelling triclads in the field. *J. Anim. Ecol.* **44**, 245–50.

Rhoades, D.F., and Cates, R.G. 1976. Towards a general theory of plant anti-herbivore chemistry. *Rec. Adv. Phytochem.* **10**, 168–213.

Richardson, B.J. 1975. *r*- and *K*-selection in kangaroos. *Nature (Lond.)*, **255**, 323–4.

Richman, S. 1958. The transformation of energy by *Daphnia pulex*. *Ecol. Monogr.* **28**, 273–91.

Ricklefs, R.E. 1973. *Ecology*. Newton, Mass., Chiron Press.

Ricklefs, R.E., and Travis, J. 1980. A morphological analysis of bird communities in scrub habitats. *Auk*. (In press.)

Ricklefs, R.E., Cochran, D., and Pianka, E.R. 1980. A morphological analysis of the structure of communities of lizards in desert habitats. *Ecology*. (In press.)

Risch, S., McClure, M., Vandermeer, J.H., and Waltz, S. 1977. Mutualism between three species of tropical *Piper* (Piperaceae) and their ant inhabitants. *Amer. Midl. Natur.* **98**, 433–44.

Roberts, A.P. 1974. The stability of a feasible random ecosystem. *Nature (Lond.)* **251**, 607–8.

Rockwood, L.L. 1973. The effect of defoliation on seed production of six Costa Rican tree species. *Ecology*, **45**, 1363–9.

Rodin, L.E. *et al.* 1975. Primary productivity of the main world ecosystems. In *Proceedings of the First International Congress of Ecology*, pp. 176–81. Wageningen, Centre for Agricultural Pub.

Rogers, D.J. 1972a. Random search and insect population models. *J. Anim. Ecol.* **41**, 369–83.

Rogers, D.J. 1972b. The ichneumon wasp *Venturia canescens*: oviposition and avoidance of superparasitism. *Ent. exp. appl.* **15**, 190–4.

Rogers, D.J., and Hassell, M.P. 1974. General models for insect parasite and predator searching behaviour: interference. *J. Anim. Ecol.* **43**, 239–53.

Rogers, D.J., and Hubbard, S. 1974. How the behaviour of parasites and predators promotes population stability. In M.B. Usher and M.H. Williamson (eds.), *Ecological Stability*, pp. 99–119. Chapman & Hall.

Rollins, H.B., and Donahue, J. 1975. Towards a theoretical basis of paleoecology: concepts of community dynamics. *Lethaia*, **8**, 255–70.

Rollins, H.B., Carothers, M., and Donahue, J. 1979. Transgression, regression, and fossil community succession. *Lethaia*, **12**, 89–104.

Root, R.B. 1967. The niche exploitation pattern of the blue-grey gnatcatcher. *Ecol. Monogr.* **37**, 317–50.

Rosen, B.R. 1979. Modules members and communes: A postscript introduction to social organisms. In J. Larwood and B.R. Rosen (eds.), *Biology and Systematics of Colonial Organisms*, pp. xiii–xxxv. London, Academic Press.

Rosenfield, A. 1980. Sociobiology stirs a controversy over limits of science. *Smithsonian*, **11(6)**, 73–80.

Rosenzweig, M.L. 1971. Paradox of enrichment: destabilization of exploitation ecosystems in ecological time. *Science*, **171**, 385–7.

Rosenzweig, M.L. 1973. Habitat selection experiments with a pair of coexisting Heteromyid rodent species. *Ecology* **54**, 111–7.

Rosenzweig, M.L., and MacArthur, R.H. 1963. Graphical representation and stability condition of predator–prey interactions. *Am. Nat.* **97**, 209–23.

Ross, H.H. 1957. Principles of natural coexistence indicated by leafhopper populations. *Evolution*, **11**, 113–29.

Ross, H.H. 1958. Further comments on niches and natural coexistence. *Evolution*, **12**, 112–13.

Ross, R. 1911. *The Prevention of Malaria* (2nd Edn.). London, Murray.

Ross, R. 1916. An application of the theory of probabilities to the study of *a priori* pathometry, number 1. *Proc. Roy. Soc. A* **92**, 204–30.

Roughgarden, J. 1972. Evolution of niche width. *Am. Nat.* **106**, 683–718.

Roughgarden, J. 1974a. Species packing and the competition function with illustrations from coral reef fish. *Theoret. Pop. Biol.* **5**, 163–86.

Roughgarden, J. 1974b. Niche width: biogeographic patterns among *Anolis* lizard populations. *Am. Nat.* **108**, 429–42.

Roughgarden, J. 1974c. The fundamental and realized niche of a solitary population. *Am. Nat.* **108**, 232–5.

Roughgarden, J. 1975a. A simple model for population dynamics in stochastic environments. *Am. Nat.* **109**, 713–36.

Roughgarden, J. 1975b. Species packing and faunal build-up: an evolutionary approach based on density-dependent natural selection. *Theor. Pop. Biol.* **7**, 1–12.

Roughgarden, J. 1976. Resource partitioning among competing species: a co-evolutionary approach. *Theor. Pop. Biol.* **9**, 388–424.

Roughgarden, J. 1979. *Theory of Population Genetics and Evolutionary Ecology: An Introduction*. New York, Macmillan.

Roughgarden, J., and Feldman, M. 1975. Species packing and predation pressure. *Ecology*, **56**, 489–92.

Rowe, J.S., and Scotter, G.W. 1973. Fire in the boreal forest. *Quarternary Res.* **3**, 444–64.

Royama, T. 1971. A comparative study of models for predation and parasitism. *Res. Popul. Ecol.* **1** (*Suppl.*) 1–91.

Rubenstein, D.I. 1980. On the evolution of alternative mating strategies. In J.E.R. Staddon (ed.), *Limits to Action: the Allocation of Individual Behaviour*. New York. Academic Press. pp. 65–100.

Rudebeck, G. 1950. The choice of prey and modes of hunting of predatory birds with special reference to their selection effect. *Oikos*, **2**, 65–88.

Rudwick, M.J.S. 1972. *The Meaning of Fossils*. New York, American Elsevier.

Rundel, P.W. 1971. Community structure and stability in the giant sequoia groves of the Sierra Nevada, California. *Am. Midl. Natur.* **85**, 478–92.

Rundel, P.W. 1972. Habitat restriction in giant sequoia: the environmental control of grove boundaries. *Am. Midl. Natur.* **87**, 81–99.

Russell, B. 1946. *The History of Western Philosophy*. London, Heinemann.

Ryan, C.A., and Green, T.R. 1974. Proteinase inhibitors in natural plant protection. *Rec. Adv. Phytochem.* **8**, 123–40.

Sabbadin, A. 1972. Results and perspectives in the study of a colonial ascidian, *Botryllus schlosseri*. *Fifth European Marine Biology Symposium*. 327–34.

Sabbadin, A. 1973. Recherches expérimentales sur l'ascidie coloniale *Botryllus schlossmi* (1). *Bulletin de la Société Zoologique de France*, **98(3)**, 417–34.

Salazar, J.A. 1976. Analysis of the damage caused by the black bean aphid (*Aphis fabae*) to the field bean (*Vicia faba*). Ph.D. Thesis, University of London.

Sale, P.F. 1974. Overlap in resource use and interspecific competition. *Oecologia*, **17**, 245–56.

Samways, M.J. 1979. Immigration, population growth and mortality of insects and mites on Cassava in Brazil. *Bull. Ent. Res.* **69**, 491–505.

Sanders, H.L. 1968. Marine benthic diversity: a comparative study. *Am. Nat.* **102**, 243–82.

Satchell, J.E. 1980. *r* worms and *K* worms; a basis for classifying lubricid earthworm strategies. *Proc. 7th Int. Soil Zoo. Coll., Int. Soc. Soil Sci. Syraccuse.* (In press.)

Saunders, P.T. 1978. Population dynamics and the length of food chains. *Nature,* **272,** 189–90.

Savory, C.J. 1978. Food consumption of red grouse in relation to the age and productivity of heather. *J. Anim. Ecol.* **47,** 269–82.

Schaefer, M.B. 1954. Some aspects of the dynamics of populations important to the management of commercial marine fisheries. *Bull. Inter-Amer. Trop. Tuna Comm.* **1,** 25–56.

Schaffer, W.M. 1974. Selection for optimal life histories: the effects of age structure. *Ecology,* **55,** 291–303.

Schall, J.J., and Pianka, E.R. 1978. Geographical trends in numbers of species. *Science,* **201,** 679–86.

Schall, J.J., and Pianka, E.R. 1980. Escape behaviour diversity. *Am. Nat.* **115,** 551–66.

Schindel, D.E., and Gould, S.J. 1977. Biological interaction between fossil species: character displacement in Bermudian land snails. *Paleobiology,* **3,** 259–69.

Schoener, A. 1974. Experimental zoogeography: colonization of marine mini-islands. *Am. Nat.* **108,** 715–38.

Schoener, T.W. 1965. The evolution of bill size differences among sympatric congeneric species of birds. *Evolution,* **19,** 189–213.

Schoener, T.W. 1968. The *Anolis* lizards of Bimini: resource partitioning in a complex fauna. *Ecology,* **49,** 704–26.

Schoener, T.W. 1969. Optimal size and specialization in constant and fluctuating environments. In *Diversity and Stability in Ecological Systems,* pp. 103–14, Springfield, Va., U.S. Department of Commerce.

Schoener, T.W. 1970. Nonsynchronous spatial overlap of lizards in patchy habitats. *Ecology,* **51,** 408–18.

Schoener, T.W. 1971. Theory of feeding strategies. *A. Rev. Ecol. Syst.* **2,** 369–404.

Schoener, T.W. 1974. Resource partitioning in ecological communities. *Science,* **185,** 27–39.

Schoener, T.W. 1975a. Competition and the form of habitat shift. *Theoret. Pop. Biol.* **5,** 265–307.

Schoener, T.W. 1975b. Presence and absence of habitat shift in some widespread lizard species. *Ecol. Monogr.* **45,** 232–58.

Schoener, T.W. 1976a. Competition and the niche. In D.W. Tinkle and W.W. Milstead (eds.), *Biology of the Reptilia.* New York, Academic Press.

Schoener, T.W. 1976b. The species-area relation within archipelagos: models and evidence from island land birds. *Proc. 16th International Ornith. Congress, Canberra, August* 1974, pp. 629–42.

Schoener, T.W., and Gorman, G.C. 1968. Some niche differences among three species of Lesser Antillean anoles. *Ecology,* **49,** 819–30.

Schoener, T.W., and Janzen, D.H. 1968. Notes on environmental determinants of tropical versus temperate insect size patterns. *Am. Nat.* **101,** 207–24.

Schoener, T.W., Huey, R.B., and Pianka, E.R. 1979. A biographical extension of the compression hypothesis: competitors in narrow sympatry. *Am. Nat.* **113,** 295–8.

Schopf, T.J.M. 1974. Permo-Triassic extinctions: relation to sea-floor spreading. *J. Geol.* **82,** 129–43.

Schopf, T.J.M. 1978. Fossilization potential of an intertidal fauna: Friday Harbor, Washington. *Paleobiology*, **4**, 261–70.

Schopf, T.J.M., and Gooch, J.C. 1972. A natural experiment to test the hypothesis that loss of genetic variability was responsible for mass extinctions of the fossil record. *J. Geol.* **80**, 481–3.

Schroder, G.D. and M.L. Rosenzweig. 1975. Perturbations analysis of competition and overlap in habitat utilization between *Dipodomys ordii* and *Dipodomys merriami*. *Oceoclogia* **19**, 9–28.

Scrimshaw, N.S., Taylor, C.E., and Gordon, J.E. 1968. Interactions of nutrition and infection. *World Health Organisation Monograph* Series 57.

Sepkoski, J.J. Jr. 1976. Species diversity in the Phanerozoic: species–area effects. *Paleobiology*, **2**, 298–303.

Sepkoski, J.J. Jr. 1978. A kinetic model of Phanerozoic taxonomic diversity I. Analysis of marine orders. *Paleobiology*, **4**, 223–51.

Sepkoski, J.J. Jr. 1979. A kinetic model of Phanerozoic taxonomic diversity II. Early Phanerozoic families and multiple equilibria. *Paleobiology*, **5**, 222–51.

Shelford, V.E. 1943. The relation of snowy owl migration to the abundance of the collared lemming. *Auk*, **62**, 592–4.

Shepherd, J.G., and Horwood, J.W. 1979. The sensitivity of exploited populations to environmental 'noise' and the implications for management. *J. Cons. Int. Explor. Mer.* **38**, 318–23.

Sherman, P.W. 1977. Nepotism and the evolution of alarm calls. *Science*, **197**, 1246–53.

Shields, W.M. 1979. Philopatry, inbreeding, and the adaptive advantages of sex. Ph.D. thesis, Ohio State University, Columbus.

Shields, W.M. 1981. Optimal inbreeding and the evolution of philopatry. In I.R. Swingland and P.J. Greenwood (eds.), *The Ecology of Animal Movement*. Oxford, Oxford University Press.

Shinozaki, K. and Kira, T. 1956. Intraspecific competition among higher plants 7. Logistic Theory of C-D effect *J. Inst. Polytech. Osaka c.y. Univ.* **7**, 35–72.

Shoemaker, C. 1973. Optimisation of agricultural pest management: II, Formulation of a control model. *Math. Biosciences*, **17**, 357–65.

Shugart, H.H. Jr., Crow, T.R., and Hett, J.M. 1973. Forest succession models: a rationale and methodology for modelling forest succession over large regions. *Forest Sci.* **19**, 203–12.

Sigurjonsdottir, H., and Reynoldson, T.B. 1979. An experimental study of competition between triclad species (Turbellaria) using the de Wit model. *Acta Zool. Fennica*, **154**, 89–104.

Siljak, D.D. 1974. Connective stability of complex systems. *Nature*, **249**, 280.

Siljak, D.D. 1975. When is a complex ecosystem stable? *Math. Biosci.* **25**, 25–50.

Simberloff, D. 1974. Permo-Triassic extinctions: effects of area on biotic equilibrium. *J. Geol.* **82**, 267–74.

Simberloff, D.S. 1976. Trophic structure determination and equilibrium in an arthropod community. *Ecology*, **57**, 395–8.

Simberloff, D.S., and Wilson, E.O. 1969. Experimental zoogeography of islands: the colonization of empty islands. *Ecology*, **50**, 278–95.

Simberloff, D.S., and Abele, L.G. 1976a. Island biogeography theory and conservation practice. *Science*, **191**, 285–6.

Simberloff, D.S., and Abele, L.G. 1976b. Island biogeography and conservation: strategy and limitations. *Science*, **193**, 1032.

Simenstad, C.A., Estes, J.A., and Kenyon, K.W. 1978. Aleuts, sea otters, and alternate stable-state communities. *Science*, **200**, 403–11.

Simmonds, H.W. 1933. The biological control of the weed *Clidemia hirta* D. Don., in Fiji. *Bull. ent. Res.* **24**, 345–8.

Simon, C.A. 1975. The influence of food abundance on territory size in the Iguanid lizard *Sceloporus jarrovi*. *Ecology*, **56**, 993–8.

Simpson, G.G. 1944. *Tempo and Mode in Evolution*. New York, Columbia University Press.

Simpson, G.G. 1949. *The Meaning of Evolution*. New Haven, Yale University Press.

Simpson, G.G. 1953. *The Major Features of Evolution*. New York, Columbia University Press.

Simpson, G.G. 1969. The first three billion years of community evolution. In *Diversity and Stability in Ecological Systems*, pp. 162–77. Springfield, Va., U.S. Department of Commerce.

Sinclair, A.R.E. 1975. The resource limitation of trophic levels in tropical grassland ecosystems. *J. Anim. Ecol.* **44**, 497–520.

Sissenwine, M.P. 1977. The effect of random fluctuations on a hypothetical fishery. *Int. Comm. Northw. Atlant. Fish. Sel. Pap.* **2**, 137–44.

Skellam, J.G. 1951. Random dispersal in theoretical populations. *Biometrika*, **38**, 196–218.

Skinozaki, K., and Kira, T. 1956. Intraspecific competition among higher plants. VII. Logistic theory of the C-D effect. *J. Inst. Polytech. Osaka Cy Univ.* **7**, 35–72.

Slatkin, M. 1974. Competition and regional coexistence. *Ecology*, **55**, 128–34.

Slobodkin, L.B. 1961. *Growth and Regulation of Animal Populations*. New York, Holt, Rinehart and Winston.

Slobodkin, L.B. 1964. The strategy of evolution. *Am. Sci.* **52**, 342–57.

Smith, C.C., and Fretwell, S.D. 1974. The optimal balance between size and number of offspring. *Am. Nat.* **108**, 499–506.

Smith, K.M. 1976. *Virus-insect Relationships*. London, Longman.

Smith, N.G. 1972. Migrations of the day-flying moth *Urania* in Central and South America. *Caribbean J. Sci.* **12**, 45–58.

Smith, R.F., and Reynolds, H.T. 1966. Principles, definitions and scope of integrated pest control. *Proc. of the FAO Symp. on Integrated Pest Control*, **1**, 11–17.

Smyth, J.D. 1977. *Introduction to Animal Parasitology* (2nd Ed.), London, Hodder and Stoughton.

Snyder, J., and Bretsky, P.W. 1971. Life habits of diminutive bivalve mollusks in the Maquoketa Formation (Upper Ordovician). *Am. J. Sci.* **271**, 227–51.

Solomon, M.E. 1949. The natural control of animal populations. *J. Anim. Ecol.* **18**, 1–35.

Soper, H.E. 1929. Interpretation of periodicity in disease-prevalence. *J. R. Statist. Soc.* **92**, 34–73.

Soulé, M.E., and Wilcox, B.A. (eds.) 1980. *Conservation Biology: an Evolutionary-Ecological Perspective*. Sunderland, Mass., Sinauer.

Southern, H.N. 1970. The annual control of a population of Tawny Owls (*Strix aluco*). *J. Zool., Lond.* **162**, 197–285.

Southwood, T.R.E. 1962. Migration of terrestrial arthropods in relation to habitat. *Biol. Rev.* **37**, 171–214.

Southwood, T.R.E. 1970. The natural and manipulated control of animal populations. In L.R. Taylor (ed.), *The Optimum Population for Britain*, pp. 87–102. *Inst. Biol. Symp.* 19. London, Academic Press.

Southwood, T.R.E. 1973. The insect/plant relationship—an evolutionary perspective. *Symp. Roy. Ent. Soc. Lond.*, **6**, 3–30.

Southwood, T.R.E. 1975. The dynamics of insect populations. In D. Pimentel (ed.), *Insects, Science and Society*, pp. 151–99. New York, Academic Press.

Southwood, T.R.E. 1976. *Ecological Methods* (Second edition). London, Chapman and Hall.

Southwood, T.R.E. 1977a. Habitat, the temple for ecological strategies? *J. Anim. Ecol.* **46**, 337–66.

Southwood, T.R.E. 1977b. The relevance of population dynamics theory to pest status. In J.M. Cherett and G.R. Sagar (eds.), *Origins of Pest, Parasite, Disease and Weed Problems*, pp. 35–54. Oxford, Blackwell, Scientific Publications.

Southwood, T.R.E. 1978. The components of diversity, In L.A. Mound and N. Waloff (eds.), *Diversity of Insect Faunas*, pp. 19–40. Oxford, Blackwell Scientific Publications.

Southwood, T.R.E., and Norton, G.A. 1973. Economic aspects of pest management strategies and decisions. In P.W. Geier, L.R. Clark, D.J. Anderson, and H.A. Nix. (eds.), *Insects: Studies in Pest Management*, pp. 168–84. Mem. Ecological Society of Australia, Canberra.

Southwood, T.R.E., May, R.M., Hassell, M.P., and Conway, G.R. 1974. Ecological strategies and population parameters. *Am. Nat.* **108**, 791–804.

Southwood, T.R.E., and Comins, H.N. 1976. A synoptic population model. *J. Anim. Ecol.* **45**, 949–65.

Stanley, S.M. 1973a. An ecological theory for the sudden origin of multicellular life in the Late Precambrian. *Proc. Natl. Acad. Sci. U.S.* **70**, 1486–9.

Stanley, S.M. 1973b. Effects of competition on rates of evolution, with special reference to bivalve mollusks and mammals. In S.J. Gould (ed.), I.C.S.E.B. Symposium on evolutionary development of form and symmetry. *Syst. Zool.* **22**, 486–506.

Stanley, S.M. 1975. A theory of evolution above the species level. *Proc. Natl. Acad. Sci. U.S.* **72**, 646–50.

Stanley, S.M. 1976. Ideas on the timing of metazoan diversification. *Paleobiology*, **2**, 209–19.

Stanley, S.M. 1979. *Macroevolution*. San Francisco, Freeman.

Starmer, W.T., Head, W.B., Miranda, M., Miller, M.W., and Phaff, H.J. 1976. The ecology of yeast flora associated with cactiphilic Drosophila and their host plants in the Sonoran Desert. *Microbiol. Ecol.* **3**, 11–30.

Steinhaus, E.A. 1958. Stress as a factor in insect disease. *Proceedings of the 10th International Congress of Entomology*, **4**, 725–30.

Stenseth, N.C. 1978. Do grazers maximise individual plant fitness? *Oikos*, **31**, 299–306.

Stephens, E.P. 1955. *The historical-developmental method of determining forest trends*. Doctoral dissertation. Harvard University, Cambridge, Mass.

Stephens, G.R. 1971. The relation of insect defoliation to mortality in Connecticut forests. *Bull. Connecticut Agr. Exp. Sta., New Haven* **723**, 1–52.

Stephens, G.R., and Waggoner, P.E. 1970. The forests anticipated from 40 years of natural transitions in mixed hardwoods. *Bull. Conn. Agric. Exp. Station, New Haven*, **707**, 1–58.

Stern, K., and Roche, L. 1974. *Genetics of Forest Ecosystems*. New York, Springer-Verlag.

Stern, V.M. 1966. Significance of the economic threshold in integrated pest control. *Proc. FAO Symp. Integrated Control*, **2**, 41–56.

Sternlicht, M. 1973. Parasitic wasps attracted by the sex pheromones of their coccid hosts. *Entomophaga*, **18**, 339–43.

Stiles, F.G. 1975. Ecology, flowering phenology, and hummingbird pollination of some Costa Rican *Heliconia* species. *Ecology*, **56**, 285–301.

Stimson, J.S. 1970. Territorial behaviour in the owl limpet, *Lottia gigantea*. *Ecology*, **51**, 113–18.

Stimson, J.S. 1973. The role of the territory in the ecology of the intertidal limpet *Lottia gigantea* (Gray). *Ecology*, **54**, 1020–30.

Stiven, A.E. 1964. Experimental studies on the epidemiology of the host-parasite system, *Hydra* and *Hydramoeba hydroxena* (Entz). II. The components of a single epidemic. *Ecological Monographs*, **34**. 119–42.

Stiven, A.E. 1967. The influence of host population space in experimental epizootics caused by *Hydraamoeba hydroxena*. *J. Invert. Path.* **9**, 536–45.

Strassman, J.E. 1979. Honey caches help female paper wasps survive Texas winters. *Science*, **204**, 207–9.

Stubbs, M. 1977. Density dependence in the life-cycles of animals and its importance in *K*- and *r*-strategies. *J. Anim. Ecol.* **46**, 677–88.

Sugihara, G. 1980. Minimal community structure: an explanation of species abundance patterns. *Am. Nat.* **116**, 770–87.

Sussman, R.W., and Raven, P.H. 1978. Pollination by lemurs and marsupials: an archaic coevolutionary system. *Science*, **200**, 731–6.

Sutherland, J.P. 1974. Multiple stable points in natural communities. *Am. Nat.* **108**, 859–73.

Sutherst, R.W., and Comins, H.N. 1979. The management of acaricide resistance in the cattle tick, *Boophilus microplus* (Canestrini) (Acari: Ixodidae). *Australia Bull. ent. Res.* **69**, 519–37.

Sutherst, R.W., Norton, G.A., Barlow, N.D., Conway, G.R., Birley, M., and Comins, H.N. 1979. An analysis of management strategies for cattle tick (*Boophilus microplus*) control in Australia. *J. Appl. Ecol.* **16**, 359–82.

Swick, K.E. 1980. Stability and bifurcation in age-dependent population dynamics. *Theor. Pop. Biol.* (In press.)

Swift, M.J., Heal, O.W., and Anderson, J.M. 1979. *Decomposition in Terrestrial Ecosystems*. Oxford, Blackwell Scientific Publications.

Swingland, I.R., and Coe, M.J. 1979. The natural regulation of giant tortoise populations on Aldabra Atoll: recruitment. *Phil. Trans. R. Soc. Lond. B.* **286**, 177–88.

Swingland, I.R., and Greenwood, P.J. (eds.) 1981. *The Ecology of Animal Movement*. Oxford, Oxford University Press.

Tanner, J.T. 1975. The stability and the intrinsic growth rates of prey and predator populations. *Ecology*, **56**, 855–67.

Tappan, H. 1971. Microplankton, ecological succession and evolution. *North Am. Paleont. Convention Chicago*, 1969. *Proc.*, *H*, 1058–103.

Taylor, W.D. 1979. Sampling data on the bactivorous ciliates of a small pond compared to neutral models of community structure. *Ecology*, **60**, 876–83.

Telfer, E.S. 1967. Comparison of moose and deer winter range in Nova Scotia. *J. Wildl. Mon.* **31**, 418–25.

Temple, S.A. 1977. Plant–animal mutualism: coevolution with Dodo leads to near extinction of plants. *Science*, **197**, 885–6.

Terborgh, J.W. 1974. Faunal equilibria and the design of wildlife preserves. In F. Golley and E. Medina (eds.), *Tropical Ecological Systems: Trends in Terrestrial and Aquatic Research*. New York, Springer.

Terborgh, J.W., and Faaborg, J. 1973. Turnover and ecological release in the avifauna of Mona Island, Puerto Rico. *Auk*, **90**, 759–79.

Terborgh, J.W. 1976. Island biogeography and conservation: strategy and limitations. *Science*, **193**, 1029–30.

Terborgh, J.W., and Winter, B. 1980. Some causes of extinction. In M.E. Soulé and B.A. Wilcox (eds.), *Conservation Biology: an Evolutionary-Ecological Perspective*, pp. 119–33. Sunderland, Mass., Sinauer.

Thayer, C.W. 1973. Taxonomic and environmental stability in the Paleozoic. *Science*, **182**, 1242–3.

Thielges, B.A. 1968. Altered polyphenol metabolism in the foliage of *Pinus sylvestris* associated with European pine sawfly attack. *Can. J. Bot.* **46**, 724–5.

Thomas, E.D., Reichelderfer, C.F., and Heimpel, A.M. 1972. Accumulation and persistence of a nuclear polyhechosis virus of the Cabbage Looper in the field. *J. Invert. Path.* **20**, 157 64.

Thompson, C.G., and Scott, D.W. 1979. Production and persistence of the nuclear polyhechosis virus of the Douglas-fir Tussode moth, *Orgyia pseudotsugata* (Lepidoptera: Lymantriidae), in the forest ecosystem. *J. Invert. Path.* **33**, 57–65.

Thompson, D.J. 1975. Towards a predator–prey model incorporating age structure: the effects of predator and prey size on the predation of *Daphnia magna* by *Ischnura elegans*. *J. Anim. Ecol.* **44**, 907–16.

Thompson, J.N. 1978. Within-patch structure and dynamics in *Pastinaca sativa* and resource availability to a specialised herbivore. *Ecology*, **59**, 443–8.

Thompson, W.A., Cameron, P.J., Wellington, W.G., and Vertinsky, I.B. 1976. Degrees of heterogeneity and the survival of an insect population. *Res. Pop. Ecol.* **18**, 1–13.

Thoreau, H.D. 1860. The succession of forest trees. In *Excursions* (1863). Boston, Houghton Mifflin & Co.

Tilden, J.W. 1951. The insect associates of *Baccharis pilularis* De Candolle. *Microent.* **16**, 149–85.

Tilman, D. 1978. Cherries, ants and tent caterpillars: timing of nectar production in relation to susceptibility of caterpillars to ant predation. *Ecology*, **59**, 686–92.

Tilman, D. 1980. Resource competition, spatial heterogeneity, and species diversity: an equilibrium approach to plant community structure. *Am. Nat.* **116**, 362–93.

Tipper, J.C. 1979. Rarefaction and rarefiction—the use and abuse of a method in paleoecology. *Paleobiology*, **5**, 423–34.

Toomey, D.F., and Cys, J.M. 1979. Community succession in small bioherms of algae and sponges in the lower Permian of New Mexico. *Lethaia*, **12**, 65–74.

Travis, C.C., and Post, W.M. 1979. Dynamic and comparative statics of mutualistic communities. *J. Theor. Biol.* **78**, 553–71.

Tregonning, K., and Roberts, A. 1979. Complex systems which evolve towards homeostasis. *Nature*, **281**, 563–4.

Trivers, R.L. 1971. The evolution of reciprocal altruism. *Q. Rev. Biol.* **46**, 35–57.

Trivers, R.L. 1974. Parent-offspring conflict. *Am. Zool.* **14**, 249–64.

Trivers, R.L., and Hare, H. 1976. Haplodiploidy and the evolution of the social insects. *Science*, **191**, 249–63.

Turelli, M. 1977. Random environments and stochastic calculus. *Theor. Pop. Biol.* **12**, 140–78.

Turelli, M. 1978. A reexamination of stability in randomly varying versus deterministic environments with comments on the stochastic theory of limiting similarity. *Theor. Pop. Biol.* **13**, 244–67.

Turnbull, A.L. 1962. Quantitative studies of the food of *Linyphia triangularis* Clerck (Araneae: Linyphiidae). *Can. Ent.* **94**, 1233–49.

Twight, P.A., and Minckler, L.S. 1972. *Ecological Forestry for the central Hardwood Forest*. Washington, National Parks and Conservation Association.

Ullyett, G.C. 1949a. Distribution of progeny by *Chelonus texanus* Cress. (Hymenoptera: Braconidae). *Can. Ent.* **81**, 25–44.

Ullyett, G.C. 1949b. Distribution of progeny by *Cryptus inornatus* Pratt (Hymenoptera: Ichneumonidae). *Can Ent.* **81**, 285–99.

Usher, M.B. 1966. A matrix approach to the management of renewable resources with special reference to selection forests. *J. Appl. Ecol.* **3**, 355–67.

Usher, M.B. 1972. Developments in the Leslie matrix model. In J.N.R. Jeffers (ed.), *Mathematical Models in Ecology*, pp. 29–60. Oxford, Blackwell Scientific Publications.

Usher, M.B. 1979. Markovian approaches to ecological succession. *J. Anim. Ecol.* **48**, 413–26.

Utida, S. 1967. Damped oscillation of population density at equilibrium. *Res. Pop. Ecol.* **9**, 1–9.

Valentine, J.W. 1970. How many marine invertebrate fossil species? A new approximation. *J. Paleontol.* **44**, 410–15.

Valentine, J.W. 1973. Phanerozoic taxonomic diversity: a test of alternate models. *Science*, **180**, 1078–9.

Valentine, J.W., Hedgecock, D., Zumwalt, G., and Ayala, F.J. 1973. Mass extinctions and genetic polymorphism in the 'killer clam', *Tridacna. Bull. Geol. Soc. Am.* **84**, 3411–14.

Valentine, J.W., and Ayala, F.J. 1974. On scientific hypotheses, killer clams, and extinctions. *Geology*, **2**, 69–71.

Valentine, J.W., Foin, T.C., and Peart, D. 1978. A provincial model of Phanerozoic marine diversity. *Paleobiology*, **4**, 55–66.

Vance, R.R. 1972. Competition and mechanism of coexistence in three sympatric species of intertidal hermit crabs. *Ecology*, **53**, 1062–74.

Vandermeer, J.H. 1972. Niche theory. *A. Rev. Ecol. Syst.* **3**, 107–32.

Vandermeer, J.H. 1980. Indirect mutualism: variations on a theme by Stephen Levine. *Am. Nat.* **116**, 441–8.

Vandermeer, J.H., and Boucher, D.H. 1978. Varieties of mutualistic interaction in population models. *J. Theor. Biol.* **74**, 549–58.

Van Valen, L. 1965. Morphological variation and the width of the ecological niche. *Am. Nat.* **100**, 377–89.

Van Valen, L. 1973. Body size and numbers of plants and animals. *Evolution*, **27**, 27–35.

Van Valen, L. 1978. Arborescent animals and other colonoids. *Nature*, **276**, 318.

Varley, G.C. 1963. The interpretation of change and stability in insect populations. *Proc. R. ent. Soc. Lond.* (*C*), **27**, 52–7.

Varley, G.C., and Gradwell, G.R. 1970. Recent advances in insect population dynamics. *An. Rev. Ent.* **15**, 1–24.

Varley, G.C., Gradwell, G.R., and Hassell, M.P. 1973. *Insect Population Ecology: an Analytical Approach*. Oxford, Blackwell Scientific Publications.

Verner, J. 1964. Evolution of polygamy in the long-billed marsh wren. *Evolution*, **18**, 252–61.

Viereck, L.A. 1966. Plant succession and soil development on gravel outwash of the Muldrow Glacier, Alaska. *Ecol. Monogr.* **36**, 181–99.

Vincent, T.L., and Anderson, L.R. 1979. Return time and vulnerability for a food chain model. *Theor. Pop. Biol.* **15**, 217–31.

Volterra, V. 1926. Variations and fluctuations of the number of individuals in animal species living together. *J. Cons. perm. int. Ent. Mer.* **3**, 3–51. (Also reprinted in Chapman, R.N. 1931. *Animal Ecology*, New York and London.)

Vrba, E.S. 1980. Evolution, species, and fossils: how does life evolve? *South African Jour. Sci.* **76**, 61–84.

Vuilleumier, F. 1970. Insular biogeography in continental regions: the northern Andes of South America. *Am. Nat.* **104**, 373–88.

Wade, M.J. 1979. The evolution of social interactions by family selection. *Am. Nat.* **113**, 399–417

Walker, K.R., and Alberstadt, L.P. 1975. Ecological succession as an aspect of structure in fossil communities. *Paleobiology*, **1**, 238–57.

Waller, D.M., and Green, D. 1980. Implications of sex for the analysis of life histories. *Am. Nat.* (In press.)

Wallner, W.E., and Walton, G.S. 1979. Host defoliation: a possible determinant of gypsy moth population quality. *Ann. Ent. Soc. America*, **72**, 62–7.

Waloff, N., and Richards, O.W. 1977. The effect of insect fauna on growth mortality and natality of broom, *Sarothamnus scoparius*. *J. appl. Ecol.* **14**, 787–98.

Wang, Y., Gutierrez, A.P., Oster, G., and Daxl, R. 1977. A population model for plant growth and development: coupling cotton-herbivore interaction. *Can. Ent.* **109**, 1359–74.

Wangersky, P.J., and Cunningham, W.J. 1957. Time lag in prey–predator population models. *Ecology*, **38**, 136–9.

Warner, R.E. 1968. The role of introduced diseases in the extinction of endemic Hawaiian Avifauna. *Condor*, **70**, 101–20.

Warren, K.S. 1973. Regulation of the prevalence and intensity of schistosomiasis in man: immunology or ecology. *J. Inf. Diseases*, **227**, 595–609.

Watkins, R., and Hurst, J.M. 1977. Community relations of Silurian crinoids at Dudley, England. *Paleobiology*, **3**, 207–17.

Watkinson, A.R. 1980. Density-dependence in single-species populations of plants. *J. Theor. Biol.* **83**, 345–7.

Watson, A., and Moss, R. 1972. A current model of population dynamics in red grouse. In K.H. Voous (ed.), *Proc. XV. Int. Orn. Cong.* 134–49.

Watt, K.E.F. 1963. Dynamic programming, 'look ahead programming', and the strategy of insect pest control. *Can. Ent.* **95**, 525–636.

Watt, K.E.F. 1968. *Ecology and Resource Management* (And references therein.) New York, McGraw-Hill.

Way, M.J. 1973. Objectives, methods and scope of integrated control. In P.W. Geier, L.R. Clark, D.J. Anderson, and H.A. Nix (eds.), *Insects: Studies in Pest Management*. Mem. 1, pp. 137–52. Ecological Society of Australia, Canberra.

Weaver, H. 1974. Effects of fire on temperate forests: Western United States. In

T.T. Kozlowski and C.E. Ahlgren (eds.), *Fire and Ecosystems*, pp. 279–319. New York, Academic Press.

Webb, L.J., Tracey, J.G., and Williams, W.T. 1972. Regeneration and pattern in the subtropical rain forest. *J. Ecol.* **60**, 675–95.

Weigal, P.D. 1969. The distribution of the flying squirrels, *Glaucomys volans* and *G. sabrinus:* an evaluation of the competitive exclusion idea. Ph.D. Thesis, Duke Univ. (from *Diss. Abstr. Int. B. Sci. Eng.*, **30**, 2966B).

Weins, J.A. 1977. On competition and variable environments. *Amer. Sci.* **65**, 590–7.

Werner, Y. 1969. Eye size in geckos of various ecological types (Reptilia: Gekkonidae and Sphaerodactlidae). *Israel J. Zool.* **18**, 291–316.

West, M.J. 1967. Foundress associations in polistine wasps: dominance hierarchies and the evolution of social behaviour. *Science*, **157**, 1584–5.

West Eberhard, M.J. 1969. The social biology of polistine wasps. *Misc. Publ. Mus. Zool. Univ. Mich.* **140**, 1–101.

West Eberhard, M.J. 1975. The evolution of social behaviour by kin selection. *Quart. Rev. Biol.* **50**, 1–33.

Whitcomb, R.F., Lynch, J.F., Opler, P.A., and Robbins, C.S. 1976. Island biogeography and conservation: strategy and limitations. *Science*, **193**, 1030–2.

White, J. 1980. Demographic factors in populations of plants. In O.T. Solbrig (ed.), *Demography and Evolution in Plant Populations*. Oxford, Blackwell Scientific Publications.

White, P.S. 1979. Pattern, process, and natural disturbance in vegetation. *Bot. Rev.* **45**, 229–99.

White, T.C.R. 1969. An index to measure weather-induced stress of trees associated with outbreaks of psyllids in Australia. *Ecology*, **50**, 905–9.

White, T.C.R. 1974. A hypothesis to explain outbreaks of looper caterpillars, with special reference to populations of *Selidosema suavis* in a plantation of *Pinus radiata* in New Zealand. *Oecologia*, **16**, 279–301.

White, T.C.R. 1976. Weather, food and plagues of locusts. *Oecologia*, **22**, 119–34.

Whitham, T.G. 1978. Habitat selection by *Pemphigus* aphids in response to resource limitation and competition. *Ecology*, **59**, 1164–76.

Whittaker, J.B. 1979. Invertebrate grazing, competition, and plant dynamics. In R.M. Anderson, B.D. Turner, and L.R. Taylor (eds.), pp. 207–22 *Symp. Brit. Ecol. Soc.* 20. Oxford, Blackwell Scientific Publications.

Whittaker, J.B., Ellistone, J., and Patrick, C.K. 1979. The dynamics of a chrysomelid beetle, *Gastrophysa viridula*, in a hazardous natural habitat. *J. Anim. Ecol.* **48**, 973–86.

Whittaker, R.H. 1957. Recent evolution of ecological concepts in relation to the eastern forests of North America. *Am. J. Bot.* **44**, 197–206.

Whittaker, R.H. 1972. Evolution and measurement of species diversity. *Taxon*, **21**, 213–51.

Whittaker, R.H. 1975a. *Communities and Ecosystems*. (Second edition). New York, Macmillan.

Whittaker, R.H. 1975b. The design and stability of some plant communities. In W.H. van Dobben and R.H. Lowe-McConnell (eds.), *Unifying Concepts in Ecology*, pp. 169–81. The Hague, Junk Pubs.

Whittaker, R.H., and Levin, S.A. 1977. The role of mosaic phenomena in natural communities. *Theoret. Pop. Biol.* **12**, 117–39.

Wickham, D.E., and Botsford, L.W. 1980. Multiple equilibria in the interaction between *Carcinonemertes errans* and its host, the Dungeness crab. *J. Anim. Ecol.* (In press.)

Wilbur, H.M. 1972. Competition, predation, and the structure of the *Ambystoma-Rana sylvatica* community. *Ecology*, **53**, 3–21.

Wilcox, B. 1980. Insular ecology and conservation. In M.E. Soulé and B.A. Wilcox (eds.), *Conservation Biology: an Evolutionary-Ecological Perspective*, pp. 95–117. Sunderland, Mass., Sinauer.

Wildbotz, T., and Meier, W. 1973. Integrated control: critical assessment of case histories in affluent economies. In P.W. Geier, L.R. Clark, D.J. Anderson, and H.A. Nix (eds.), *Insects: Studies in Pest Management*. Mém. 1, pp. 137–52, Ecological Society of Australia, Canberra.

Williams, G.C. 1975. *Sex and Evolution*. Princeton, Princeton University Press.

Williams, W.T., Lance, G.N., Webb, L.J., Tracey, J.G., and Dale, M.B. 1969. Studies in the numerical analysis of complex rain-forest communities III. The analysis of successional data. *J. Ecol.* **57**, 515–35.

Williamson, M. 1972. *The Analysis of Biological Populations*. London, Edward Arnold.

Williamson, M. 1973. Species diversity in ecological communities. In M.S. Bartlett and R.W. Hiorns (eds.), *The Mathematical Theory of the Dynamics of Biological Populations*, pp. 325–35. New York, Academic Press.

Willis, E.O. 1974. Populations and local extinction of birds on Barro Colorado Island, Panama. *Ecol. Monographs*, **44**, 153–69.

Wilson, A.G.L., Hughes, R.D., and Gilbert, N.E. 1972. The response of cotton to pest attack. *Bull. ent. Res.* **61**, 405–14.

Wilson, D.S. 1980. *The Natural Selections of Populations and Communities*. Menlo Park, Cal., Benjamin/Cummings.

Wilson, E.O. 1969. The new population biology. *Science*, **163**, 1184–5.

Wilson, E.O. 1971. *The Insect Societies*. Cambridge, Mass., Harvard University Press.

Wilson, E.O. 1973. Group selection and its significance for ecology. *Bio. Science*, **23**, 631–8.

Wilson, E.O. 1975. *Sociobiology: The New Synthesis*. Cambridge, Mass., Harvard University Press.

Wilson, E.O. 1978. *On Human Nature*. Cambridge, Mass., Harvard University Press.

Wilson, E.O. *et al.* 1973. *Life on Earth*. Stamford, Conn., Sinauer.

Wilson, E.O., and Willis, E.O. 1975. Applied biogeography. In M.L. Cody and J.M. Diamond (eds.), *Ecology and Evolution of Communities* pp. 522–34. Cambridge, Mass., Harvard University Press.

Winterhalder, B.P. 1980. Canadian fur beaver cycles and Cree-Ojibwa hunting and trapping practices. *Am. Nat.* **115**, 870–9.

Wolda, H. 1978. Fluctuations in abundance of tropical insects. *Am. Nat.* **112**, 1017–45.

Wood, B.J. 1971. Development of integrated control programs for pests of tropical perennial crops in Malaysia. In C.B. Huffaker (ed.), *Biological Control*, pp. 422–57. New York, Plenum.

Wood, D.L., Browne, L.E., Bedard, W.D., Tilden, P.E., Silverstein, R.M., and Rodin, J.O. 1968. Response of *Ips confusus* to synthetic sex pheromones in Nature. *Science*. **159**, 1373–4.

Woods, K.D. 1979. Reciprocal replacement and maintenance of codominance in a beech-maple forest. *Oikos*, **33**, 31–9.

Wratten, S.D. 1973. The effectiveness of the coccinellid beetle, *Adalia bipunctata* (L.), as a predator of the lime aphid, *Eucallipterus tiliae* L. *J. Anim. Ecol.* **42**, 785–802.

Wright, H.E., Jr. 1974. Landscape development, forest fires, and wilderness management. *Science*, **186**, 487–95.

Wright, H.E., and Heinselman, M.L. 1973. The ecological role of fire in natural conifer forests in western and northern North America—Introduction. *Quarternary Res.* **3**, 319–28.

Wright, S. 1945. Book review: Tempo and Mode in Evolution, by G.G. Simpson. *Ecology*, **26**, 415–19.

Wynne-Edwards, V.C. 1962. *Animal Dispersion in Relation to Social Behaviour*, Edinburgh, Oliver & Boyd.

Yoda, K., Kira, T., Ogawa, H., and Hozumi, K. 1963. Self-thinning in over-crowded pure stands under cultivated and natural conditions. *J. Biol. Osaka Cy. Univ.* **14**, 107–29.

Yodzis, P. 1980. The connectance of real ecosystems. *Nature*, **284**, 544–5.

Yorke, J.A., and London, W.P. 1973. Recurrent outbreaks of measles, chicken pox and mumps. II. Systematic differences in contact rates and stochastic effects. *Am. J. Epidmiol.* **98**, 469–82.

Yorke, J.A., Heathcote, H.W., and Nold, A. 1978. Dynamics and control of the transmission of Gonorrhea. *Sex. Trans. Dis.* **5**, 51–6.

Yorke, J.A., Nathanson, N., Pianigiani, G., and Martin, J. 1979. Seasonality and the requirements for perpetuation and eradication of viruses in populations. *Am. J. Epidemiol.* **109**, 103–23.

Yoshiyama, R.M., and Roughgarden, J. 1977. Species packing in two dimensions. *Am. Nat.* **111**, 107–21.

Young, A.M. 1974. On the biology of *Hamadryas februa* (Lepidoptera: Nymphalidae) in Guanacuste, Costa Rica. *Z. ang. Ent.* **76**, 380–93.

Young, A.M., and Muyshondt, A. 1972. Biology of *Morpho polyphemus* (Lepidoptera: Morphidae) in El Salvado. *J. New York ent. Soc.* **80**, 18–42.

Index